Biogenic Amines in Food
Analysis, Occurrence and Toxicity

Food Chemistry, Function and Analysis

Series editors:
Gary Williamson, *University of Leeds, UK*
Alejandro G. Marangoni, *University of Guelph, Canada*
Juliet A. Gerrard, *University of Auckland, New Zealand*

Titles in the series:

1: Food Biosensors
2: Sensing Techniques for Food Safety and Quality Control
3: Edible Oil Structuring: Concepts, Methods and Applications
4: Food Irradiation Technologies: Concepts, Applications and Outcomes
5: Non-extractable Polyphenols and Carotenoids: Importance in Human Nutrition and Health
6: Cereal Grain-based Functional Foods: Carbohydrate and Phytochemical Components
7: Steviol Glycosides: Cultivation, Processing, Analysis and Applications in Food
8: Legumes: Nutritional Quality, Processing and Potential Health Benefits
9: Tomato Chemistry, Industrial Processing and Product Development
10: Food Contact Materials Analysis: Mass Spectrometry Techniques
11: Vitamin E: Chemistry and Nutritional Benefits
12: Anthocyanins from Natural Sources: Exploiting Targeted Delivery for Improved Health
13: Carotenoid Esters in Foods: Physical, Chemical and Biological Properties
14: Eggs as Functional Foods and Nutraceuticals for Human Health
15: Rapid Antibody-based Technologies in Food Analysis
16: DNA Techniques to Verify Food Authenticity: Applications in Food Fraud
17: Advanced Gas Chromatography in Food Analysis
18: Handbook of Food Structure Development
19: Mitigating Contamination from Food Processing
20: Biogenic Amines in Food: Analysis, Occurrence and Toxicity

How to obtain future titles on publication:
A standing order plan is available for this series. A standing order will bring delivery of each new volume immediately on publication.

For further information please contact:
Book Sales Department, Royal Society of Chemistry, Thomas Graham House, Science Park, Milton Road, Cambridge, CB4 0WF, UK
Telephone: +44 (0)1223 420066, Fax: +44 (0)1223 420247,
Email: booksales@rsc.org
Visit our website at www.rsc.org/books

Biogenic Amines in Food
Analysis, Occurrence and Toxicity

Edited by

Bahruddin Saad
Universiti Teknologi PETRONAS, Malaysia
Email: bahruddin.saad@utp.edu.my

and

Rosanna Tofalo
Università degli Studi di Teramo, Italy
Email: rtofalo@unite.it

ROYAL SOCIETY OF CHEMISTRY

Food Chemistry, Function and Analysis No. 20

Print ISBN: 978-1-78801-436-6
PDF ISBN: 978-1-78801-581-3
EPUB ISBN: 978-1-78801-919-4
Print ISSN: 2398-0656
Electronic ISSN: 2398-0664

A catalogue record for this book is available from the British Library

The Royal Society of Chemistry is a charity, registered in England and Wales, Number 207890, and a company incorporated in England by Royal Charter (Registered No. RC000524), registered office: Burlington House, Piccadilly, London W1J 0BA, UK, Telephone: +44 (0) 20 7437 8656.

Visit our website at www.rsc.org/books

Printed in the United Kingdom by CPI Group (UK) Ltd, Croydon, CR0 4YY, UK

Preface

Microbes surrounding us have different traits which can affect different aspects of our health. Biogenic amines (BAs) are generally produced by microbial decarboxylation of amino acids in foods and beverages. Some BAs, such as histamine, tyramine and phenylethylamine, can have toxic effects on organisms and the consumption of foods and beverages containing high amounts of these compounds can negatively impact our health. The degree of BAs intoxication depends on the amount and type of BAs ingested and on the activity of monoamine oxidase (MAO) and diamine oxidase (DAO) enzymes in the liver as a detoxification system. The first part of this book reports the toxicology and regulation of BAs in foods in different countries (Chapter 1) and the modulation of microbial activities related to BAs production (Chapter 2). The main strategies used to reduce BAs occurrence in both fermented and non-fermented foods are reported in Chapter 3. An overview of BAs presence in different foods and beverages is also reported (Chapters 4–8 and 10). Particular attention is given to baby foods (Chapter 9) and to the formation of heterocyclic amines during meat processing and their impact on increasing the risk of cancer (Chapter 11).

As regards consumers and food producers, as well as regulatory agencies, the content of BAs in foods and beverages is of great interest. A wealth of analytical methods that describe the details of the analysis are available in the literature. Nevertheless, the choice of analytical method for the intended purpose can be a daunting task for the novice due to the diversity of these methods. Thus, we have tried to address the challenges in the analysis of BAs in a simple and straightforward manner for it to be appreciated by broad-based readers. We started off by highlighting the importance of preparing the sample for the eventual analysis in Chapter 12. Here, the basics of the extraction from the conventional liquid–liquid extraction

Food Chemistry, Function and Analysis No. 20
Biogenic Amines in Food: Analysis, Occurrence and Toxicity
Edited by Bahruddin Saad and Rosanna Tofalo

Published by the Royal Society of Chemistry, www.rsc.org

all the way to the more futuristic applications offered by the various microextraction methods are outlined. These are aided by diagrams to illustrate the techniques. For many methods of analysis, the BAs need to be converted into a more suitable chemical form before the analysis. In this regard, derivatization becomes essential (discussed in Chapter 13), where schemes for gas chromatography and high-performance liquid chromatography derivatizations are outlined. Pertinent information on these derivatization reagents are also presented. After the sample preparation (may also include derivatizations), the samples can be analysed by suitable analytical methods, among which chromatographic methods (especially HPLC) are the most widely used (Chapter 14). The potential of capillary electrophoresis in BAs analysis is also discussed. Chemical sensors offer interesting possibilities for BAs detection. This and other miscellaneous methods are treated in Chapter 15.

This book would not be possible without the valuable contributions of those involved – in particular, the chapter contributors for their willingness to sacrifice their professional time to embark on this interesting project and to give us an update in this field, as well as the staff of the Royal Society of Chemistry for their professionalism in guiding and assisting us throughout the course of preparation of the book. Bahruddin expresses his sincere thanks to past postgraduate students who were directly involved in BAs research (Dr Mardiana Saaid, Dr Ahmad Makahleh, Dr Elbaleeq Gubartallah, Dr Abdussalam Hafiz). Special thanks are also due to my wife Halijah for her continuous support, sacrifices and silent prayers, not only during this book project but also throughout my academic career. Rosanna thanks her research group (Dr Giorgia Perpetuini, Dr Fabrizia Tittarelli, Dr Noemi Battistelli) and Prof. Giovanna Suzzi, who transferred to her the interest in studying the occurrence of BAs in fermented foods. We each owe special thanks to our families for their support and understanding.

Bahruddin Saad and Rosanna Tofalo

Contents

Food Chemistry, Function and Analysis No. 20
Biogenic Amines in Food: Analysis, Occurrence and Toxicity
Edited by Bahruddin Saad and Rosanna Tofalo

Published by the Royal Society of Chemistry, www.rsc.org

CHAPTER 1

Biogenic Amines Formation, Toxicity, Regulations in Food

YESIM ÖZOGUL AND FATIH ÖZOGUL*

Department of Seafood Processing Technology, Faculty of Fisheries, Cukurova University, 01330, Balcali, Adana, Turkey
*Email: fozogul@cu.edu.tr

1.1 Introduction

Biogenic amines (BAs) are nitrogenous low molecular weight organic bases and they have an aliphatic, aromatic, or a heterocyclic structure. Biogenic amines are produced by the decarboxylation of their respective free precursor amino acids, through the catalytic action of substrate-specific microbial decarboxylases that remove the α-carboxyl group of amino acids to give the corresponding amines.[1] They are present in a wide range of food products, including fish, fish products, meat products, eggs, cheeses, fermented vegetable, fruits, soybean products, beers, wines, nuts, and chocolate.[2–5] The most significant biogenic amines occurring in foods are histamine (HIS), tyramine (TYR), putrescine (PUT), cadaverine (CAD), tryptamine (TRY), and β-phenethylamine (PHE) and they originate from the precursor amino acids histidine, tyrosine, ornithine, lysine, tryptophane, and phenylalanine, respectively. The demand for safer foods has led to more research into biogenic amines over the last decades. The presence of biogenic amines in these foods is associated with toxicological reason and quality indicators.[6,7] Improper storage of food can lead to the formation of BAs. A majority of BAs are malodorous, toxic, and carcinogenic, causing dizziness, faintness,

Food Chemistry, Function and Analysis No. 20
Biogenic Amines in Food: Analysis, Occurrence and Toxicity
Edited by Bahruddin Saad and Rosanna Tofalo

Published by the Royal Society of Chemistry, www.rsc.org

headaches, cardiac palpitations, mucosal burns, and irritation to the eyes *etc.*[8,9] The purpose of this chapter is to provide basic information on BAs, including chemical structure, properties, their toxicology, and regulations on their levels in food for different countries.

1.2 Biogenic Amine Formation

1.2.1 Chemical Structure of Amino Acid and Biogenic Amines

α-Amino acids are basic units of proteins and they consist of α-carbon atom covalently attached to a hydrogen atom (H), an amino group (NH_2), a carboxyl group (COOH), and a side-chain R group.

Amines are basic nitrogenous compounds in which one, two, or three atoms of hydrogen in ammonia are replaced by alkyl or aryl groups. The formation of these low molecular nitrogen compounds occurs through four enzymatic reactions: (i) decarboxylation, (ii) transamination, (iii) reductive amination, and (iv) degradation of certain precursor amino compounds.[10] Decarboxylation of amino acid is the most common way of synthesis of amines in foods. Since these amines are formed by the action of living organisms by the decarboxylation process of amino acids, they are called biogenic. Amino acid decarboxylation occurs by removal of the α-carboxyl group to give the corresponding amine. The names of many biogenic amines correspond to the names of their originating amino acid; for instance, histamine from histidine, tyramine from tyrosine, agmatine from arginine, β-phenylethylamine from phenylethylalanine, and tryptamine from tryptophan (Figure 1.1). An alternative pathway for plants and some microorganisms takes place to form putrescine from arginine *via* agmatine. Lysine is decarboxylated by lysine decarboxylase to produce cadaverine, and it can also be produced by ornithine decarboxylase when the content of ornithine is low, but lysine content is high.[11]

The chemical structure of biogenic amines can be three forms: aliphatic (putrescine, cadaverine, spermine, and spermidine); aromatic (tyramine and phenylethylamine); heterocyclic (histamine and tryptamine). They can also be classified into monoamines (phenylethylamine and tyramine), diamines (cadaverine and putrescine), or polyamines (spermidine and spermine) based on the number of amine groups.[12] Putrescine, spermidine, spermine, and cadaverine are essential components of living cells, regulating nucleic acid function, protein synthesis, and the stabilization of membranes.[13] Dietary BAs are metabolized by acetylation and oxidation reactions enhanced by the enzymes monoamine oxidase (MAO) or diamine oxidase (DAO).[14] Preconditions for biogenic amine formation by microorganisms are: (i) availability of free amino acids, (ii) presence of decarboxylase-positive microorganisms, and (iii) conditions that allow bacterial growth, decarboxylase synthesis, and decarboxylase activity.[15]

Proteolysis that occurs either autolytic or bacterial may play an important role in the release of free amino acids from tissue proteins which offer a

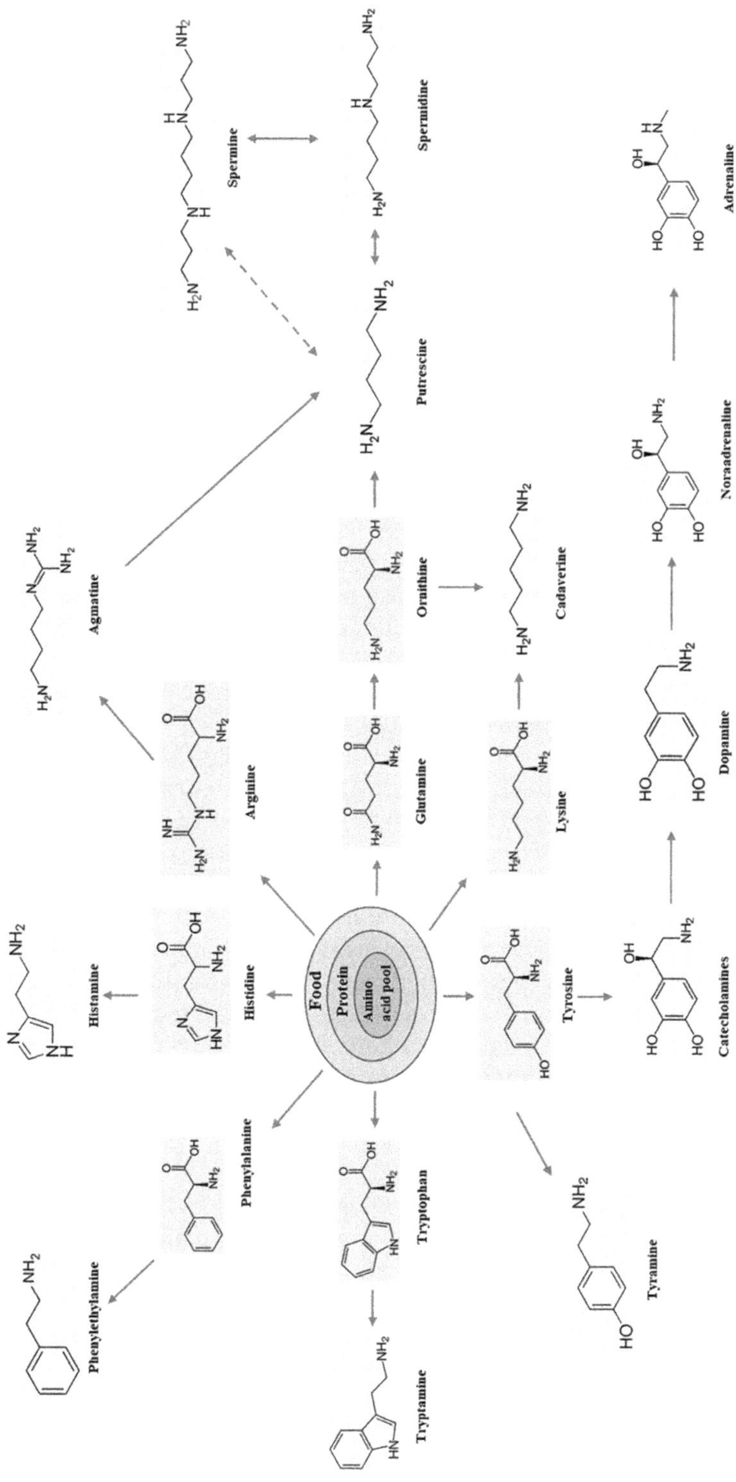

Figure 1.1 Metabolic pathway of biogenic amines formation.

substrate for decarboxylases reactions.[2] It is reported that two mechanisms of reaction for amino acid decarboxylation take place, which is a pyridoxal phosphate dependent reaction and a non-pyridoxal phosphate dependent reaction.[16]

1.2.2 Physiological Significance of Biogenic Amines

Biogenic amines are nitrogen sources and precursors for the synthesis of hormones, alkaloids, nucleic acids, and proteins. BAs have many important biological functions in the human nervous system (*e.g.* as neurotransmitter) and the cardiovascular system (controlling blood pressure), and play a significant role in the regulation of body temperature and digestion.[2,17–19] BAs can be of endogenous origin at low concentrations in fresh food products. However, BAs presence in foods is closely associated with microbial activity. High level of BAs in food may be harmful to the nervous and cardiovascular systems of humans and also change food flavour. In addition, BAs are well known as potential precursors of carcinogenic nitrosamines.[20]

BAs can be used as spoilage indicators for different meat products. In particular, the biological amines index (BAI = histamine + putrescine + cadaverine + tyramine) and quality index (QI) = (histamine + putrescine + cadaverine)/(1 + spermidine + spermine) have been used to evaluate the freshness of meat products.[21,22] BAs formation are species specific. Putrescine and cadaverine have been detected in significant concentrations in fermented meat and fish.[23–25] Cadaverine was reported to be a reliable spoilage indicator of poultry meat,[26,27] whereas histamine has been regarded as an index of the fish quality, particularly dark-muscle fish.[28] Tyramine is reported to cause food intoxication commonly associated with ripened cheeses,[29,30] affecting health due to its capacity to potentiate sympathetic cardiovascular activity by releasing noradrenaline, called "cheese reaction".[31] Histamine, putrescine, cadaverine, tyramine, 2-phenylethylamine, and tryptamine are often detected in fermented products.[25,32] In addition, the level of BAs in alcoholic beverages (*i.e.* wine and beer) has received much attention since ethanol and acetaldehyde can increase the risk to human health by retarding the enzymes responsible for detoxification.[33,34]

Spermine and spermidine are mainly found bound to polyanionic molecules such as DNA, RNA, ATP, and phospholipids.[35] They are involved in the regulation of cell growth and proliferation, in regulating DNA transcription, RNA translation, protein biosynthesis, the modulation of kinase activity, the activity of ion channels, apoptosis, and in adjusting the immune response.[36–38] They play a role in the growth, maturation, and regeneration of intestinal cells, and possess potent antioxidant activity at physiological concentrations, preventing the damage of cell membranes and DNA.[36] However, spermine and spermidine were reported to be harmful to health. In laboratory animals, the administration of spermine and spermidine resulted in an acute decrease in blood pressure, respiratory symptoms, and nephrotoxicity.[36] They are also involved in carcinogenesis, tumour invasion, and metastasis.[37] The catabolic enzyme spermine/spermidine N^1-acetyltransferase, which reduces cell spermine and spermidine contents, also seems to reduce cell growth, migration, and invasion in hepatocellular carcinoma and colorectal cancer.[39] Although high levels of spermine and spermidine are associated with these potential health risks, legal limits of spermine and spermidine in foods have not been established yet.

del Rio *et al.*[40] reported that spermine or spermidine in food is not harmful to healthy people. However, patients with cancer are not advised to consume spermine/spermidine-rich foods.[38] Patients with chronic kidney failure show high plasma polyamine oxidase activity, which causes an increase in spermine and spermidine catabolism and the accumulation of toxic acrolein. In such patients, a high polyamine (PA) might be harmful.[41] In addition, under acidic conditions, spermidine reacts with nitrous acid preservatives in food, forming the nitrosamines *N*-nitrosopiperidine and *N*-nitrosodimethylamine, which are carcinogens.[42]

Cadaverine, putrescine, spermine, and spermidine have been classified within the group of polyamines. All of them are involved in DNA, RNA, and protein synthesis and they are necessary for normal growth, renovation, and metabolism of every organ in the body, as well as maintaining the metabolic activity of normal functions and the immunological system of the gut. Since polyamines are preferred by tissues with high demands, they could be beneficial for post-operation patients or wound healing, and the development of a neonate's digestive system.[43] Polyamines have also been demonstrated to inhibit the oxidation of polyunsaturated fatty acids, and the antioxidative effect is related to the number of amine groups in the polyamine.[11]

Biogenic amines also play a role as neurotransmitters in physiological functions. They are synthesized in the nervous system and are divided into catecholamines, indoleamines, and imidazoleamines. It was indicated that psychoactive amines (dopamine and serotonin) are neurotransmitters in the central nervous system. Biogenic amines located in the central nervous system are crucial to life. A lack of these neurotransmitters can cause mental disorders, such as infantile autism, sleep disorders, depression, and Parkinson's disease.[44–46] A better understanding of the conformational binding of those bio-active compounds in confined biological environments is essential in the treatment of those diseases.[47] Determination of specific amine roles is difficult due to their complicated interactions and the complexity of the nervous system. Thus, many amine functions have not been completely explained.[48] Biogenic amines, such as tyramine and β-phenylethylamine, have been proposed as the initiators of a hypertensive crisis in certain patients and of dietary-induced migraines.[49] Tyramine and histamine also act as hormonal mediators in humans and animals.

1.3 Toxicity of Biogenic Amines

1.3.1 Histamine (Scombroid) Poisoning

Histamine poisoning is a foodborne chemical intoxication caused by the ingestion of foods containing high amounts of histamine. Histamine fish poisoning, also known as scombroid poisoning, is the most common cause of ichythyotoxicosis worldwide. This disease was first described in 1799 in Britain and re-emerged in the medical literature in the 1950s when outbreaks were recorded in Japan.[47] Although commonly associated with the consumption of scombroid-type fish, other foods such as cheese have also been associated with outbreaks of histamine poisoning. The term "scombroid" derives from a type of fish (*i.e.* Scombridae) such as tuna and mackerel. Scombrid fish are characterized by high levels of free histidine in their muscle tissues.[50] Other (non-scombroid) fish species are also reported in scombroid poisoning, including anchovies, yellowtail steak, mahi-mahi, tuna, herring, and sardine bluefish, and fish products such as sauce and dried fish.[51–55] These species have relatively high levels of histidine in their flesh.[8,54] Histidine levels range from 1 g kg^{-1} in herring to 15 g kg^{-1} in tuna.[56] Fresh fish contains negligible quantities of histamine, usually $<$0.1 mg per 100 g.[57] Scombroid poisoning results from consumption of mishandled fish. Histamine (2-(1*H*-imidazol-4-yl) ethanamine) and other decomposition products are produced in raw fish by bacterial, enzymatic conversion of free histidine. Figure 1.2 shows a schematic summary of histamine poisoning in food.

The metabolism of histidine basically follows two ways: the major route of catabolism of histidine passes through its transformation to glutamic acid, which begins with the degradation of histidine to urocanic acid by action of the enzyme histidase. The glutamate product is converted to α-ketoglutarate, which is an intermediate in the citric acid cycle (Krebs cycle). The second is the decarboxylation (loss COO–) for action of the enzyme histidine decarboxylase with formation of histamine.[58] The process of decarboxylation is initiated mainly by enzymes produced by gram-negative enteric bacteria such as *Morganella morganii*, *Escherichia coli*, *Klebsiella* spp., and *Pseudomonas aeruginosa* which are found in the intestine and skin of the fish.[59] These bacteria are present in the marine environment and also in the intestine and gills of the fish, causing no disease. Bacteria capable of amine production include *Escherichia*, *Enterobacter*, *Salmonella*, *Shigella*, *Streptococcus*, *Lactobacillus*, *Leuconostoc* spp., and *Clostridium perfringens*.[60] To prevent poisoning, the fish should be iced during the food chain until consumption. If the preservation techniques are not good, the degradation of histidine to histamine can occur due to bacteria growth. Storage temperature is the most important factor controlling BAs formation.[61] Other parameters such as pH, water activity, NaCl concentration, and additives may affect microbiota composition and result in differences in BAs content.[62]

Fermented foods such as wine, dry sausage, sauerkraut, miso, and soy sauce can also contain histamine along with other biogenic amines.

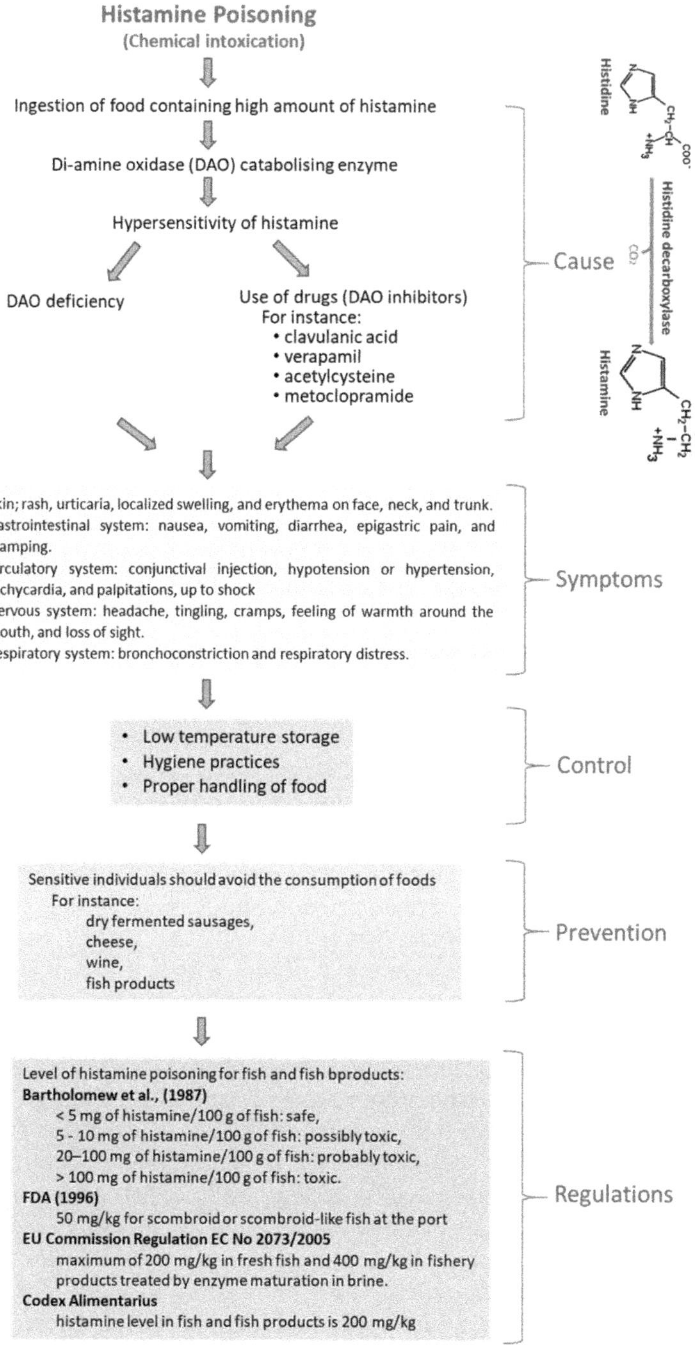

Figure 1.2 The schematic summary of histamine poisoning in food.

Microorganisms with histidine decarboxylase enzyme, which converts histidine to histamine, are responsible for the formation of histamine in foods. The toxicity of histamine is enhanced by the presence of other biogenic amines (putrescine and cadaverine) in foods, inhibiting histamine-metabolizing enzymes in the intestine.[49] DAO is the major histamine catabolizing enzyme in the intestinal tract. The hypersensitivity to histamine could occur due to DAO deficiency or the use of drugs which are DAO inhibitors (DAOI), such as clavulanic acid, verapamil, acetylcysteine, and metoclopramide.[63] The consumption of products with histamine could cause adverse effects in a person treated with DAOI. Therefore, sensitive individuals should avoid the consumption of foods such as dry fermented sausages, cheese, wine, and fish products that are also potential sources of biogenic amines.[64]

Estimating occurrence of histamine poisoning is hard because most countries do not regulate histamine levels in foods and they do not require warning when an incident of histamine poisoning takes place. In addition to this, histamine poisoning induces allergy-like symptoms which closely resembles a food allergy.[65] Thus, it may often be misdiagnosed.

Ohnuma *et al.*[66] categorized the symptoms as follows:

- *Skin:* Rash, urticaria, localized swelling, and erythema on the face, neck, and trunk.
- *Gastrointestinal system:* Nausea, vomiting, diarrhoea, epigastric pain, and cramping.
- *Circulatory system:* Conjunctival injection, hypotension or hypertension, tachycardia, and palpitations, up to shock.
- *Nervous system:* Headache, tingling, cramps, feeling of warmth around the mouth, and loss of sight.
- *Respiratory system:* Bronchoconstriction and respiratory distress.

The onset of the symptoms generally occurs within a few minutes after ingestion of the implicated food, and the duration of symptoms ranges from a few hours to 24 h. More severe symptoms (*e.g.* respiratory distress, swelling of the tongue and throat, and blurred vision) can occur and require medical treatment with antihistamines. The United States Federal Drug Administration (FDA) and the European Union (EU) suggest that safe levels of histamine in fish meat should not exceed 50 ppm and 100 ppm, respectively.[67,68] The New Zealand cut-off is 200 mg kg^{-1} in fish or fish products, which is the same as in South Korea. When histamine levels exceed these standards in fish or fish products before shipment or distribution, they are discarded. The only effective method for prevention of scombroid fish poisoning is consistent temperature control of fish at 4.4 °C between catching and consumption.[69] The bacterial spoilage and the production of histamine can occur at any stage of the food chain such as fishing and fish landing, processing, distribution systems, as caterers, or at home. The rapid cooling of the fish and a maintenance of the cold chain until the consumption of the product are very important to avoid bacterial proliferation, activation of the enzyme histidine decarboxylase, and the conversion of histidine to histamine. In future, scombroid outbreaks can be prevented if fishermen, public health officials, workers in restaurant, and medical professionals work together to set international safety standards and increase awareness of this poisoning.

After fish, fermented meats, vegetables, dairy products, and alcoholic beverages may accumulate high levels of histamine.[6,70] In some types of cheese, histamine can even exceed 1000 mg kg^{-1}, causing histamine poisoning.[71] In spite of the cold chain, cheeses with high concentrations of histamine are on the market.[72,73] Although a number of microorganisms in cheese, including moulds, yeasts, and Gram-negative bacteria, are able to produce histamine,[74] lactic acid bacteria (LAB), which produce histidine decarboxylase enzyme, are thought to be mainly responsible for histamine accumulation.[71,75] LAB may form the milk microbiota, or can be introduced by contamination before, during, or after processing, or may even be part of the starter culture. Diaz *et al.*[76] reported that *Lactobacillus parabuchneri*, responsible for histamine accumulation in cheese, can grow and produce histamine at 4 and 8 °C. Moreover, *L. parabuchneri* produced toxic histamine concentrations in refrigerated cheeses from only 14 days. Thus, refrigeration delays but does not prevent the accumulation of toxic histamine levels in cheese.

1.3.2 Tyramine Toxicity

Fermented dairy products, especially cheese, are the main sources of biogenic amines, mainly histamine, tyramine, putrescine, and cadaverine. Many factors have been reported to affect the production of biogenic amines in cheeses, including the presence of microorganisms decarboxylating free amino acids, the presence of spoilage microorganisms, and the synergistic effects of microorganisms and free amino acids as an outcome of proteolysis levels.[77,78] In addition, pH, salt-in-moisture levels, and ripening temperature also play an important role.[79,80] Since the content of BAs is influenced by ripening and quality of storage conditions, attention in the amount of BAs in food is always being paid, not only due to the potential toxicity but also being a parameter for ripening and the quality of storage. TYR is produced from amino acid tyrosine by tyrosine decarboxylase enzyme and metabolized by monoamine oxidase. It is also present in many foods and beverages such as meat, fish and fish products, alcoholic beverages, and dairy products.[81] TYR is the major biogenic amine in cheese and is responsible for the "cheese effect" which is characterized by hypertension, headache, and migraine. It was also reported that TYR can cause necrosis of human intestinal cells and this cytotoxicity was even stronger than that of histamine.[6] Normally, ingested TYR does not have any adverse effects on human health since the intestinal wall and liver include monoamine oxidase, an enzyme metabolizing the TYR to less active *p*-hydroxylphenylacetic acid before being involved in the general circulation. On the other hand, when ingested in high amounts in food or in patients treated with monoamine oxidase inhibitor drugs, TYR cannot be detoxified adequately and passes into the circulation, resulting in the release of noradrenaline from the nervous system.[82] TYR reacts with monoamine oxidase inhibitor (MAOI) drugs causing a hypertensive crisis.[2] The enzyme, monoamine oxidase, oxidatively deaminates amines derived from food and plays a role in their degradation before they reach the blood. The use of MAOI drugs for treatment of mental depression eliminates this detoxification process and a high concentration of tyramine can be accumulated in the blood resulting in a hypertensive crisis for the patient.[83] Halasz *et al.*[84] reported that high intake of TYR can cause adverse toxicological effects and diseases such as high blood pressure, brain haemorrhage, palpitations, nausea, and diarrhoea.

1.4 Regulations

In general, histamine and tyramine are thought to be the main cause of numerous cases of food intoxication.[85] Putrescine and cadaverine have been reported as potentiators of toxic effect of other amines due to the inhibition of detoxifying enzymes.[2,86] People with a lack of natural mechanisms for detoxifying BAs caused by genetic reasons and taking antidepressant medicines show that they are more sensitive to BAs poisoning.[28] The toxicological level of BAs is very difficult to set as it depends on individual characteristics and the presence of other amines. However, a maximum total BAs level of 750–900 mg kg^{-1} has been proposed by Ladero *et al.*[87]

Histamine poisoning is concerned with public health and safety issues, and also the global fish trade.[88] Processing methods such as cooking, canning, or freezing cannot eliminate histamine because it is heat stable.[89] Therefore, it is important to accurately determine its level in fish and fish products.

According to Bartholomew *et al.*,[90] levels of histamine poisoning are as follows:

1. <5 mg of histamine per 100 g of fish: safe,
2. 5–10 mg of histamine per 100 g of fish: possibly toxic,
3. 20–100 mg of histamine per 100 g of fish: probably toxic,
4. >100 mg of histamine per 100 g of fish: toxic.

The FDA in 1996 executed the Hazard Analysis and Critical Control Point (HACCP) program to prevent seafood processing hazards which cause foodborne illness. The level of 50 mg kg^{-1} was established for scombroid or scombroid-like fish at the port. According to Commission Regulation EC No 2073/2005 (Regulation 2073/2005/EC),[91] the limit for histamine was established in fish species with a high amount of histidine, including families of Scombridae, Scombresosidae, Clupeidae, Engraulidae,

Coryfenidae, and Pomatomidae, both fresh and treated by enzyme maturation in brine. Regulatory limits set by European legislation are a maximum of 200 mg kg^{-1} in fresh fish and 400 mg kg^{-1} in fishery products treated by enzyme maturation in brine. According to Codex Alimentarius, the histamine level in fish and fish products is 200 mg kg^{-1}. Brazil and Mercosur limit histamine to 100 mg kg^{-1} [89,92,93] whereas European regulation limits histamine content in fish and fish products to 100 and 200 mg kg^{-1}.[94]

The main aim of the guideline established by the FDA[89] is to inhibit the growth of spoilage bacteria which produce histamine during the handling and chilling of fish. There are a number of factors for time required to decrease the internal temperature of fish after capture, including the harvest and chilling methods, as well as fish size. The scombrotoxin-forming fish should be stored as close as possible to the freezing point until it is consumed.

Tao *et al.*[54] determined histamine content in fresh scombroid fish (tuna, mackerel *etc.*) and non-scombroid fish (mahi–mahi, sardine, herring *etc.*), and fish products (sauce, dried fish) from Fiji, Germany, the Netherlands, Norway, Thailand, Cambodia, the Philippines, Japan, and China. Low level of histamine (12–31 ppm) was found in Fiji's fish samples. Histamine was determined in 2 of 12 samples from Germany (184 ppm in sardine and 68 ppm in herring). These concentrations were higher than the FDA regulation of 50 ppm histamine. Histamine was found in two of nine samples from the Netherlands (39 ppm in horse mackerel and 1439 ppm in tuna). Histamine level in the tuna (1439 ppm) exceeded the EC regulation of histamine (100 ppm–200 ppm) from the Netherlands although the tuna meat appeared to very fresh at that time. Histamine was found in 4 of 11 fish samples from Thailand in the range of 56 ppm and 1964 ppm. Histamine was determined in the three dried freshwater fish samples from Cambodia in the range of 25–148 ppm. Histamine was detected in 60% of the dried bonito samples from the Philippines and its level was in the range of 19 ppm to 1530 ppm. Histamine and biogenic amines were not detected in the samples from Japan. Histamine of 15 ppm was found in one sample from China. Putrescine (13–25 ppm) was detected in two samples (horse mackerel and saury), also cadaverine (6 ppm) was detected in one sample of horse mackerel. There is a need for monitoring of histamine to ensure the safety of commercial fish products in the world's fish markets.

Evaluation of alimentary exposure of consumers to individual biogenic amines and also maximum tolerable levels of these biogenic amines in different foods were carried out. For instance, Poulsen *et al.*[95] estimated maximum tolerable levels of tyramine in foods in Austria. They reported that for individuals with no susceptibility to tyramine, it is reasonable to suggest a NOAEL (no observed adverse effect level) of 200 mg tyramine per meal. Maximum tolerable tyramine levels in cheese, fermented sausage, fish/fishery products, and sauerkraut could be 1000 mg kg^{-1}, 2000 mg kg^{-1},

950 mg kg^{-1}, and 800 mg kg^{-1}, respectively. It was reported that normal consumption of tyramine is 100–800 mg kg^{-1} whereas levels higher than 1080 mg kg^{-1} are regarded as toxic. However, individuals under MAOI treatment are more sensitive to tyramine consumption and values of 60 mg kg^{-1} in the diet could cause a mild crisis, while 100–250 mg kg^{-1} is associated with severe headache with intracranial haemorrhage and sequelae.[96]

Due to the rapid metabolism of tryptamine, acute effects after oral intake are the main concern. Tryptamine contents in foods are in the range of a few mg kg^{-1} or lower.[97] The highest reported tryptamine value in the EFSA Scientific Opinion was 362 mg kg^{-1} in fish and fish products and 10.1 mg kg^{-1} in fermented fish meat.[98] The highest reported tryptamine level in cheese in the EFSA Scientific Opinion, which summarized data of over 2000 cheese samples from Europe, and in an Austrian semi-hard cheese, was 312 mg kg^{-1} However, 4% of the total EFSA's cheese samples gave a tryptamine level above the limit of detection (LOD).[98,99] In a study by Wüst *et al.*,[97] dietary exposure through fish/seafood, beer, cheese, and meat products was estimated for Austrian consumers, based on 543 food samples. For fresh/cooked fish, preserved fish, cheese, raw sausage, condiments, sauerkraut, and fermented tofu, maximum tolerable levels were found as 1650, 3200, 2840, 4800, 14 120, 1740, and 2400 mg kg^{-1}, respectively. For raw sausages and beer, tryptamine contents were below the proposed maximum tolerable tryptamine level. Although they did not take into account combined effects of ingested biogenic amines and increased susceptibility to tryptamine, *e.g.* due to reduced MAO activity, they concluded that dietary intake of tryptamine should not result in adverse effects on healthy individuals.

The physicochemical control of six Italian-type salami brands sold in the city of Niteroi was evaluated by dos Santos *et al.*[100] (Rio de Janeiro, Brazil). The quality control of salami should be essential, particularly moisture and protein contents since only half of the brands met the requirements of Brazilian law. Two brands presented histamine contents above the legal limit (100 mg kg^{-1}). They concluded that the levels of biogenic amines found would cause an adverse effect in sensitive consumers, which depends on the amount and frequency of the intake of these products. Rauscher-Gabernig *et al.*[101] suggested tolerable levels for putrescine and cadaverine in cheese, fermented sausages, fish, sauerkraut, and seasonings based on toxicological information and consumption habits in Austria. For putrescine, proposed maximum tolerable levels for sauerkraut, fish, cheese, fermented sausages, and seasonings are 140, 170, 180, 360, and 510 mg kg^{-1}, respectively, whereas for cadaverine, maximum tolerable levels are 430, 510, 540, 1080, and 1540 mg kg^{-1}, respectively in sauerkraut, fish, cheese, fermented sausages, and seasonings. In addition, studies showed that tyramine and histamine have synergistic cytotoxicity towards intestinal cell cultures. It was also indicated that histamine, below the legal limit, can raise the cytotoxicity of tyramine at a concentration frequently present in foods.[6,40]

References

1. R. Bermúdez, J. M. Lorenzo, S. Fonseca, I. Franco and J. Carballo, *Front. Microbiol.*, 2012, **3**, 1–6.
2. A. R. Shalaby, *Food Res. Int.*, 1996, **29**, 675–690.
3. F. Özogul and Y. Özogul, *Eur. Food Res. Technol.*, 2007, **225**, 385–394.
4. C. Ruiz-Capillas and F. Jiménez-Colmenero, *Crit. Rev. Food Sci. Nutr.*, 2004, **44**, 489–499.
5. D. M. Linares, M. C. Martín, V. Ladero, M. A. Álvarez and M. Fernández, *Crit. Rev. Food Sci. Nutr.*, 2011, **51**, 691–703.
6. D. M. Linares, B. del Rio, B. Redruello, V. Ladero, M. C. Martin, M. Fernandez and M. A. Alvarez, *Food Chem.*, 2016, **197**, 658–663.
7. G. I. Mohammed, A. S. Bashammakh, A. A. Alsibaai, H. Alwael and M. S. El-Shahawi, *TrAC, Trends Anal. Chem.*, 2016, **78**, 84–94.
8. S. L. Taylor, *Crit. Rev. Toxicol.*, 1986, **17**(2), 91–128.
9. H. S. Woo, C. W. Na, I. D. Kim and J. H. Lee, *Nanotechnology*, 2012, **23**, 245501.
10. P. Sebastian, P. Herr and U. Fischer, *S. Afr. J. Enol. Vitic.*, 2011, **32**, 300–309.
11. W. Lovaas, *J. Am. Oil Chem. Soc.*, 1991, **68**, 357.
12. G. Spano, P. Russo, A. Lonvaud-Funel, P. Lucas, H. Alexandre, C. Grandvalet, E. Coton, M. Coton, L. Barnavon, B. Bach, F. Rattray, A. Bunte, C. Magni, V. Ladero, M. Alvarez, M. Fernández, P. Lopez, P. F. de Palencia, A. Corbi, H. Trip and J. S. Lolkema, *Eur. J. Clin. Nutr.*, 2010, **64**, 95–100.
13. M. H. S. Santos, *Int. J. Food Microbiol.*, 1996, **29**, 213–231.
14. S. Bardocz, *Trends Food Sci. Technol.*, 1995, **6**, 341–346.
15. B. ten Brink, C. Damink, H. M. L. J. Joosten and J. H. J. Huis in't Veld, *Int. J. Food Microbiol.*, 1990, **11**, 73–84.
16. R. R. Eitenmiller and S. C. De Souza, in *Seafood Toxins*, ed. E. P. Ragelis, American Chemical Society, Washington, DC, 1984, ch. 36, pp. 431–442.
17. J. E. Stratton, R. W. Hutkins and S. L. Taylor, *J. Food Prot.*, 1991, **54**, 460–470.
18. A. Bouchereau, P. Guénot and F. Larher, *J. Chromatogr. B: Biomed. Sci. Appl.*, 2000, **747**, 49–67.
19. S. C. Jansen, M. van Dusseldorp, K. C. Bottema and A. E. J. Dubois, *Ann. Allergy, Asthma, Immunol.*, 2003, **91**, 233–241.
20. S. Lu, H. Ji, Q. Wang, B. Li, K. Li, C. Xu and C. Jiang, *Food Control*, 2015, **50**, 869–875.
21. M. T. Veciana-Nogues, A. Marine-Font and M. C. Vidal-Carou, *J. Agric. Food Chem.*, 1997, **45**, 2036–2041.
22. J. L. Mietz and E. Karmas, *J. Food Sci.*, 1997, **42**, 155–158.
23. B. Y. Byun, X. Bai and J. H. Mah, *Food Sci. Biotechnol.*, 2013, **22**, 55–62.
24. Y. Ozogul, M. Durmus, E. K. Boga, Y. Ucar and F. Ozogul, *J. Food Sci.*, 2018, **83**, 318–325.

25. L. Li, X. X. Wen, Z. Y. Wen, S. W. Chen, L. Wang and X. T. Wei, *Front. Microbiol.*, 2018, **9**, 1–9.
26. C. C. Balamatsia, E. K. Paleologos, M. G. Kontominas and I. N. Savvaidis, *Antonie van Leeuwenhoek*, 2006, **89**, 9–17.
27. W. Wojnowski, J. Płotka-Wasylka, K. Kalinowska, T. Majchrzak, T. Dymerski, P. Szweda and J. Namieśnik, *Monatsh. Chem.*, 2018, **149**, 1521–1525.
28. L. Prester, *Food Addit. Contam., Part A*, 2011, **28**(11), 1547–1560.
29. A. Scavnicar, I. Rogelj, D. Kocar, S. Kose and M. Pompe, *J. AOAC Int.*, 2018, **101**, 1542–1547.
30. H. K. Mayer and G. Fiechter, *Food Control*, 2018, **93**, 9–16.
31. M. B. H. Youdim, D. Edmondson and K. F. Tipton, *Nat. Rev. Neurosci.*, 2006, 7, 295–309.
32. A. Pircher, F. Bauer and P. Paulsen, *Eur. Food Res. Technol.*, 2007, **226**, 225–231.
33. B. Bach, S. Le Quere, P. Vuchot, M. Grinbaum and L. Barnavon, *Anal. Chim. Acta*, 2012, **732**, 114–119.
34. C. Almeida, J. O. Fernandes and S. C. Cunha, *Food Control*, 2012, **25**, 380–388.
35. K. Igarashi and K. Kashiwagi, *Biochem. Biophys. Res. Commun.*, 2000, **271**(3), 559–564.
36. A. E. Pegg, *Chem. Res. Toxicol.*, 2013, **26**(12), 1782–1800.
37. D. Ramani, J. P. De Bandt and L. Cynober, *Clin. Nutr.*, 2014, **33**(1), 14–22.
38. P. Kalac, *Food Chem.*, 2014, **161**, 27–39.
39. Y. Wang, J. Zhang, M. Li, M. Li, S. Xie and C. Wang, *Chem. Biol. Drug Des.*, 2017, **89**, 670–680.
40. B. del Rio, B. Redruello, D. M. Linares, V. Ladero, P. Ruas-Madiedo, M. Fernandez, M. C. Martin and M. A. Alvarez, *Food Chem.*, 2018, **269**, 321–326.
41. K. Igarashi, S. Ueda, K. Yoshida and K. Kashiwagi, *Amino Acids*, 2006, **31**, 477–483.
42. G. Drabik-Markiewicz, B. Dejaegher, E. De Mey, T. Kowalska, H. Paelinck and Y. Vander Heyden, *Food Chem.*, 2011, **126**, 1539–1545.
43. P. Kalac and P. Krausova, *Food Chem.*, 2005, **90**, 219–230.
44. S. Ramboz, R. Oosting and D. A. Amara, *Proc. Natl. Acad. Sci. U. S. A.*, 1998, **95**, 14476–14481.
45. B. Kobilka, *Angew. Chem. Int. Ed.*, 2013, **52**, 6380–6388.
46. G. Arena, A. Pappalardo and S. Pappalardo, *J. Therm. Anal. Calorim.*, 2015, **121**, 1073–1079.
47. C. Feng, S. Teuber and M. E. Gershwin, *Clin. Rev. Allergy Immunol.*, 2016, **50**(1), 64–69.
48. K. Dora, Bachelor's Thesis, *Functions of biogenic amines as neurotransmitters in the central nervous system*, Faculty of Science, Department of Biology, University of Zagreb, 2016.
49. E. S. Jayne, W. H. Rovert and L. T. Steve, *J. Food Prot.*, 1991, **54**, 460–470.

50. C. Ruiz-Capillas and A. Moral, *Amino Acids*, 2004, **26**, 125–132.
51. A. Alfonzo, R. Gaglio, N. Francesca, M. Barbera, F. Saiano, A. Santulli, M. Matraxia, F. Rallo and G. Moschetti, *Food Control*, 2018, **92**, 301–311.
52. C. R. Kang, Y. Y. Kim, J. I. Lee, H. D. Joo, S. W. Jung and S. I. Cho, *J. Korean Med. Sci.*, 2018, **33**(38), 1–6.
53. T. Dole, S. Koltun, S. M. Baker, R. M. Goodrich-Schneider, M. R. Marshall and P. J. Sarnoski, *J. Aquat. Food Prod. Technol.*, 2016, **26**(7), 781–789.
54. Z. Tao, M. Sato, H. Zhang, T. Yamaguchi and T. Nakano, *Food Control*, 2011, **22**, 430–432.
55. V. Simat and P. Dalgaard, *LWT--Food Sci. Technol.*, 2011, **44**(2), 399–406.
56. P. Ijomah, M. N. Clifford, R. Walker, J. Wright, R. Hardy and C. K. Murray, Further volunteer studies on scombrotoxicosis, in *Pelagic Fish: The Resource and its Exploitation*, ed. J. R. Burt, R. Hardy and K. J. Whittle, Fishing News Books, Oxford, 1992, pp. 194–199.
57. H. A. Frank, D. H. Yoshinaga and W. K. Nip, *Mar. Fish. Rev.*, 1981, **43**(10), 9–14.
58. V. Tortorella, P. Masciari, M. Pezzi, A. Mola, S. P. Tiburzi, M. C. Zinzi, A. Scozzafava and M. Verre, *Case Rep. Emerg. Med.*, 2014, **2014**, 482531.
59. S. L. Taylor, J. E. Stratton and J. A. Nordlee, *J. Toxicol., Clin. Toxicol.*, 1989, **27**, 225–240.
60. O. Pinho, A. I. E. Pintado, A. M. P. Gomes, M. M. E. Pintado, F. X. Malcata and I. M. P. L. V. O. Ferreira, *J. Food Prot.*, 2004, **67**(12), 2779–2785.
61. C. Y. Chong, F. Abu Bakar, A. R. Russly, B. Jamilah and N. A. Mahyudin, *Int. Food Res. J.*, 2011, **18**, 867–876.
62. G. Suzzi and F. Gardini, *Int. J. Food Microbiol.*, 2003, **88**, 41–54.
63. J. Sattler, D. Hafner, H. J. Klotter, W. Lorenz and P. K. Wagner, *Agents Actions*, 1988, **23**, 361–365.
64. M. L. Latorre-Moratalla, T. Veciana-Nogués, S. Bover-Cida, M. Garriga, T. Aymerich, E. Zanardi, A. Ianieri, J. Fraqueza, L. Patarata, E. H. Drosinos, A. Lauková, R. Talon and M. C. Vidal-Carou, *Food Chem.*, 2008, **107**, 912–921.
65. J. M. Hungerford, *Toxicon*, 2010, **56**, 231–243.
66. S. Ohnuma, M. Higa, S. Hamanaka, K. Matsushima and W. Yamamuro, *Int. Med.*, 2001, **40**, 833–835.
67. FDA, Scombrotoxin (histamine) formation, in *Fish and Fishery Products Hazards and Controls Guidance*, Food and Drug Administration, Center for Food Safety and Applied Nutrition, Rockville, MD, 4th edn, 2001, pp. 113–152.
68. EFSA Panel on Biological Hazards (BIOHAZ), Scientific opinion on risk based control of biogenic amine formation in fermented foods, *EFSA J.*, 2011, **9**, 2301–2393.

69. Centers for Disease Control and Prevention (CDC), Scombroid fish poisoning associated with tuna steaks–Louisiana and Tennessee, *Morb. Mortal Wkly. Rep.*, 2007, **56**(32), 817.
70. M. A. Alvarez and M. V. Moreno-Arribas, *Trends Food Sci. Technol.*, 2014, **39**, 146–155.
71. M. Fernández, D. M. Linares, B. del Rio, V. Ladero and M. A. Alvarez, *J. Dairy Res.*, 2007, **74**, 276–282.
72. M. Schirone, R. Tofalo, P. Visciano, A. Corsetti and G. Suzzi, *Front. Microbiol.*, 2012, **3**, 171.
73. J. Liu, M. Y. Su, Z. Y. Xu, C. P. You and Z. M. Liu, *Int. Dairy J.*, 2017, **85**, 263–269.
74. D. M. Linares, B. del Rio, V. Ladero, N. Martinez, M. Fernandez, M. C. Martin and M. A. Alvarez, *Front. Microbiol.*, 2012, **3**, 1–10.
75. M. Diaz, B. del Rio, V. Ladero, B. Redruello, M. Fernandez, M. C. Martin and M. A. Alvarez, *Int. J. Food Microbiol.*, 2015, **215**, 117–123.
76. M. Diaz, B. del Rio, E. Sanchez-Llana, V. Ladero, B. Redruello, M. Fernández, M. C. Martin and M. A. Alvarez, *Int. J. Food Sci. Technol.*, 2018, **53**, 2342–2348.
77. M. Martuscelli, F. Gardini, S. Torriani, D. Mastrocola, A. Serio, C. Chaves-Lopez, M. Schirone and G. G. Suzzi, *Int. Dairy J.*, 2005, **15**, 571–578.
78. M. Bernardeau, J. P. Vernoux, S. Henri-Dubernet and M. Guéguen, *Int. J. Food Microbiol.*, 2008, **126**, 278–285.
79. F. Gardini, M. Martuscelli, M. C. Caruso, F. Galgano, M. A. Crudele, F. Favati, M. E. Guerzoni and G. Suzzi, *Int. J. Food Microbiol.*, 2001, **64**, 105–117.
80. M. C. Gennaro, V. Gianotti, E. Marengo, D. Pattono and R. M. Turi, *Food Chem.*, 2003, **82**, 545–551.
81. H. D. Duong and J. I. Rhee, *Talanta*, 2007, **72**, 1275–1282.
82. T. A. Smith, *Food Chem.*, 1981, **6**, 169–200.
83. C. K. Smith and D. T. Durack, *Ann. Intern. Med.*, 1978, **88**(4), 520–521.
84. A. Halasz, A. Barath, L. Simon-Sarkadi and W. Holzapfel, *Trends Food Sci. Technol.*, 1994, **5**, 42–49.
85. A. I. Ordonez, F. C. Ibanez, P. Torre and Y. Barcina, *J. Food Prot.*, 1997, **60**, 1371–1375.
86. K. Valsamaki, A. Michaelidou and A. Polychroniadou, *Food Chem.*, 2000, **71**, 259–266.
87. V. Ladero, M. Fernandez, I. Cuesta and M. A. Alvarez, *Food Microbiol.*, 2010, **27**, 933–939.
88. W. P. Evangelista, T. M. Silva, L. R. Guidi, P. A. S. Tette, R. M. D. Byrro, P. Santiago-Silva, C. Fernandes and M. B. A. Gloria, *Food Chem.*, 2016, **211**, 100–106.
89. FDA, *Fish and fisheries products hazards and controls guide*, Office of Seafood, Food and Drug Administration, Washington, DC, 4th edn, 2011.

90. B. A. Bartholomew, P. R. Berry, J. C. Rodhouse, R. J. Gilbert and C. K. Murray, *Epidemiol. Infect.*, 1987, **99**, 775–782.
91. Commission Regulation EC No 2073/2005, Microbiological criteria for foodstuffs, OJ. L 338, pp. 1–26.
92. Brasil, *Manual de garantia da qualidade analítica*, Secretaria de Defesa Agropecuária, Ministério da Agricultura, Brasília, DF, Brasil, 1 ed, 2011.
93. FAO, Codex Alimentarius Commission, Joint FAO/WHO Food Standards programme, Codex Committee on Fish and Fishery Products, 32 session discussion paper histamine, 2012, pp. 1–14.
94. J. Stroka, K. Bouten, C. Mischke, A. Breidbach and F. Ulberth, Joint Research Centre Institute for Reference Materials and Measurements, Report EUR 26605 EN. Equivalence testing of histamine methods – Final report, Administrative Arrangement N SANCO/2011/G4/JRC32515/SI2.611754 between DG Health and Consumers (DG SANCO) and Joint Research Centre (JRC), European Union, 2014.
95. P. Paulsen, R. Grossgut, F. Bauer and E. Rauscher Gabernig, *J Food Nutr. Res.*, 2012, **51**, 52–59.
96. FAO/WHO, *Public health risks of histamine and other biogenic amines from fish and fishery products*, Food and Agriculture Organization, Rome, Italy, 2013.
97. N. Wüst, E. Rauscher-Gabernig, J. Steinwider, F. Bauer and P. Paulsen, *Food Addit. Contam., Part A*, 2017, **34**, 404–420.
98. EFSA European Food Safety Authority, Scientific opinion on risk based control of biogenic amine formation in fermented foods, *EFSA J.*, 2011, **9**, 2393–2486.
99. H. K. Mayer, G. Fiechter and E. Fischer, *J. Chromatogr. A*, 2010, **1217**, 3251–3257.
100. L. F. L. dos Santos, E. T. Mársico, C. A. Lázaro, R. Teixeira, L. Doro and C. A. C. Júnior, *Ital. J. Food Saf.*, 2015, **4**(3), 4048.
101. E. Rauscher-Gabernig, R. Gabernig, W. Brueller, R. Grossgut, F. Bauer and P. Paulse, *Eur. Food Res. Technol.*, 2012, **235**, 209–220.

CHAPTER 2

Microorganisms Able to Produce Biogenic Amines and Factors Affecting Their Activity

PASQUALE RUSSO,[a] MARIAGIOVANNA FRAGASSO,[a] CARMEN BERBEGAL,[a,b] FRANCESCO GRIECO,[c] GIUSEPPE SPANO[a] AND VITTORIO CAPOZZI*[a]

[a] Department of the Sciences of Agriculture, Food and Environment, University of Foggia, Via Napoli 25, 71122 Foggia, Italy; [b] Enolab. Estructura de Recerca Interdisciplinar en Biotecnología i Biomedicina (ERIBioTecMed), Universitat de València, c/Dr. Moliner 50, E46100, Burjassot-València, Spain; [c] Istituto di Scienze delle Produzioni Alimentari, Consiglio Nazionale delle Ricerche, Unità di Lecce, Via Prov. Lecce-Monteroni, 73100 Lecce, Italy
*Email: vittorio.capozzi@gmail.com, vittorio.capozzi@unifg.it

2.1 Introduction

Microbial diversity is able to influence the different facets of the quality of every food matrices: hygienic quality, sensory quality, nutritional quality and functional quality.[1,2] In fact, among the other effects, microorganisms in food can: (i) improve vitamin content;[3] (ii) modulate the content of volatile organic compounds;[4] (iii) positively influence the fermentation process;[5] (iv) exert biocontrol activity;[6] (v) inhibit the negative effects of toxic compounds;[7] (vi) introduce probiotic properties;[8] (vii) influence the risks for human health;[9] and (viii) help to face the negative effects of climate

Food Chemistry, Function and Analysis No. 20
Biogenic Amines in Food: Analysis, Occurrence and Toxicity
Edited by Bahruddin Saad and Rosanna Tofalo

Published by the Royal Society of Chemistry, www.rsc.org

change on food quality.[10,11] A current challenge is the management of the microbial diversity to preserve food attributes, reducing adverse effects and enhancing desired characteristics.[2,12–16] In the framework of this large picture, the understanding of 'micro-biodiversity' in relation to the production of biogenic amines (BAs) in food and beverages deserves high interest because of the polyphasic influence of this chemical class on the final features of associated foodstuffs. With concern about food quality, BAs are organic and basic nitrogenous compounds of low molecular weight, mainly obtained via decarboxylation of amino acids and they are responsible for sensorial changes and negatively impact food safety.[17,18] Moreover, they are an index for freshness evaluation in several food chains, especially in the fish and meat industries.[19,20] The toxicological effects of dietary BAs are deeply connected with their physiological role in humans, such as neurotransmitters, hormones and modulators of numerous physiological processes. As a consequence, the ingestion of foods/beverages containing high concentrations of BAs can lead to adverse reactions such as nausea, headaches, rashes and changes in blood pressure.[21] This chapter supplies an overview of the diversity of BA-producer microbes, underlining the presence of microorganisms with different aptitudes with respect to different foods and beverages and how their main chemical, physical and biological determinants can modulate microbial activities. The aim of this chapter is to encourage microbial management in order to minimize quality depreciation of foods and beverages associated with the presence of BAs.

2.2 BAs in Food: Chemical Diversity, Microbial Diversity and the Different Significance of Microbial Producers

Biogenic amines are polar or semi-polar compounds with aliphatic (*e.g.* putrescine, cadaverine), aromatic (*e.g.* tyramine, phenylethylamine) or heterocyclic (*e.g.* histamine) structures.[22] Genetic and phenotypic traits associated with microbial BAs production are widespread among prokaryotes, with a general presence across both Gram-positive and Gram-negative bacteria. Table 2.1 summarizes exemplificative cases of BAs production by bacterial species, highlighting the corresponding matrices of isolation.

The physiological role of BAs in prokaryotes helps to explain the number of species/strains able to produce nitrogenous organic compounds: when BAs are depleted from the cell, a wide number of biological processes have shown to be affected.[23,24] In particular, the relevant effect of polyamines on cell proliferation is due to the stimulation of the translation of transcription factors related to growth in prokaryotes, while it is addressable to a putative role in cell cycle progression among eukaryotes.[23] Among Gram-negative bacteria that can be found in foods, strains belonging to *Escherichia coli*, *Hafnia alvei*, *Klebsiella pneumoniae*, *Morganella moorganii*, *Pseudomonas* spp.

Table 2.1 Exemplificative cases of biogenic amine (BA) production among bacterial species, highlighting the corresponding matrices of isolation. Abbreviations: histamine (HIS), tryptamine (TRPT), phenylethylamine (PHEN), tyramine (TYR), putrescine (PUT), isoamylamine (ISO) and cadaverine (CAD).

Microbial species	Produced BAs	Food matrices
Aerococcus viridans[97]	HIS	Cheese
Aeromonas hydrophila[98,99]	HIS, CAD, PUT	Fish, vegetable
Acinetobacter baumannii[39]	HIS	Fish
Bacillus coagulans[100]	HIS	Fish
Bacillus megaterium[100]	HIS	Fish
Citrobacter freundii[96,101,102]	CAD, HIS, ISO, PHEN, PUT, TYR, TRPT	Cheese, meat
Enterobacter aerogenes[40,103]	CAD, HIS, PUT, TYR	Cheese, fish
Enterobacter cloacae[40,96,99,103,104]	CAD, HIS, PUT, TYR	Cheese, fish, vegetables
Enterobacter gergoviae[103]	CAD, HIS, PUT, TYR	Cheese
Enterobacter sakazakii[97,99]	CAS, HIS, PUT, TYR	Cheese, vegetables
Enterococcus casseliflavus[97]	HIS, TYR	Cheese
Enterococcus durans[97]	TYR	Cheese
Enterococcus faecalis[25,97]	TYR, PUT	Cheese
Enterococcus faecium[97,105]	TYR	Cheese, wine
Escherichia coli[101,103]	CAD, HIS, PUT, TYR	Cheese, meat
Escherichia fergusonii[101]	PUT, CAD	Meat
Escherichia vulnaris[101]	CAD	Meat
Hafnia alvei[40,99,101,102,106]	CAD, ISO, HIS, PUT, TYR	Cheese, fish, meat, vegetables
Halomonas venusta[102]	CAD, HIS, PHEN, PUT, TYR, TRPT	Cheese
Klebsiella oxytoca[102,103]	CAD, HIS, ISO, PHEN, PUT, TYR	Cheese
Klebsiella pneumoniae[99]	CAD, HIS, PUT	Vegetables
Klebsiella terrigena[99]	CAD, PUT	Vegetables
Lactobacillus brevis[65,97]	TYR	Cheese, wine
Lactobacillus buchneri[97]	HIS	Cheese
Lactobacillus casei[97]	HIS, TYR	Cheese
Lactobacillus collinoides[28]	TYR	Cider
Lactobacillus delbrueckii[97]	HIS, TYR	Cheese
Lactobacillus diolivorans[28]	HIS	Cider
Lactobacillus hilgardii[107]	HIS	Wine
Lactococcus lactis[26,97]	HIS, PUT, TYR	Cheese
Leuconostoc mesenteroides[97]	TYR	Cheese
Lactobacillus plantarum[97]	TYR	Cheese
Lactobacillus rhamnosus[108]	HIS	Wine
Lactobacillus saerimneri[109]	CAD, HIS, PUT	Reference strain for BA studies
Morganella morganii[40,71,99,101,102]	CAD, HIS, PHEN, PUT, TYR, TRPT	Cheese, fish, meat, vegetables
Morganella psychrotolerans[38,110]	HIS	Fish
Pantoea agglomerans[104]	HIS	Fish
Pediococcus damnosus[97]	TYR	Cheese

(*continued*)

Table 2.1 (*Continued*)

Microbial species	Produced BAs	Food matrices
Photobacterium phosphoreum[38]	HIS	Fish
Proteus alcalifaciens[101]	PUT	Meat
Proteus mirabilis[101]	HIS, PUT	Meat
Proteus penneri[101]	PUT	Meat
Pseudomonas putida[102]	CAD, HIS, PHEN, PUT, TYR, TRPT	Cheese
Pseudomonas lundensis[102]	CAD, HIS, PUT, TYR	Cheese
Raoultella ornithinolytica[40,106]	HIS	Fish
Raoultella planticola[106]	HIS	Fish
Serratia grimesii[101]	PUT	Meat
Serratia liquefaciens[99,102,103]	CAD, HIS, PHEN, PUT, TYR, TRPT	Cheese, vegetables
Serratia marcescens[40,99]	CAD, HIS, PUT	Fish, vegetables
Staphylococcus xylosus[111]	HIS	Fish

and *Serratia* spp. are often BA producers. On the other hand, several Gram-positive bacteria, especially lactic acid bacteria (LAB) are capable of producing BAs. Even, in some cases, BA biosynthesis is a species-level characteristic. It is the case of tyramine production by *Enterococcus faecalis*, *Enterococcus faecium* and *Enterococcus durans*, putrescine synthesis by *E. faecalis*[25] and putrescine production by *Lactococcus lactis*.[26]

However, numerous evidence highlights the importance of the strain-dependent and, not only of the species-dependent character, of BA production by bacteria.[27,28] This trend is well testified reviewing the information in Tables 2.2 and 2.3, that report the intraspecific variability in the production of cadaverine, histamine, putrescine and tyramine within different species belonging to *Enterobacteriaceae* and LAB, respectively. Concerning this finding, several investigations suggest a concrete role of horizontal gene events to explain the dissemination of the character among strains.[29,30]

Considering the food systems and the different significance of microorganisms able to produce BAs, various diverse edible matrices have been found involved in cases of noteworthy accumulation of BAs: non-fermented foods such as meat, fish and seafood,[31–33] and fermented foods such as meat, cheeses, vegetables and fermented beverages.[34–37] This non-fermented/fermented dichotomy is propaedeutic to discuss the different managements that can be adopted to minimize the incidence of food-related microorganisms able to produce BAs. In fact, in the first case, the objective to minimize BA content corresponds to the avoidance of microbial growth. In the second case, the purpose to reduce BA concentration can be achieved only through a tailored microbial management in the framework of fermentative processes required.

Table 2.2 Data reported in the table belong to the study of Pircher *et al.*[96] *Enterobacteriae* isolates from different species producing amines according to the following ranges <10 mg l^{-1} (−); 10–100 mg l^{-1} (+); >100–1000 mg l^{-1} (++); >1000 mg l^{-1} (+++). Adapted from ref. 96 with permission from Springer Nature, Copyright 2006.

		Putrescine	Cadaverine	Histamine	Tyramine
Species	N	−/+/++/+++	−/+/++/+++	−/+/++/+++	−/+/++/+++
Hafnia alvei	61	17/4/18/22	6/2/24/29	57/3/1/0	60/1/0/0
Serratia liquefaciens	31	2/0/12/17	2/1/15/13	27/2/1/1	29/2/0/0
Escherichia coli	29	3/4/8/14	2/3/8/16	27/2/0/0	29/0/0/0
Enterobacter cloacae	7	2/0/2/3	2/3/2/0	6/1/0/0	6/1/0/0
Citrobacter freundii	6	0/0/1/5	0/6/0/0	6/0/0/0	4/2/0/0
Enterobacter aerogenes	4	1/0/0/3	0/1/1/2	3/1/0/0	4/0/0/0
Klebsiella oxytoca	3	1/0/0/2	0/0/2/1	3/0/0/0	3/0/0/0
Klebsiella terrigena	2	0/2/0/0	0/0/0/2	2/0/0/0	2/0/0/0
Klebsiella ornithinolytica	1	0/0/1/0	0/0/1/0	1/0/0/0	1/0/0/0

Table 2.3 Data reported in the table belong to the study of Pircher *et al.*[96] Lactic acid bacteria (LAB) isolates from different species producing amines according to the following ranges <10 mg l^{-1} (−); 10–100 mg l^{-1} (+); >100–1000 mg l^{-1} (++); >1000 mg l^{-1} (+++). Adapted from ref. 96 with permission from Springer Nature, Copyright 2006.

		Putrescine	Cadaverine	Histamine	Tyramine
Species	N	−/+/++/+++	−/+/++/+++	−/+/++/+++	−/+/++/+++
Lactobacillus paracasei ssp. *paracasei*	77	57/20/0/0	69/8/0/0	67/8/2/0	32/27/11/7
Leuconostoc mesenteroides ssp. *mesenteroides*	76	46/29/1/0	66/9/1/0	66/6/4/0	46/21/7/2
Lactobacillus lactis ssp. *lactis*	26	15/11/0/0	23/3/0/0	24/2/0/0	15/10/0/1
Lactobacillus curvatus	25	16/9/0/0	23/2/0/0	21/3/1/0	8/7/5/5
Leuconostoc lactis	13	6/7/0/0	11/2/0/0	7/2/2/0	3/6/1/3
Lactobacillus fermentum	12	9/3/0/0	10/2/0/0	12/0/0/0	8/2/2/0
Lactobacillus plantarum	12	7/5/0/0	11/1/0/0	12/0/0/0	6/4/2/0
Lactobacillus rhamnosus	10	6/4/0/0	8/2/0/0	6/4/0/0	4/4/1/1
Enterococcus faecalis	75	65/10/0/0	75/0/0/0	73/2/0/0	4/1/34/36
Enterococcus faecium	62	56/6/0/0	62/0/0/0	59/3/0/0	2/1/37/22

Gram-negative bacteria are often found to be associated with BA-related outbreaks, beyond that to be considered as spoilage microorganisms. As possibly responsible for histamine fish poisoning events associated with cold-smoked tuna consumption reported in Denmark during 2004, two strains were isolated belonging to the species *Morganella psychrotolerans* and *Photobacterium phosphoreum*, respectively.[38] Histamine-producing *Acinetobacter baumannii* strains were isolated from a fish sample with a histamine content higher than the hazard action level, in association with a food-borne poisoning causing illness in more than 300 people in 2007 (Taiwan).[39] In a similar more recent event (2014, Taiwan), strains belonging to the species *Enterobacter aerogenes*, *Enterobacter cloacae*, *Serratia marcescens* and *H. alvei* were indicated as possibly being responsible of an illness that involved more than 35 victims.[40] These species, with the exception of *Morganella morganii*, were considered as human pathogens.

Gram-positive BA producers represent an insidious problem in fermentative applications, particularly in the case of LAB. *Lactobacillus* appears to be the more representative genus among Gram-positive bacteria able to synthetize BAs. The long and safe history of use of LAB in food fermentations has resulted in the classification of several LAB species as GRAS (General Regarded As Safe) or QPS (Qualified Presumption of Safety). Hence, in several cases, LAB strains able to produce BAs belong to species with a GRAS/QPS status and often they are part of the protechnological microflora associated to food/beverage fermentations. This aspect helps us to clarify the reason why, in the overview we propose in the next section, some parameters that might modulate BAs release in association with growth.

Remaining in the field of BA-producing microbes, a particular aspect deals with the eukaryotic BA producers. A general influence of the different *Saccharomyces cerevisiae* yeast strains on BA content has been depicted,[41] a pronounced interest has been deserved to the potential contribution of strains belonging to the heterogeneous class of non-*Saccharomyces*. In a model matrix such as wine, several authors have reported the inability of wine yeasts to produce BAs[42–44] and to decrease BA contents during both spontaneous and induced commercial must fermentations.[45] However, numerous investigations have described a straight function of wine yeasts in rising BA amounts during vinification experiments on laboratory- and large-scale[41,46] (Table 2.4).

Furthermore, the outcome of two independent studies have demonstrated that different strains belonging to the species *S. cerevisiae*, *Kloeckera apiculata*, *Candida stellata*, *Metschnikowia pulcherrima* and *Brettanomyces bruxellensis* were capable of producing putrescine, phenylethylamine and ethanolamine.[45,47] Recently, Tristezza and co-workers[48] experimentally demonstrated the ability of strains of *Issatchenkia terricola, M. pulcherrima* and *Pichia manshurica* to synthesize histamine, a phenotype never reported in strains isolated from wine belonging to these species.

Table 2.4 Eukaryotic species involved in biogenic amine (BA) production found in association with food and beverages. Abbreviations: histamine (HIS), tryptamine (TRPT), tyramine (TYR), putrescine (PUT), cadaverine (CAD), agmatine (AGM), ethanolamine (ETH) and spermidine (SPE).

Eukaryotic species	Produced BAs	Food matrices
Brettanomyces bruxellensis[45,47]	AGM, PHEN	Wine
Candida stellate[45,47]	AGM, ETH, CAD	Wine
Candida versatilis[112]	CAD, HIS, TYR, SPE	Soy sauce
Debaryomyces hansenii[113,114]	HIS, PUT	Cheese
Geotrichum candidum[97]	HIS	Cheese
Issatchenkia terricola[48]	CAD, HIS	Wine
Kloeckera apiculata[45,47]	AGM, PHE	Wine
Metschnikowia pulcherrima[47,48]	AGM, HIS, PHE	Wine
Pichia manshurica[48,115]	CAD, HIS	Wine
Yarrowia lipolytica[113,114]	CAD, PHE, TYR	Cheese
Zygosaccharomyces rouxii[112]	CAD, HIS, TYR, SPE	Soy sauce
Pichia manshurica[48]	CAD, HIS	Wine

2.3 Abiotic and Biotic Factors Affecting the Activity of Microorganisms that Produce BAs

Several physical, chemical and biological parameters of different food and beverage matrices can influence the life of microbial BA producers and modulate the intensity of this peculiar phenotype.[49]

In general, BA production has been suggested as a strategy to supply alternative metabolic energy under suboptimal and/or stress conditions and to increase the pH of acidic environments. It is well-established that BA formation occurs if (i) high levels of free amino acids are available; (ii) microorganisms carried out the genetic determinants for their synthesis; and (iii) microorganisms faced suboptimal and/or stressing environmental conditions. Therefore, the control of the factors that can determine and/or impact one of these conditions is critical to avoid and/or limit the occurrence of these toxic compounds into the food chain.[50] For this reason, we provide an overview of the effects of the main abiotic (*e.g.* time effect and growth phases, changes in pH, temperature, precursors content, atmosphere compositions, salt concentrations) and biotic factors.

Time effect is one of the main determinants in influencing BA content, considering that microorganisms can proliferate and produce metabolic compounds during both transformation and storage.[51,52] The effect is particularly pronounced in matrices rich in BA precursors. In general, a positive relationship is reported among BA levels and ripening time, since the proteolytic reaction increased the availability of free amino acids.[53] For example, the presence of tyramine (64.4 mg kg^{-1}) has been detected in fermented meat after only 15 days of fermentation.[51] Obliviously, this time effect is also deeply related to growth aspects. In fact, BA formation is generally considered as growth phase dependent, with a general association with the stationary phase, probably because of maximum energy depletion and acidification that characterizes this growth stage.[54] However, controversial evidence suggests that the relationship with the growth phase can be considered a strain-dependent character. For example, in *Lactobacillus hilgardii*, the *hdc* (histidine decarboxylase) gene has been found up-regulated during exponential growth phase (even if an inhibitory effect of both glucose and fructose has also been reported),[55] while, in *L. hilgardii*, *Lactobacillus brevis* and *E. faecalis*, tyramine accumulation is associated to the presence of fructose or fructose plus L-malic acid when sugars are not yet consumed.[56,57] However, the link among metabolic sources and the occurrence of BAs in wine is debated, since several authors detected a reduction of the expression of decarboxylase genes or their enzymatic activity in the presence of glucose, fructose, organic acids or phenolic compounds.[55,58–60]

In fermented dry sausages production, addition of sugars encourages the growth of the starter cultures and is correlated with lower levels of BAs, probably due to a consequent faster acidification of the meat.[61,62] On the one hand, a low pH hinders the growth of undesired microorganisms, and on the other, creates favourable conditions for the production of BAs. Indeed, amino acids decarboxylation and functional substrate/product transmembrane exchangers results in a proton motive force which generates alkalinization of the cytoplasm.[35] Accordingly, transcriptional analysis confirmed that genes responsible for tyramine and putrescine production were induced at low pH.[63] However, although decarboxylase enzymes in cell-free extracts were active in a wide acidic pH range, the optimum activity was observed between pH 4.5 and 5.5.[64,65] Moreover, acidic conditions could encourage proteolysis during the ripening of cheeses and sausages, increasing the availability of substrates for BA production.[66] Likewise, oenological practices enhancing the levels of amino acids as well as maceration, employment of pectolytic enzymes and ageing with lees, potentially contribute to the risk of higher concentrations of BAs in wines.[67–70]

The impact of the temperature on BA biosynthesis is often related to the different behaviour of BA-producing microorganisms with respect to the principal range of temperatures (*e.g.* thermophiles, mesophiles, psychrophiles). For instance, the mesophilic *M. morganii* is able to release toxic concentrations of histamine in seafood at or above 7 to 10 °C.[71] Oppositely, the psychrotolerant *M. psychrotolerans* is likely responsible for histamine formation in cold-smoked tuna and found its optimal growth temperature at 2 °C, not at 37 °C.[38] Similarly, the storage of raw milk under refrigeration conditions increased the levels of putrescine, histamine and β-phenylethylamine, probably due to the occurrence of spoilage psychrotrophic microorganisms.[72] In general, low temperatures should be recommended in the food industry to control the microbial load and indirectly BA formation. Accordingly, higher ripening temperature increased the BA content, including tyramine (until 500 mg kg^{-1}), putrescine and cadaverine in several cheeses.[73–76] Similarly, BA accumulation was faster and higher at a temperature above the environment during ripening of Spanish dry fermented sausages.[66]

The content of the different precursors in the food and beverage matrices is a crucial factor in determining the potential in BA accumulation. Also within the same foodstuff typologies (in terms of human consumption), it is possible to highlight consistent differences. As is the case of two kinds of fish often consumed as smoked fish, *i.e.* filets of Atlantic salmon (*Salmo salar*) and tuna (*Thunnus albacares*), that were found to contain concentrations of free histidine in muscle tissues of 100–200 mg kg^{-1} and 10 000 mg kg^{-1}, respectively.[77,78]

Different oxygen levels in the atmosphere associated to the food matrices can affect BA synthesis depending on the effects exerted on the producing microbiota. In the case of chicken meat, Balamatsia and coworkers[52] compared the BAs concentration in aerobically produced preparations with those obtained under modified atmospheric packaging (MAP; 30%, CO_2, 70% N_2), thus observing lower levels of putrescine and cadaverine in the MAP-produced meat. Yassoralipour *et al.*[79] offer another kind of perspective of gradually increasing the gas mixtures. The different concentrations were M1 (100% CO_2), M2 (75% CO_2/25% N_2), M3 (50% CO_2/50% N_2), M4 (25% CO_2/75% N_2) and air-packed (control). At the end of the storage period, histamine concentration in M1, M2, M3, M4 and control was 169.9, 180.8, 211.7, 240.4 and 473.6 mg kg^{-1}, respectively. The total BA formation was inversely correlated to CO_2 concentration; however, there was no significant difference in the total BAs between the samples kept in 75 and 100% CO_2.

In effect/control on the microbial growth, NaCl concentration was found able to modulate BA content produced in association to food shelf life. In the histamine fish poisoning events associated with cold-smoked tuna consumption reported in Denmark during 2004, Emborg and Dalgaard[38] isolated *M. psychrotolerans* and *P. phosphoreum* in the presence of 2.2% ± 0.6% and 1.3% ± 0.1% NaCl (concentration in the water phase associated to the packaged tuna), respectively. In the following challenge tests conceived at 5 °C, the same authors assessed how the NaCl concentration of 4.4% represents a sufficient hurdle to inhibit histamine formation by *P. phosphoreum*, but not for *M.* psychrotolerans.[38] On the other hand, at 6.9% of NaCl concentration in the water phase, no significant histamine formation was observed in both cases.[38] Addition of sodium chloride is usually employed in sausages and dairy production to control undesired microorganisms potentially able to produce BAs.[80] However, in a starter-induced fermentation of white cabbage performed by *Lactobacillus plantarum* and *Leuconostoc mesenteroides*, the BA content of sauerkraut was lower by adding 0.5% NaCl rather than 1.5%.[81] Similarly, *E. durans* and *L. lactis* produced higher levels of tyramine in a growth medium containing 20 g l^{-1} than 10 g l^{-1} of NaCl.[82] It is probable that an increasing of the osmotic pressure might modulate the decarboxylase activity in order to regulate the intracellular pH.[83] Electrolyzed oxidizing water and ice treatments have been reported as treatments able to reduce histamine-producing bacteria on fish skin and the food contact surface.[84] Similarly, methods including irradiation, hydrostatic pressures and controlled atmosphere could be useful to avoid and/or limit the proliferation of BA producers during food processing.[85]

Mah and Hwang[86] evaluated the effect of a panel of food additives on BA formation using a salted and fermented anchovy as a model matrix. The authors found the most pronounced effect in the case of glycine utilization, leading to a reduction of about 32%, 78%, 93%, 100% and 100% for the contents of putrescine, cadaverine, histamine, tyramine and spermidine, respectively.

Finally, biotic factors can also be exploited in order to reduce the influence of BA producers. A first aspect to underline is that when we consider fermented foods/beverages, bacteria involved in the fermentative processes exert an inhibitory/competitive effect against BA-producing Gram-negative bacteria.[87,88] However, the control of Gram-positive bacteria, and especially LAB, can be more arduous in the field of fermented matrices.[89] In fact, BA-producing LAB can be part of the autochthonous microbes associated with spontaneous processes or may even 'contaminate' starter cultures.[90,91] With the purpose to avoid these insidious events, more effective management is the use of starter cultures designed that start from strains characterized in the absence of the BA-producing phenotype.[92] Furthermore, recent evidence suggests the possible exploitation of selected LAB with a biogenic amine-degrading aptitude.[90,93]

2.4 Future Perspectives

The continuous exploration of microbial diversity is needed to improve our understanding of phenomena susceptible to influence the microbial contribution to food quality. The isolation of novel microbial species/strains able to produce BAs is also relevant if we consider the renewed interest in spontaneous fermentations, and the BA-producer microbes present in the spontaneous microbiota, particularly in the case of artisanal fermented foods produced through uncontrolled fermentative processes that can lead to increased risks for consumer health.[94] Furthermore, the isolation and characterization of a more representative microbial diversity responsible for the presence of these chemical contaminants allows the design of more accurate challenge tests.

Even the microbial production of BAs in foods and beverages represents a field of deservedly applicative relevance, as with other applicative scientific issues, this topic takes noteworthy advantages of molecular investigation delving into cellular mechanisms involved in biosynthesis and metabolism. In fact, the molecular approaches are fundamental in order to improve the understanding of the influence of single determinants among the several physico-chemical and biological features associated with food matrices. These kind of studies can be useful in understanding the separation between the effects of biotic and abiotic stressors on the growth of

Table 2.5 Chronological list of molecular studies elucidating the molecular basis associated to biogenic amine (BA) phenotypes and the factor affecting BA production. Abbreviations: histidine decarboxylase (HDC), ornithine decarboxylase (ODC) and tyrosine decarboxylase (TDC).

Studied organism	Molecular approach	Outcome
Lactobacillus hilgardii[116]	Genomic studies, heterologous expression	• A four-gene cluster involved in histamine production is encoded on an unstable plasmid
Lactobacillus hilgardii[117]	Proteomic	• HDC and ODC biosynthesis are shown to be strictly reliant on the presence of high concentrations of free amino acids in the growth medium • Histamine production is only induced during the stationary phase
Lactobacillus brevis[118]	Genomic studies, heterologous expression	• The molecular basis of agmatine deiminase and tyrosine catabolic pathways relies on a genomic region connected to acid resistance, subjected to horizontal gene transfers among LAB
Lactobacillus hilgardii, Pediococcus parvulus[55]	Northern blotting, biochemical studies	• The HDC gene expression is reduced by glucose, fructose, malic acid and citric acid • Ethanol improves HDC enzyme activity • The conditions associated to malolactic fermentation could support histamine production • SO_2 precludes bacterial growth, but does not reduce HDC enzyme activity
Enterococcus faecalis[56]	Proteomic	• TDC catalyses both tyrosine and phenylalanine decarboxylation • Tyrosine is recognized as the first substrate and completely transformed into tyramine, while phenylalanine is converted to beta-phenylethylamine only when tyrosine was completely depleted • Tyrosine decarboxylation seems to not compete with the other energy-providing routes
Lactococcus lactis[26]	Northern blotting, RT-PCR	• The putrescine-producing cluster is composed of a putative regulatory gene, *aguR*, followed by the genes (*aguB*, *aguD*, *aguA* and *aguC*) encoding the catabolic enzymes • This operon is induced in the presence of agmatine
Lactobacillus brevis[119]	Quantitative real-time RT-PCR	• The bacterium is able to produce both tyramine and putrescine during wine fermentation • Increased cell viability is also observed in wine supplemented with tyrosine or agmatine

(continued)

Table 2.5 (*Continued*)

Studied organism	Molecular approach	Outcome
Lactobacillus saerimneri[109]	Whole genome sequencing	• qRT-PCR analysis suggests a strong influence of substrate availability on the expression of genes coding for tyrosine decarboxylase and agmatine deiminase • Draft genome sequence of *L. saerimneri* 30a (formerly *Lactobacillus* sp. strain 30a) and the principal findings from its annotation reveal the genes encoding the enzymatic machineries that are involved in the production of BAs

BA-producing bacteria and/or on the regulation/activity of genes/proteins/cellular machineries involved in BA production. The exemplificative outcomes reported in Table 2.5 testify to the crucial role of molecular biology studies in order to enhance the knowledge necessary to propose management conceived to minimize microbial synthetized BAs in foods and beverages.

In our opinion, one last aspect that deserves to be underlined deals with the understanding of the complex dynamics at the community level and of the existing exchanges of BA-producing microbial ecosystems associated with the food environment and with other environments connected to food chains/consumption. For example, recent evidence highlights the presence of a wide diversity of prokaryotes associated with the human gut that are able to produce and degrade BAs.[95] The bacterial diversity within this niche and the possible relationship with other food niches represents an interesting field to investigate.

All the examined thematic issues will benefit from the adoption of holistic investigations (*e.g.* complementary integrate analysis, systems biology) and of approaches capable of better exploiting available large databases and analytical techniques capable of producing a huge quantity of data.

References

1. G. Suzzi, From wild strain to domesticated strain: The philosophy of microbial diversity in foods, *Front. Microbiol.*, 2011, **2**, 169.
2. M. E. Guerzoni, Human food chain and microorganisms: A case of co-evolution, *Front. Microbiol.*, 2010, **1**, 106.
3. A. Yépez, P. Russo, G. Spano, I. Khomenko, F. Biasioli, V. Capozzi, *et al.*, *In situ* riboflavin fortification of different kefir-like cereal-based beverages using selected Andean LAB strains, *Food Microbiol.*, 2019, 77, 61–68.

4. C. Garofalo, M. Tristezza, F. Grieco, G. Spano and V. Capozzi, From grape berries to wine: Population dynamics of cultivable yeasts associated to 'Nero di Troia' autochthonous grape cultivar, *World J. Microbiol. Biotechnol.*, 2016, **32**, 59.
5. G. Perpetuini, G. Caruso, S. Urbani, M. Schirone, S. Esposto, A. Ciarrocchi, *et al.*, Changes in polyphenolic concentrations of table olives (cv. Itrana) produced under different irrigation regimes during spontaneous or inoculated fermentation, *Front. Microbiol.*, 2018, **9**, 1287.
6. P. Russo, M. P. Arena, D. Fiocco, V. Capozzi, D. Drider and G. Spano, *Lactobacillus plantarum* with broad antifungal activity: A promising approach to increase safety and shelf-life of cereal-based products, *Int. J. Food Microbiol.*, 2017, **247**, 48–54.
7. R. Prete, R. Tofalo, E. Federici, A. Ciarrocchi, G. Cenci and A. Corsetti, Food-associated *Lactobacillus plantarum* and yeasts inhibit the genotoxic effect of 4-nitroquinoline-1-oxide, *Front. Microbiol.*, 2017, **8**, 2349.
8. C. Gheziel, P. Russo, M. P. Arena, G. Spano, H.-I. Ouzari, O. Kheroua, *et al.*, Evaluating the probiotic potential of *Lactobacillus plantarum* strains from algerian infant feces: Towards the design of probiotic starter cultures tailored for developing countries, *Probiotics Antimicrob. Proteins*, 2019, **11**, 113–123.
9. V. Capozzi, M. Fragasso, R. Romaniello, C. Berbegal, P. Russo and G. Spano, Spontaneous food fermentations and potential risks for human health, *Fermentation*, 2017, **3**, 49.
10. N. Teslić, F. Patrignani, M. Ghidotti, G. P. Parpinello, A. Ricci, R. Tofalo, *et al.*, Utilization of 'early green harvest' and non-*Saccharomyces cerevisiae* yeasts as a combined approach to face climate change in winemaking, *Eur. Food Res. Technol.*, 2018, **244**, 1301–1311.
11. C. Berbegal, N. Peña, P. Russo, F. Grieco, I. Pardo, S. Ferrer, *et al.*, Technological properties of *Lactobacillus plantarum* strains isolated from grape must fermentation, *Food Microbiol.*, 2016, **57**, 187–194.
12. P. D. Cotter, C. Hill and R. P. Ross, Food microbiology: Bacteriocins: Developing innate immunity for food, *Nat. Rev. Microbiol.*, 2005, **3**, 777–788.
13. L. Dijkshoorn, P. De Vos and T. Dedeurwaerdere, Understanding patterns of use and scientific opportunities in the emerging global microbial commons, *Res. Microbiol.*, 2010, **161**, 407–413.
14. T. Mattila-Sandholm, P. Myllärinen, R. Crittenden, G. Mogensen, R. Fondén and M. Saarela, Technological challenges for future probiotic foods, *Int. Dairy J.*, 2002, **12**, 173–182.
15. G. Spano and V. Capozzi, Food microbial biodiversity and 'microbes of protected origin', *Front. Microbiol.*, 2011, **2**, 237.

16. A. H. Havelaar, S. Brul, A. de Jong, R. de Jonge, M. H. Zwietering and B. H. ter Kuile, Future challenges to microbial food safety, *Int. J. Food Microbiol.*, 2010, **139**, S79–S94.
17. P. Russo, G. Spano, M. P. Arena, V. Capozzi, D. Fiocco, F. Grieco, *et al.*, Are consumers aware of the risks related to biogenic amines in food?, in *Current Research, Technology and Education Topics in Applied Microbiology and Microbial Biotechnology*, ed. A. Méndez-Vilas, 2010, pp. 1087–1095.
18. S. Chytiri, E. Paleologos, I. Savvaidis and M. G. Kontominas, Relation of biogenic amines with microbial and sensory changes of whole and filleted freshwater rainbow trout (*Onchorynchus mykiss*) stored on ice, *J. Food Prot.*, 2004, **67**, 960–965.
19. S. Baixas-Nogueras, S. Bover-Cid, M. T. Veciana-Nogués, A. Mariné-Font and M. C. Vidal-Carou, Biogenic amine index for freshness evaluation in iced Mediterranean hake (*Merluccius merluccius*), *J. Food Prot.*, 2005, **68**, 2433–2438.
20. G. Vinci and M. L. Antonelli, Biogenic amines: Quality index of freshness in red and white meat, *Food Control*, 2002, **13**, 519–524.
21. V. Ladero, M. Calles-Enriquez, M. Fernandez and M. A. Alvarez, Toxicological effects of dietary biogenic amines, *Curr. Nutr. & Food Sci.*, 2010, **6**, 145–156.
22. M. Papageorgiou, D. Lambropoulou, C. Morrison, E. Kłodzińska, J. Namieśnik and J. Płotka-Wasylka, Literature update of analytical methods for biogenic amines determination in food and beverages, *TrAC, Trends Anal. Chem.*, 2018, **98**, 128–142.
23. L. Miller-Fleming, V. Olin-Sandoval, K. Campbell and M. Ralser, Remaining mysteries of molecular biology: The role of polyamines in the cell, *J. Mol. Biol.*, 2015, **427**, 3389–3406.
24. P. Shah and E. Swiatlo, A multifaceted role for polyamines in bacterial pathogens, *Mol. Microbiol.*, 2008, **68**, 4–16.
25. V. Ladero, M. Fernández, M. Calles-Enríquez, E. Sánchez-Llana, E. Cañedo, M. C. Martín, *et al.*, Is the production of the biogenic amines tyramine and putrescine a species-level trait in enterococci?, *Food Microbiol.*, 2012, **30**, 132–138.
26. V. Ladero, F. P. Rattray, B. Mayo, M. C. Martín, M. Fernández and M. A. Alvarez, Sequencing and transcriptional analysis of the biosynthesis gene cluster of putrescine-producing *Lactococcus lactis*, *Appl. Environ. Microbiol.*, 2011, **77**, 6409–6418.
27. A. Marcobal, B. de las Rivas, M. V. Moreno-Arribas and R. Muñoz, Identification of the ornithine decarboxylase gene in the putrescine-producer *Oenococcus oeni* BIFI-83, *FEMS Microbiol. Lett.*, 2004, **239**, 213–220.
28. G. Garai, M. T. Dueñas, A. Irastorza and M. V. Moreno-Arribas, Biogenic amine production by lactic acid bacteria isolated from cider, *Lett. Appl. Microbiol.*, 2007, **45**, 473–478.

29. E. Coton and M. Coton, Evidence of horizontal transfer as origin of strain to strain variation of the tyramine production trait in *Lactobacillus brevis*, *Food Microbiol.*, 2009, **26**, 52–57.
30. Á. Marcobal, B. de las Rivas, M. V. Moreno-Arribas and R. Muñoz, Evidence for horizontal gene transfer as origin of putrescine production in *Oenococcus oeni* RM83, *Appl. Environ. Microbiol.*, 2006, **72**, 7954–7958.
31. I. Al Bulushi, S. Poole, H. C. Deeth and G. A. Dykes, Biogenic amines in fish: Roles in intoxication, spoilage, and nitrosamine formation–a review, *Crit. Rev. Food Sci. Nutr.*, 2009, **49**, 369–377.
32. C. Ruiz-Capillas and F. Jiménez-Colmenero, Biogenic amines in meat and meat products, *Crit. Rev. Food Sci. Nutr.*, 2004, **44**, 489–499.
33. K. B. Biji, C. N. Ravishankar, R. Venkateswarlu, C. O. Mohan and T. K. S. Gopal, Biogenic amines in seafood: A review, *J. Food Sci. Technol.*, 2016, **53**, 2210–2218.
34. G. Suzzi and F. Gardini, Biogenic amines in dry fermented sausages: A review, *Int. J. Food Microbiol.*, 2003, **88**, 41–54.
35. D. M. Linares, B. del Río, V. Ladero, N. Martínez, M. Fernández, M. C. Martín, *et al.*, Factors influencing biogenic amines accumulation in dairy products, *Front. Microbiol.*, 2012, **3**, 180.
36. G. Spano, P. Russo, A. Lonvaud-Funel, P. Lucas, H. Alexandre, C. Grandvalet, *et al.*, Biogenic amines in fermented foods, *Eur. J. Clin. Nutr.*, 2010, **64**(Suppl 3), S95–S100.
37. L. Beneduce, A. Romano, V. Capozzi, P. Lucas, L. Barnavon, B. Bach, *et al.*, Biogenic amine in wines, *Ann. Microbiol.*, 2010, **60**, 573–578.
38. J. Emborg and P. Dalgaard, Formation of histamine and biogenic amines in cold-smoked tuna: An investigation of psychrotolerant bacteria from samples implicated in cases of histamine fish poisoning, *J. Food Prot.*, 2006, **69**, 897–906.
39. H.-C. Chen, Y.-R. Huang, H.-H. Hsu, C.-S. Lin, W.-C. Chen, C.-M. Lin, *et al.*, Determination of histamine and biogenic amines in fish cubes (*Tetrapturus angustirostris*) implicated in a food-borne poisoning, *Food Control*, 2010, **21**, 13–18.
40. Y.-C. Lee, H.-F. Kung, C.-H. Wu, H.-M. Hsu, H.-C. Chen, T.-C. Huang, *et al.*, Determination of histamine in milkfish stick implicated in food-borne poisoning, *J. Food Drug Anal.*, 2016, **24**, 63–71.
41. D. T. Goñi and C. A. Azpilicueta, Influence of yeast strain on biogenic amines content in wines: Relationship with the utilization of amino acids during fermentation, *Am. J. Enol. Vitic.*, 2001, **52**, 185–190.
42. P. Herbert, M. J. Cabrita, N. Ratola, O. Laureano and A. Alves, Free amino acids and biogenic amines in wines and musts from the Alentejo region. Evolution of amines during alcoholic fermentation and relationship with variety, sub-region and vintage, *J. Food Eng.*, 2005, **66**, 315–322.

43. A. Marcobal, P. J. Martín-Alvarez, M. C. Polo, R. Muñoz and M. V. Moreno-Arribas, Formation of biogenic amines throughout the industrial manufacture of red wine, *J. Food Prot.*, 2006, **69**, 397–404.
44. J. M. Landete, S. Ferrer and I. Pardo, Biogenic amine production by lactic acid bacteria, acetic bacteria and yeast isolated from wine, *Food Control*, 2007, **18**, 1569–1574.
45. L. Granchi, P. Romano, S. Mangani, S. Guerrini and M. Vincenzini, Production of biogenic amines by wine microorganisms, *Bull OIV*, 2005, **78**, 595–609.
46. D. Torrea and C. Ancín, Content of biogenic amines in a Chardonnay wine obtained through spontaneous and inoculated fermentations, *J. Agric. Food Chem.*, 2002, **50**, 4895–4899.
47. M. Caruso, C. Fiore, M. Contursi, G. Salzano, A. Paparella and P. Romano, Formation of biogenic amines as criteria for the selection of wine yeasts, *World J. Microbiol. Biotechnol.*, 2002, **18**, 159–163.
48. M. Tristezza, C. Vetrano, G. Bleve, G. Spano, V. Capozzi, A. Logrieco, *et al.*, Biodiversity and safety aspects of yeast strains characterized from vineyards and spontaneous fermentations in the Apulia region, Italy, *Food Microbiol.*, 2013, **36**, 335–342.
49. V. Capozzi, D. Fiocco, M. L. Amodio, A. Gallone and G. Spano, Bacterial stressors in minimally processed food, *Int. J. Mol. Sci.*, 2009, **10**, 3076–3105.
50. M. L. Mohedano, P. López, G. Spano and P. Russo, 12 - Controlling the formation of biogenic amines in fermented foods, in *Advances in Fermented Foods and Beverages*, ed. W. Holzapfel, Woodhead Publishing, Cambridge, 2015, pp. 273–310.
51. J. D. Coïsson, C. Cerutti, F. Travaglia and M. Arlorio, Production of biogenic amines in 'Salamini italiani alla cacciatora PDO', *Meat Sci.*, 2004, **67**, 343–349.
52. C. C. Balamatsia, E. K. Paleologos, M. G. Kontominas and I. N. Savvaidis, Correlation between microbial flora, sensory changes and biogenic amines formation in fresh chicken meat stored aerobically or under modified atmosphere packaging at 4 °C: Possible role of biogenic amines as spoilage indicators, *Antonie van Leeuwenhoek*, 2006, **89**, 9–17.
53. F. Pasini, F. Soglia, M. Petracci, M. F. Caboni, S. Marziali, C. Montanari, *et al.*, Effect of fermentation with different lactic acid bacteria starter cultures on biogenic amine content and ripening patterns in dry fermented sausages, *Nutrients*, 2018, **10**, 1497.
54. E. Pessione and S. Cirrincione, Bioactive molecules released in food by lactic acid bacteria: Encrypted peptides and biogenic amines, *Front. Microbiol.*, 2016, 7, 876.
55. J. M. Landete, I. Pardo and S. Ferrer, Regulation of *hdc* expression and HDC activity by enological factors in lactic acid bacteria, *J. Appl. Microbiol.*, 2008, **105**, 1544–1551.

56. E. Pessione, A. Pessione, C. Lamberti, D. J. Coïsson, K. Riedel, R. Mazzoli, *et al.*, First evidence of a membrane-bound, tyramine and beta-phenylethylamine producing, tyrosine decarboxylase in *Enterococcus faecalis*: A two-dimensional electrophoresis proteomic study, *Proteomics*, 2009, **9**, 2695–2710.
57. M. V. Moreno-Arribas, M. C. Polo, F. Jorganes and R. Muñoz, Screening of biogenic amine production by lactic acid bacteria isolated from grape must and wine, *Int. J. Food Microbiol.*, 2003, **84**, 117–123.
58. M. E. Arena, D. Fiocco, M. C. Manca de Nadra, I. Pardo and G. Spano, Characterization of a *Lactobacillus plantarum* strain able to produce tyramine and partial cloning of a putative tyrosine decarboxylase gene, *Curr. Microbiol.*, 2007, **55**, 205–210.
59. M. E. Arena, J. M. Landete, M. C. M. de Nadra, I. Pardo and S. Ferrer, Factors affecting the production of putrescine from agmatine by *Lactobacillus hilgardii* X1B isolated from wine, *J. Appl. Microbiol.*, 2008, **105**, 158–165.
60. M. R. Alberto, M. E. Arena and M. C. Manca de Nadra, Putrescine production from agmatine by *Lactobacillus hilgardii*: Effect of phenolic compounds, *Food Control*, 2007, **18**, 898–903.
61. S. Bover-Cid, M. Izquierdo-Pulido and M. Carmen Vidal-Carou, Changes in biogenic amine and polyamine contents in slightly fermented sausages manufactured with and without sugar, *Meat Sci.*, 2001, **57**, 215–221.
62. C. González-Fernández, E. M. Santos, I. Jaime and J. Rovira, Influence of starter cultures and sugar concentrations on biogenic amine contents in chorizo dry sausage, *Food Microbiol.*, 2003, **20**, 275–284.
63. M. P. Arena, P. Russo, V. Capozzi, L. Beneduce and G. Spano, Effect of abiotic stress conditions on expression of the *Lactobacillus brevis* IOEB 9809 tyrosine decarboxylase and agmatine deiminase genes, *Ann. Microbiol.*, 2011, **61**, 179–183.
64. M. Fernández, D. M. Linares, A. Rodríguez and M. A. Alvarez, Factors affecting tyramine production in *Enterococcus durans* IPLA 655, *Appl. Microbiol. Biotechnol.*, 2007, **73**, 1400–1406.
65. V. Moreno-Arribas and A. Lonvaud-Funel, Tyrosine decarboxylase activity of *Lactobacillus brevis* IOEB 9809 isolated from wine and *L. brevis* ATCC 367, *FEMS Microbiol. Lett.*, 1999, **180**, 55–60.
66. M. L. Latorre-Moratalla, S. Bover-Cid and M. C. Vidal-Carou, Technological conditions influence aminogenesis during spontaneous sausage fermentation, *Meat Sci.*, 2010, **85**, 537–541.
67. P. J. Martín-Álvarez, Á Marcobal, C. Polo and M. V. Moreno-Arribas, Influence of technological practices on biogenic amine contents in red wines, *Eur. Food Res. Technol.*, 2006, **222**, 420–424.
68. A. Y. Smit, W. J. du Toit, M. Stander and M. du Toit, Evaluating the influence of maceration practices on biogenic amine formation in wine, *LWT--Food Sci. Technol.*, 2013, **53**, 297–307.

69. C. Ancín-Azpilicueta, A. González-Marco and N. Jiménez-Moreno, Comparative study of the amine concentration in wines obtained from the traditional fermentation and from a more anaerobic fermentation method, *LWT--Food Sci. Technol.*, 2010, **43**, 771–776.
70. M. García-Marino, Á Trigueros and T. Escribano-Bailón, Influence of oenological practices on the formation of biogenic amines in quality red wines, *J. Food Compos. Anal.*, 2010, **23**, 455–462.
71. S. H. Kim, R. J. Price, M. T. Morrissey, K. G. Field, C. I. Wei and H. An, Histamine production by Morganella morganii in mackerel, albacore, mahi-mahi, and salmon at various storage temperatures, *J. Food Sci.*, 2002, **67**, 1522–1528.
72. S. Novella-Rodríguez, M. T. Veciana-Nogués, A. X. Roig-Sagués, A. J. Trujillo-Mesa and M. C. Vidal-Carou, Evaluation of biogenic amines and microbial counts throughout the ripening of goat cheeses from pasteurized and raw milk, *J. Dairy Res.*, 2004, **71**, 245–252.
73. V. Pachlová, F. Buňka, R. Flasarová, P. Válková and L. Buňková, The effect of elevated temperature on ripening of Dutch type cheese, *Food Chem.*, 2012, **132**, 1846–1854.
74. L. Bunková, F. Bunka, G. Mantlová, A. Cablová, I. Sedlácek, P. Svec, *et al.*, The effect of ripening and storage conditions on the distribution of tyramine, putrescine and cadaverine in Edam-cheese, *Food Microbiol.*, 2010, **27**, 880–888.
75. T. Komprda, E. Rejchrtová, P. Sládková, L. Zemánek and L. Vymlátilová, Effect of some external factors on the content of biogenic amines and polyamines in a smear-ripened cheese, *Dairy Sci. Technol.*, 2012, **92**, 367–382.
76. O. Pinho, I. M. P. L. V. O. Ferreira, E. Mendes, B. M. Oliveira and M. Ferreira, Effect of temperature on evolution of free amino acid and biogenic amine contents during storage of Azeitão cheese, *Food Chem.*, 2001, **75**, 287–291.
77. J. Emborg, B. G. Laursen and P. Dalgaard, Significant histamine formation in tuna (*Thunnus albacares*) at 2 degrees C–effect of vacuum- and modified atmosphere-packaging on psychrotolerant bacteria, *Int. J. Food Microbiol.*, 2005, **101**, 263–279.
78. M. Espe, E. Lied and K. R. Torrissen, Changes in plasma and muscle free amino acids in Atlantic salmon (*Salmo salar*) during absorption of diets containing different amounts of hydrolysed cod muscle protein, *Comp. Biochem. Physiol., Part A: Mol. Integr. Physiol.*, 1993, **105**, 555–562.
79. A. Yassoralipour, J. Bakar, R. A. Rahman and F. Abu Bakar, Biogenic amines formation in barramundi (*Lates calcarifer*) fillets at 8 °C kept in modified atmosphere packaging with varied CO_2 concentration, *LWT--Food Sci. Technol.*, 2012, **48**, 142–146.
80. C. Roseiro, C. Santos, M. Sol, L. Silva and I. Fernandes, Prevalence of biogenic amines during ripening of a traditional dry fermented pork

sausage and its relation to the amount of sodium chloride added, *Meat Sci.*, 2006, **74**, 557–563.
81. E. Peñas, J. Frias, B. Sidro and C. Vidal-Valverde, Impact of fermentation conditions and refrigerated storage on microbial quality and biogenic amine content of sauerkraut, *Food Chem.*, 2010, **123**, 143–150.
82. L. Buňková, F. Buňka, E. Pollaková, T. Podešvová and V. Dráb, The effect of lactose, NaCl and an aero/anaerobic environment on the tyrosine decarboxylase activity of *Lactococcus lactis* subsp. *cremoris* and *Lactococcus lactis* subsp. *lactis*, *Int. J. Food Microbiol.*, 2011, **147**, 112–119.
83. W. A. M. Wolken, P. M. Lucas, A. Lonvaud-Funel and J. S. Lolkema, The mechanism of the tyrosine transporter TyrP supports a proton motive tyrosine decarboxylation pathway in *Lactobacillus brevis*, *J. Bacteriol.*, 2006, **188**, 2198–2206.
84. S. Phuvasate and Y.-C. Su, Effects of electrolyzed oxidizing water and ice treatments on reducing histamine-producing bacteria on fish skin and food contact surface, *Food Control*, 2010, **21**, 286–291.
85. A. Naila, S. Flint, G. Fletcher, P. Bremer and G. Meerdink, Control of biogenic amines in food–existing and emerging approaches, *J. Food Sci.*, 2010, **75**, R139–R150.
86. J.-H. Mah and H.-J. Hwang, Effects of food additives on biogenic amine formation in Myeolchi-jeot, a salted and fermented anchovy (*Engraulis japonicus*), *Food Chem.*, 2009, **114**, 168–173.
87. M. R. Adams and L. Nicolaides, Review of the sensitivity of different foodborne pathogens to fermentation, *Food Control*, 1997, **8**, 227–239.
88. E. Caplice and G. F. Fitzgerald, Food fermentations: Role of microorganisms in food production and preservation, *Int. J. Food Microbiol.*, 1999, **50**, 131–149.
89. D. M. Linares, M. C. Martín, V. Ladero, M. A. Alvarez and M. Fernández, Biogenic amines in dairy products, *Crit. Rev. Food Sci. Nutr.*, 2011, **51**, 691–703.
90. M. A. Alvarez and M. V. Moreno-Arribas, The problem of biogenic amines in fermented foods and the use of potential biogenic amine-degrading microorganisms as a solution, *Trends Food Sci. Technol.*, 2014, **39**, 146–155.
91. A. Costantini, E. Vaudano, V. Del Prete, M. Danei and E. Garcia-Moruno, Biogenic amine production by contaminating bacteria found in starter preparations used in winemaking, *J. Agric. Food Chem.*, 2009, **57**, 10664–10669.
92. EFSA Panel on Biological Hazards (BIOHAZ), Scientific Opinion on the risk posed by pathogens in food of non-animal origin. Part 1 (outbreak data analysis and risk ranking of food/pathogen combinations), *EFSA J.*, 2013, **11**, 3025.
93. V. Capozzi, P. Russo, V. Ladero, M. Fernandez, D. Fiocco, M. A. Alvarez, *et al.*, Biogenic amines degradation by *Lactobacillus*

plantarum: Toward a potential application in wine, *Front. Microbiol.*, 2012, **3**, 122.
94. P. Russo, G. Spano and V. Capozzi, Safety evaluation of starter cultures, in *Starter Cultures in Food Production*, ed. B. Speranza, A. Bevilacqua, R. Corbo, and M. Sinigaglia, John Wiley & Sons, Ltd, Hoboken (NY), 2017, pp. 101–128.
95. B. Pugin, W. Barcik, P. Westermann, A. Heider, M. Wawrzyniak, P. Hellings, *et al.*, A wide diversity of bacteria from the human gut produces and degrades biogenic amines, *Microb. Ecol. Health Dis.*, 2017, **28**, 1353881.
96. A. Pircher, F. Bauer and P. Paulsen, Formation of cadaverine, histamine, putrescine and tyramine by bacteria isolated from meat, fermented sausages and cheeses, *Eur. Food Res. Technol.*, 2007, **226**, 225–231.
97. A. X. Roig-Sagués, A. P. Molina and M. Hernández-Herrero, Histamine and tyramine-forming microorganisms in Spanish traditional cheeses, *Eur. Food Res. Technol.*, 2002, **215**, 96–100.
98. B. L. Middlebrooks, P. M. Toom, W. L. Douglas, R. E. Harrison and S. Mcdowell, Effects of storage time and temperature on the microflora and amine-development in Spanish mackerel (*Scomberomorus maculatus*), *J. Food Sci.*, 1988, **53**, 1024–1029.
99. T. Lavizzari, M. Breccia, S. Bover-Cid, M. C. Vidal-Carou and M. T. Veciana-Nogués, Histamine, cadaverine, and putrescine produced *in vitro* by *enterobacteriaceae* and *pseudomonadaceae* isolated from spinach, *J. Food Prot.*, 2010, **73**, 385–389.
100. Y.-H. Tsai, C.-Y. Lin, L.-T. Chien, T.-M. Lee, C.-I. Wei and D.-F. Hwang, Histamine contents of fermented fish products in Taiwan and isolation of histamine-forming bacteria, *Food Chem.*, 2006, **98**, 64–70.
101. F. Durlu-Özkaya, K. Ayhan and N. Vural, Biogenic amines produced by *Enterobacteriaceae* isolated from meat products, *Meat Sci.*, 2001, **58**, 163–166.
102. M. Coton, C. Delbés-Paus, F. Irlinger, N. Desmasures, A. Le Fleche, V. Stahl, *et al.*, Diversity and assessment of potential risk factors of Gram-negative isolates associated with French cheeses, *Food Microbiol.*, 2012, **29**, 88–98.
103. M. Marino, M. Maifreni, S. Moret and G. Rondinini, The capacity of *Enterobacteriaceae* species to produce biogenic amines in cheese, *Lett. Appl. Microbiol.*, 2000, **31**, 169–173.
104. Y.-H. Tsai, C.-Y. Lin, S.-C. Chang, H.-C. Chen, H.-F. Kung, C.-I. Wei, *et al.*, Occurrence of histamine and histamine-forming bacteria in salted mackerel in Taiwan, *Food Microbiol.*, 2005, **22**, 461–467.
105. V. Capozzi, V. Ladero, L. Beneduce, M. Fernández, M. A. Alvarez, B. Benoit, *et al.*, Isolation and characterization of tyramine-producing *Enterococcus faecium* strains from red wine, *Food Microbiol.*, 2011, **28**, 434–439.

106. H.-F. Kung, T.-Y. Wang, Y.-R. Huang, C.-S. Lin, W.-S. Wu, C.-M. Lin, *et al.*, Isolation and identification of histamine-forming bacteria in tuna sandwiches, *Food Control*, 2009, **20**, 1013–1017.
107. R. Mazzoli, C. Lamberti, J. D. Coisson, M. Purrotti, M. Arlorio, M. G. Giuffrida, *et al.*, Influence of ethanol, malate and arginine on histamine production of *Lactobacillus hilgardii* isolated from an Italian red wine, *Amino Acids*, 2009, **36**, 81–89.
108. K. Henríquez-Aedo, D. Durán, A. Garcia, M. B. Hengst and M. Aranda, Identification of biogenic amines-producing lactic acid bacteria isolated from spontaneous malolactic fermentation of chilean red wines, *LWT--Food Sci. Technol.*, 2016, **68**, 183–189.
109. A. Romano, H. Trip, H. Campbell-Sills, O. Bouchez, D. Sherman, J. S. Lolkema, *et al.*, Genome sequence of *Lactobacillus saerimneri* 30a (formerly *Lactobacillus* sp. Strain 30a), a reference lactic acid bacterium strain producing Biogenic amines, *Genome Announc.*, 2013, **1**, e00097-12.
110. J. Emborg, P. Dalgaard and P. Ahrens, *Morganella psychrotolerans* sp. nov., a histamine-producing bacterium isolated from various seafoods, *Int. J. Syst. Evol. Microbiol.*, 2006, **56**, 2473–2479.
111. H.-F. Kung, C.-Y. Huang, C.-M. Lin, L.-H. Liaw, Y.-C. Lee and Y.-H. Tsai, The histamine content of dried flying fish products in Taiwan and the isolation of halotolerant histamine-forming bacteria, *J. Food Drug Anal.*, 2015, **23**, 335–342.
112. W. Qi, L.-H. Hou, H.-L. Guo, C.-L. Wang, Z.-C. Fan, J.-F. Liu, *et al.*, Effect of salt-tolerant yeast of *Candida versatilis* and *Zygosaccharomyces rouxii* on the production of biogenic amines during soy sauce fermentation, *J. Sci. Food Agric.*, 2014, **94**, 1537–1542.
113. F. Gardini, R. Tofalo, N. Belletti, L. Iucci, G. Suzzi, S. Torriani, *et al.*, Characterization of yeasts involved in the ripening of Pecorino Crotonese cheese, *Food Microbiol.*, 2006, **23**, 641–648.
114. M.-T. Wyder, H.-P. Bachmann and Z. Puhan, Role of selected yeasts in cheese ripening: An evaluation in foil wrapped raclette cheese, *LWT--Food Sci. Technol.*, 1999, **32**, 333–343.
115. J. S. Saez, C. A. Lopes, V. E. Kirs and M. Sangorrín, Production of volatile phenols by *Pichia manshurica* and *Pichia membranifaciens* isolated from spoiled wines and cellar environment in Patagonia, *Food Microbiol.*, 2011, **28**, 503–509.
116. P. M. Lucas, W. A. M. Wolken, O. Claisse, J. S. Lolkema and A. Lonvaud-Funel, Histamine-producing pathway encoded on an unstable plasmid in *Lactobacillus hilgardii* 0006, *Appl. Environ. Microbiol.*, 2005, **71**, 1417–1424.
117. E. Pessione, R. Mazzoli, M. G. Giuffrida, C. Lamberti, E. Garcia-Moruno, C. Barello, *et al.*, A proteomic approach to studying biogenic amine producing lactic acid bacteria, *Proteomics*, 2005, **5**, 687–698.

118. P. M. Lucas, V. S. Blancato, O. Claisse, C. Magni, J. S. Lolkema and A. Lonvaud-Funel, Agmatine deiminase pathway genes in *Lactobacillus brevis* are linked to the tyrosine decarboxylation operon in a putative acid resistance locus, *Microbiology*, 2007, **153**, 2221–2230.
119. M. P. Arena, A. Romano, V. Capozzi, L. Beneduce, M. Ghariani, F. Grieco, *et al.*, Expression of *Lactobacillus brevis* IOEB 9809 tyrosine decarboxylase and agmatine deiminase genes in wine correlates with substrate availability, *Lett. Appl. Microbiol.*, 2011, **53**, 395–402.

CHAPTER 3

Controlling Biogenic Amine Formation in Food

JOSÉ M. LORENZO,*[a] PAULO E. MUNEKATA,[a] BELÉN GÓMEZ,[a] ARISTIDE MAGGIOLINO,[b] DANIEL FRANCO,[a] PASQUALE DE PALO[b] AND FRANCISCO J. BARBA[c]

[a] Centro Tecnológico de la Carne de Galicia, Adva. Galicia n° 4, Parque Tecnológico de Galicia, San Cibrao das Viñas, 32900 Ourense, Spain; [b] Department of Veterinary Medicine – University of Bari A. Moro, Italy, S.P. per Casamassima, km 3, 70010, Valenzano, Bari, Italy; [c] Nutrition and Food Science Area, Preventive Medicine and Public Health, Food Sciences, Toxicology and Forensic Medicine Department, Faculty of Pharmacy, Universitat de València, Avda. Vicent Andrés Estellés, s/n, 46100 Burjassot, València, Spain

*Email: jmlorenzo@ceteca.net

3.1 Introduction

Biogenic amines (BAs) occur in food as a consequence of microbial activity, thus converting specific amino acids into organic bases, mainly by a decarboxylation reaction. The main factors associated with formation of BAs are as follows: the availability of amino acids; growth of decarboxylase-positive microorganisms; food characteristics that induce the activity of microorganisms and decarboxylases; and the processing and storage conditions. The most relevant BAs in food are histamine, cadaverine, β-phenylethylamine, putrescine, spermidine, spermine, tyramine, and

Food Chemistry, Function and Analysis No. 20
Biogenic Amines in Food: Analysis, Occurrence and Toxicity
Edited by Bahruddin Saad and Rosanna Tofalo

Published by the Royal Society of Chemistry, www.rsc.org

tryptamine.[1] Each BA has a precursor amino acid that is released by proteases during processing and/or storage: histidine (histamine), lysine (cadaverine), ornithine (putrescine), and tyrosine (tyramine) (Figures 3.1–3.3).[2]

The presence of BAs in food has been reported in several studies, which revealed the distribution in the modern human diet: fermented food such as sausages, cheese and fish.[3–5] It is worth noting that the occurrence of BAs has been reported even in ready-to-eat baby food even though at very low levels: up to 2.4 mg g^{-1} for histamine and 1.7 mg g^{-1} for tyramine.[6] The consumption of food contaminated with BAs, mainly histamine, can lead to several symptoms that include alteration of blood pressure, headache, itching, skin oedema, dizziness, faintness, a burning sensation in the mouth, inability to swallow, diarrhoea, and vomiting.[7,8] The development of symptoms is also concentration dependent in food. For histamine, a safe limit was recommended by European legislation, particularly for fish and

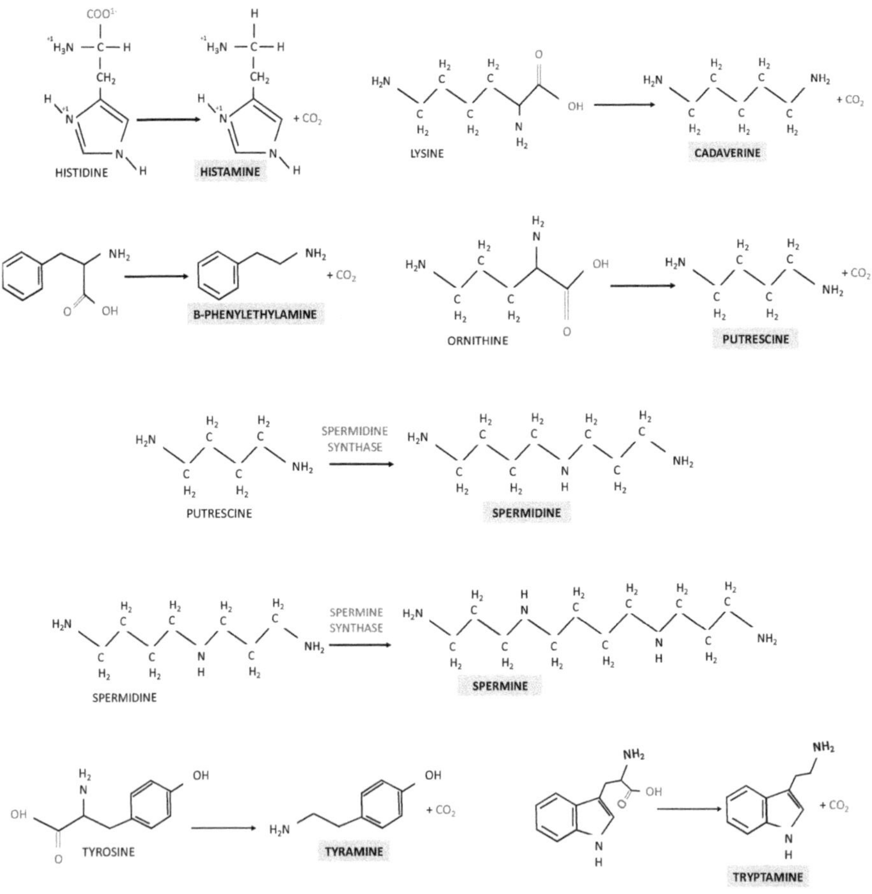

Figure 3.1 Relevant biogenic amines (BAs) in food.

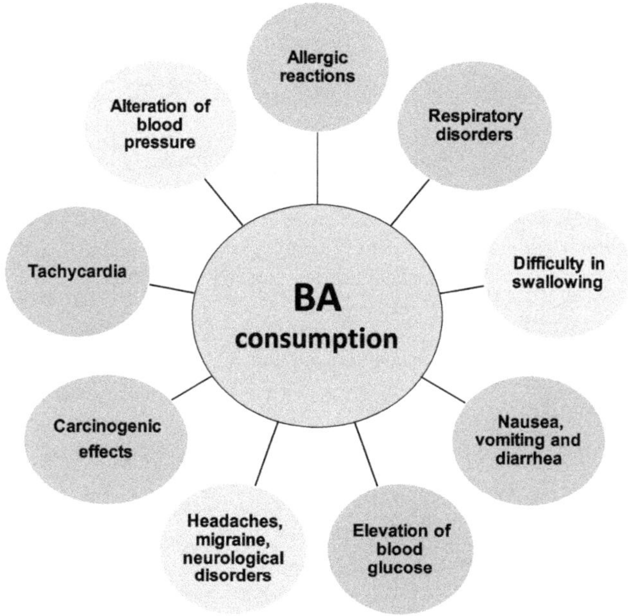

Figure 3.2 Symptoms of consumption of food contaminated with biogenic amines (BAs).

fish products: 200 and 400 mg kg^{-1}, respectively. However, the same authority also indicated that at lower doses, histamine can cause effects in health that vary according to individual susceptibility.[9]

Cases reported in the scientific literature indicate the variety of scenarios that lead to BA poisoning: incorrect storage and cooking of tilapia induced a fast response (5 min) which included flushing in the face, chest, and back along with erythemas;[10] consumption of spoilage canned tuna with altered appearance and taste led to a severe case of recurrent admission due to expected symptoms of histamine contamination and other symptoms associated to prior asthma which led to difficulties in breathing and low blood oxygenation;[11] and the consumption of cooked spoilage crabs, bought from a local market, induced intense abdominal pain, vomiting, and dizziness, followed by no signs of improvement after proper medication, and led to death.[12]

Food processing and storage has a determinant role in reducing the risk associated with BA poisoning. The simultaneous occurrence of one or more factors that can induce BA formation and accumulation does not necessarily indicate that BA will be formed but preventive actions are still necessary. However, food processing is complex and relies on several parameters that must be properly evaluated to achieve safety levels. This chapter will discuss the impact of food processing in the reduction of BA poisoning risk by means of conventional and emerging technologies as well as innovative strategies.

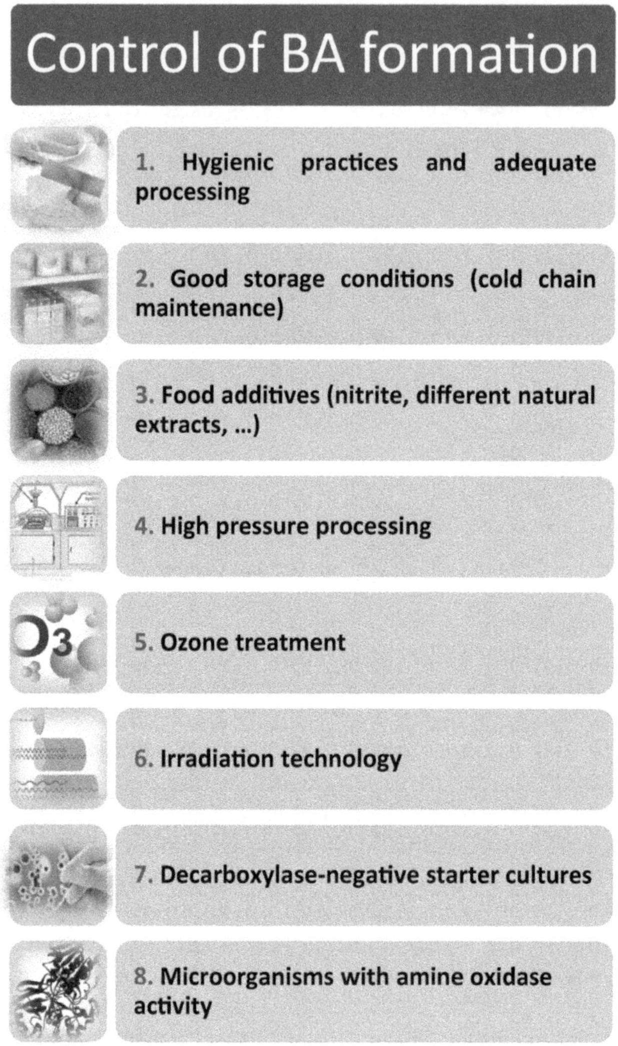

Figure 3.3 Reported factors to control the accumulation of biogenic amines (BAs).

3.2 Factors that Control the Occurrence of BAs

Any food that contains precursor amino acids is subjected to the accumulation of BAs, which is not just restricted to the main protein sources in the human diet (*e.g.* fish, meat, cheese, and milk), but has also been observed in wine, beer, juices, and yogurt.[13] Changes in protein structure, increasing acidity, loss of moisture, and presence of sodium chloride can favour the release of amino acids.[2]

Several microbial strains have been characterized by decarboxylase activity. The occurrence of decarboxylase-positive strains is not limited to microorganisms related to food spoilage but also to strains commonly used on food fermentation: *Bacillus*, *Carnobacterium*, *Citrobacter*, *Clostridium*, *Escherichia*, *Enterococcus*, *Klebsiella*, *Kocuria*, *Lactobacillus*, *Lactococcus*, *Leuconostoc*, *Pediococcus*, *Photobacterium*, *Proteus*, *Pseudomonas*, *Salmonella*, *Shigella*, *Staphylococcus*, and *Streptococcus*.[2,14] Important aspects used to prevent the contamination of food by carboxylase-positive bacteria are hygiene practices. The contamination during processing and storage can induce the formation of BAs.[15,16] The activity of decarboxylase can be influenced by food pH, sodium chloride content, and dimension, storage temperature and time, and food additives (mainly sodium nitrite).[2]

Hygienic practices and adequate processing and storage conditions play a central role in the prevention of BA formation in food. Several studies in the scientific literature indicate that contamination during processing and storage of food leads to BA food poisoning such as those reported by Emborg and Dalgaard for cold smoked tuna;[17] Tsai *et al.* for canned mackerel;[18] and by Chang *et al.* for swordfish fillets.[19] The control of each main factor related to the production of BAs in food is crucial to plan and evaluate the impact of strategies to reduce the accumulation in food.

3.3 Practical Ways to Control BA Formation

The use of strategies and technologies to control the formation of BAs in food is mainly focused in the control and reduction of carboxylase-positive microorganisms. Moreover, some technologies and strategies can also reduce the BA content in food by directly degrading BAs. The use of thermal technologies is a common practice worldwide to reduce microbial load in food. When food processing is properly executed, the main microorganisms associated with production of BAs are affected and the risk of food poisoning can be reduced to a safe level. The study carried out by Novella-Rodríguez *et al.* indicated that pasteurization of milk did not influence the cadaverine, putrescine, and spermidine and spermine content in cow's milk but reduced the formation of BAs during cheese ripening when pasteurized milk was used.[20] Similarly, a study carried out by Veciana-Nogués, Mariné-Font, and Vidal-Carou revealed non-significant differences on individual and total BA content between fresh and sterilized canned tuna.[21]

Cold chain management during processing and storage is another current strategy to prevent BA formation in food. The effect of frozen and refrigerated storage followed by 20 days of storage at either 4 or 15 °C of raw materials on BA content of spontaneously fermented sausages was evaluated by Bover-Cid, Izquierdo-Pulido, and Vidal-Carou. The authors observed that frozen storage followed by either 4 or 15 °C prevented the formation of histamine and octopamine. However, refrigerated storage followed by either 4 or 15 °C did not display the same effect and relevant increases in BA content were observed. According to the authors, increasing temperature was a factor associated with higher BA content even with the use of good hygienic quality of raw materials.[22] A similar outcome was obtained in the experiment carried out by Komprda *et al.* This study revealed that storage at 22 °C is not advised due to a significant increase in total BA content of Czech dry-fermented sausage as a consequence of microbial activity. Since conventional thermal processing and cold chain management are suitable approaches to prevent BA formation, the technologies and strategies discussed in this chapter are important alternatives that can be equal, and even more effective, in preventing both BA formation and accumulation.[23]

3.3.1 Food Additives

Food additives are also associated with an inhibitory effect on decarboxylase-positive microorganisms. The traditional use of nitrite as an antimicrobial in food was also associated with reduced accumulation of BAs.[24] The authors observed that increasing concentrations of nitrite prevented the accumulation of cadaverine, histamine, and putrescine. However, the same study revealed that nisin, a natural antimicrobial used in food, induced the formation of cadaverine, putrescine, and spermidine. The use of natural extracts can also contribute to prevent BA accumulation in food. The study carried out by Fan *et al.* observed that the combined use of bamboo and tea polyphenols in pork sausages prevented the accumulation of both total and individual BA content during 42 days of storage. The authors argued that final BA contents could be attributed to the antimicrobial activity of bamboo and tea polyphenols.[25]

A study carried out by Lu *et al.* evaluated the addition of a complex mixture of tea polyphenols with cinnamon, cloves, ginger, and anise essential oils in Chinese smoked horsemeat sausage. The use of such a natural mixture prevented the accumulation of cadaverine, histamine, 2-phenylethylamine, putrescine, and tyramine after 28 days of ripening in comparison with the control sausage.[26] Wang *et al.* studied the effect of tea polyphenols, grape seed extract, gingerol, and α-tocopherol on BA content of dry-cured bacons. This study indicated that tea polyphenols were the most effective additive to prevent BA accumulation.[27]

3.3.2 High Pressure Processing

High pressure processing (HPP) is one of the main non-thermal technologies applied to improve food safety and the shelf life of food, particularly those with thermolabile compounds. The impact of HPP is dependent on processing parameters, mainly pressure (between 100 and 100 MPa), temperature (from <0 to >100 °C), and time (usually up to 20 min). The inactivation of microorganisms and enzymes is one of the main advantages of HPP, which improves product stability. Regarding chemical bounds, HPP is suggested to impact the tertiary and quaternary structure of complexes molecules (usually with pressures higher than 200 MPa) while covalent bonds remains unaltered during processing.[28]

In microbial inactivation, some mechanisms have been proposed to explain its effectiveness, which include: loss of protein conformity and consequent partial or even total denaturation; change in membrane fluidity impairing its functionality along with possible membrane damage; and inactivation of crucial organelles such as ribosomes and related pathways.[29] By the occurrence of such changes in microorganisms, its normal functionality is inhibited which also reduces the production of BAs.

HPP can reduce the formation of BAs during the processing of fermented food and storage of non-fermented food (Table 3.1). The study carried out by Calzada *et al.* evaluated the effect of HPP during the ripening of raw ewe's milk cheese (400 and 600 MPa for 5 min after 3 and 5 weeks of ripening). The highest levels of inhibition during 240 days of storage were obtained by 600 MPa treatments after both 3 and 5 weeks of ripening, mainly for histamine, putrescine, and tyramine (main BAs). On the other hand, the 400 MPa treatments did not display the same effect for most BAs, except for cadaverine, which was reduced by all HPP treatments.[30]

In seafood, HPP has shown promising results. In raw squid, the reduction in total BA content was shown to be pressure-dependent (200, 300, and 400 MPa). A similar effect was reported for trimethylamine wherein increasing pressure inhibited its formation during 20 days of storage.[34] Křížek *et al.* evaluated the effect of pressure and temperature to preserve Pike (*Esox lucius*) fillets by HPP technology. This study revealed that temperature induced a more expressive reduction on the content of cadaverine, putrescine, histamine, and tyramine on samples subjected to HPP at 3.5 °C than at 12 °C during 70 days of storage. For most BAs, non-significant differences were observed. However, the formation of both spermidine and spermine were not influenced by either pressure or temperature during HPP treatment.[36]

In meat batter (raw meat, sodium nitrite, NaCl, and sodium tripolyphosphate), the influence of HPP in the formation of BAs was evaluated by Ruiz-Capillas *et al.* Increasing temperature was an important factor, as indicated by the authors, since samples treated at 7 °C (300 MPa) displayed the lower agmatine, spermidine, spermine, and tyramine contents in raw batter in comparison with other temperatures (20 °C and 4 °C). However, the lowest temperature also induced the formation of putrescine in raw batter. A similar outcome was observed for cooked batter after HPP treatment. During storage, an opposite trend was observed wherein samples treated at 40 °C displayed lower tyramine, histamine, and putrescine contents while a mild effect was observed on agmatine, cadaverine, spermidine, and spermine after 54 days.[38]

Latorre-Moratalla *et al.* assessed the effect of HPP (200 MPa for 10 min) in two traditional Spanish sausages: chorizo and fuet. The authors observed that cadaverine and putrescine production were inhibited during 21 days of storage for chorizo. Tyramine formation was not influenced by HPP in chorizo and no effect was observed in fuet for none of the BAs. The authors also argued that differences observed between chorizo and fuet sausages could

Table 3.1 High pressure processing to prevent biogenic amine (BA) formation and accumulation.[a]

Pressure	Matrix	Effect on BA	Ref.
400 and 600 MPa (5 min)	Raw ewe's milk cheese	600 MPa (3 weeks) was the most efficient on CAD, HIS, PHE, PUT, TRP, and TYR content after 120 days	30
400 and 600 MPa (5 min)	Raw cow's milk cheese	600 MPa (14 days) was the most efficient on PUT and TYR content after 240 days	31
50 (72 h) and 400 (5 min) MPa	Goat's milk cheese	400 MPa reduced TYR and total BA content; no effect on other BAs during ripening	32
500 (2 min)	Goat's milk (cheese)	Similar effect on cheese BA profile as obtained for pasteurized milk cheese	33
200, 300, and 400 MPa (20 min)	Raw squid	Reduction of TMA and total BA formation	34
500 MPa (0, 5, and 10 min)	Semidried squid	500 MPa for 10 min prevented DMA and TMA; no clear effect on other BA	35
300 and 500 MPa (10 min)	Pike (*Esox lucius*) fillets	Reduction of CAD, PUT, HIS, and TYR content by both pressures at 3.5 °C	36
300 and 500 MPa (10 min)	Trout fillets	500 MPa was the most efficient to prevent CAD, PUT, and TYR formation; no effect on SPD and SPM	37
400 MPa (30 min)	Frankfurters	No effect on BA content	38
200 MPa (10 min)	Spanish fuet and chorizo	Reduced CAD and PUT formation in chorizo; no effect on fuet	39
400 MPa (10 min)	Cooked ham	Delayed CAD, HIS, PUT, and TYR formation; no effect on AGM, SPD, and SPM	40
500 MPa (10 min)	Hungarian fermented sausages	Reduced the formation of CAD and PUT; induced the formation of SPM and TYR	41
100 MPa[b]	Bovine (caciotta cheese) and ovine (pecorino cheese) milk	Reduced CAD and SPD; delayed HIS formation, no effect on PHE, PUT, SPM, and TYR	42

[a]Abbreviations: AGM, agmatine; CAD, cadaverine; HIS, histamine; PHE, β-phenylethylamine; PUT, putrescine; SPD, spermidine; SPM, spermine; TMA, trimethylamine; TRP, tryptamine; TYR, tyramine; DMA, dimethylamine; TMA, trimethylamine.
[b]High pressure homogenization.

be explained by the higher microbial activity observed in chorizo in comparison with fuet.[39]

An interesting experiment was carried out by Lanciotti *et al.* about the effect of high pressure homogenization (HPH) in bovine and ovine milk prior to cheese production. The authors observed that cadaverine and spermidine were reduced in both caciotta (from bovine milk) and pecorino (from goat's milk) cheese. The formation of histamine was delayed but no effect on β-phenylethylamine, putrescine, spermine, and tyramine were observed during 27 days of storage. Although poorly explored, HPH seems to be an interesting technology to treat fluid raw materials and prevent/delay the formation of BAs.[42]

3.3.3 Ozone Treatment

Ozone technology promotes the decontamination of food by interacting with membrane, cytoplasmic, and nucleic components. At the exterior layer, the components of the bacterial cell envelope (*e.g.* polyunsaturated fatty acids and membrane binding proteins) are oxidized and its integrity is compromised. Eventually, the control of membrane permeability is reduced which results in lysis and the leakage of cytoplasmic components. Once ozone is absorbed, enzymes responsible for normal activity are oxidized and important cellular processes, such as the respiratory chain, are inactivated. The structure of nucleic acids can be affected which diminishes the transcription activity. Spores can also be affected by ozone, where the degradation of the protein coat occurs and leads to inactivation.[43]

The use of ozone to reduce microbial load is an emerging technology that can also influence BA accumulation in food. To the best of our knowledge, only a single study has been conducted so far regarding the effect of ozone gas on BA accumulation. In this study, poultry meat was treated with a single dose of ozone (0.82–1.2 ppm) which reduced the levels of putrescine and cadaverine (32.4 and 132.3 mg kg^{-1}, respectively) in comparison with untreated samples (53.6 and 175.2 mg kg^{-1}, respectively). Although this promising study revealed the impact of ozone on BA formation, additional studies are still necessary to evaluate the impact in other matrixes and other related food characteristics.[44]

3.3.4 Irradiation

Irradiation is one of the non-thermal technologies applied in food processing to increase food safety and prolong shelf life. The use of irradiation is dose-dependent and can prevent sprouting in vegetables, reduce insect and parasite contamination, and reduce/eliminate microorganisms. This technology is particularly relevant for packaged (prevent microbial growth and recontamination) and heat-sensible food (preservation of compounds associated with aroma and sensory attributes). Healthy and regulatory authorities support the use of irradiation to process food and recognize that a 10 kGy irradiation dose does not present hazardous effect. Moreover, the lack of specific effects on the nutritional quality of irradiated food also strengthens the use of irradiation.[45,46]

The protective effect attributed to irradiation against BA accumulation has been attributed to two mechanisms: reduction of decarboxylase-positive bacteria load and direct degradation of BAs by exposure to irradiation. In microorganisms, irradiation induces several mechanisms ultimately leading to cell death. The main effects involve the irreversible damage to nucleic acids wherein free radicals produced by radiolysis of water play a central role in this mechanism.[45] As a consequence, the production of BAs is reduced. Likewise, ionizing radiation affects the integrity of BAs leading to degradation after exposure.[3,47] Several studies support the use of irradiation to prevent BA formation and accumulation in food (Table 3.2).

An interesting study was carried out by Kim *et al.*, who evaluated the stability of several BAs (as aqueous solutions) to direct exposure to irradiation. All BAs tested were affected by irradiation doses of 2.5 kGy. Particularly for putrescine and spermine, this dose induced a reduction of 80 and 90%, respectively. The authors also observed that at 10 kGy putrescine, spermidine and spermine in solution were completely destroyed and reductions in the range of 70–98% for cadaverine, tryptamine, β-phenylethylamine, histamine, tyramine, and agmatine. All BAs were destroyed when 25 kGy doses were applied to solutions. In addition, the authors argued that the use of irradiation can induce radiolytic degradation of BAs and it can be explored as a technology in the food industry to reduce the risk of food poisoning related to BA.[47]

The use of irradiation in food processing to reduce the accumulation of BAs in food has been explored in many studies wherein promising results support the use of such technology. Kim *et al.* studied the effect of irradiation doses on BAs in soybean paste during the fermenting stage. After 12 weeks of fermentation, the final content of most BAs was reduced

Table 3.2 Irradiation to prevent biogenic amine (BA) formation and accumulation.[a]

Irradiation dose (kGy)	Matrix	Effect on BA	Ref.
2.5, 5, 10, 20, or 25 kGy	Distilled water	>95% reduction at 10 kGy; completely degraded by 25 kGy dose	47
5, 10, and 15 kGy	Fermented soybean paste	PUT, CAD, PHE, SPD, TRP, HIS, and TYR at 15 kGy; no effect on AGM and SPM	48
1 and 3 kGy	Atlantic horse mackerel	AGM, CAD, HIS, PUT, TMA, and TYR reduced by 1 and 3 kGy doses	49
5, 10, and 20 kGy	Pepperoni sausage	PUT, SPD, SPM, and TYR; no effect on PHE	50
2, 4, and 6 kGy	Egyptian fermented sausage	Individual and total BA content by irradiation; highest inhibition at 6 kGy dose	51
0.5, 1, and 2 kGy	Chicken breast and thigh	Cut-dependent effect for individual BA; no clear effect for total BA content	52
5 kGy	Chinese Rugao ham	Degradation of PUT, SPM, and TYR; increasing of CAD, PHE, SPD, and TRP	53

[a]Abbreviations: AGM, agmatine; CAD, cadaverine; HIS, histamine; PHE, β-phenylethylamine; PUT, putrescine; SPD, spermidine; SPM, spermine; TMA, trimethylamine; TRP, tryptamine; TYR, tyramine; DMA, dimethylamine; TMA, trimethylamine.

by the use of irradiation except for spermine and agmatine. The study revealed that putrescine was the most affected by irradiation, with a reduction from 207 (non-irradiated) to 132.8, 130.6, and 105.2 mg per 100 mg (5, 10, and 15 kGy, respectively). The authors argued that irradiation reduced microbial load and its capacity to produce BAs during the fermentation of soy paste.[48]

Irradiation does of 1 and 3 kGy protected Atlantic horse mackerel against the accumulation of several BAs. The protective effect induced by irradiation was observed after 8 days of refrigerated storage when a sharp increase in all tested BAs was observed. In comparison, treated samples displayed lower agmatine, cadaverine, histamine, putrescine, trimethylamine, and tyramine contents. Only spermidine content was not influenced by irradiation doses. The irradiation treatment increased the shelf life of Atlantic horse mackerel from 8 (control) to 12 days.[49]

Similar outcomes have been observed for other products of animal origin. Kim *et al.* observed that BA degradation was dose-dependent in pepperoni sausages irradiated with 5, 10, and 20 kGy. The highest dose led to complete degradation of putrescine after 4 weeks of storage in samples packaged in both vacuum and air. The content of other BAs (spermidine, spermine, and tyramine) remained the same or were reduced after 4 weeks in the same package systems. Interestingly, the use of modified atmosphere technology by the combination of CO_2/N_2 gases induced the formation of BAs.[50]

However, some contrasting results were also observed in the literature. Min *et al.* observed that total BA content in chicken breast and thigh inoculated with different *Bacillus cereus*, *Enterobacter cloacae*, or *Alcaligenes faecalis* was mildly affected by irradiation. In addition, the microbial content was reduced in a concentration dependent manner. The authors also compared the effect of irradiation treatment with organic acids (2%) and reported similar results between these two strategies in BA content.[52] Interestingly, the study carried out by Wei *et al.* indicated that the irradiation dose of 5 kGy prevented the accumulation of tyramine, putrescine, and spermine during ripening (265 days) and post-ripening (between 265 and 325 days) periods of Chinese Rugao ham. The authors observed that irradiation induced the formation of phenylethylamine, cadaverine, and tryptamine, which was attributed to continuous microbial activity (process was carried out according to traditional practices) during ripening.[53]

3.3.5 Starter Cultures with Negative-decarboxylase Activity

The selection of decarboxylase-negative microorganisms can be of great advantage to reduce the production of BA in food. The use of such a strategy can increase safety and improve the control of BA forming microorganisms during processing and storage of food.[54] The effect of selecting microbial strains with carboxylase-negative activity is indicated in Table 3.3.

Table 3.3 Selected decarboxylase-negative autochthonous starter cultures to prevent biogenic amine (BA) accumulation in food.[a]

Product	Starter culture	Effect on BA	Sensory attributes	Ref.
Chorizo	*Pediococcus acidilactici* and *Staphylococcus vitulus*	Dependent of strain	Effect on hardness, softness, saltiness, and fibrousness	55
Salchichón	*Pediococcus acidilactici* and *Staphylococcus vitulus*	Reduced CAD and total BA content	No effect on overall acceptability	56
Dry-fermented sausage	*Pediococcus acidilactici* and *Staphylococcus vitulus*	Reduced individual and total BA content	Reduction on hardness, gumminess, and chewiness; no effect on acceptability	57
Dacia sausage	*Lactobacillus sakei*, *Lactobacillus acidophilus*, and *Staphylococcus equorum*	Reduced total BA content by the combination of three bacteria	Improved flavour intensity, quality, and overall acceptability	58
Salsiccia sarda	*Lactobacillus curvatus*, *Lactobacillus plantarum*, and *Staphylococcus xylosus*	Reduced total BA content on *Staphylococcus xylosus*, *Lactobacillus* and *Curvatus* treatment	Effect on texture and colour, and ripened flavour was dependent of treatment	59
Dry-fermented sausage	*Lactobacillus plantarum* and *Staphylococcus xylosus*	Reduced all BAs except SPM	n.d.	60
Silver carp sausage	*Lactobacillus plantarum*	Reduced PUT and CAD	n.d.	61
Fish sauce	*Tetragenococcus halophilus*	Reduced HIS	n.d.	62
Ewe's milk cheese	*Lactococcus lactis*, *Lactobacillus plantarum*, and *Enterococcus raffinosus*	Reduced individual and total BA content	n.d.	63
Harbin dry sausages	*Lactobacillus plantarum* and *Staphylococcus xylosus*	Reduced individual and total BA content	Improved all sensory attributes	64
Chinese sauerkraut	*Lactobacillus plantarum* and *Zygosaccharomyces rouxii*	Reduced CAD, PUT, and TYR	n.d.	66

[a]Abbreviations: CAD, cadaverine; HIS, histamine; PUT, putrescine; SPM, spermine; TYR, tyramine; n.d., not determined.

The study carried out by Casquete *et al.* revealed that combining *Pediococcus acidilactici* and *Staphylococcus vitulus* reduced the accumulation of total BA content in comparison with spontaneous fermented in chorizo (a Spanish dry-fermented sausage). Each combination of strains had a specific effect on individual BAs. While the combination P200S34 (*Pediococcus acidilactici* MS200 and *Staphylococcus vitulus* RS34) was the most effective on preventing the accumulation of cadaverine, histamine was greatly reduced by the P198S34 (*Pediococcus acidilactici* MS198 and *Staphylococcus vitulus* RS34) combination. The authors also highlighted the importance of ripening time on aminogenesis since chorizo sausages (65 or 90 days) ripened for a longer time displayed higher levels of total BA content than those ripened for short periods.[55] In another study, Casquete *et al.* evaluated the use of a single combination of *Pediococcus acidilactici* (MS198) and *Staphylococcus vitulus* (RS34) in salchichón (a traditional Spanish dry-fermented sausage) to prevent BA formation. Differently, only cadaverine and total BA content was significantly reduced in comparison with spontaneously fermented salchichón after 86 days of ripening.[56]

Simion *et al.* studied the effect of combinations of *Lactobacillus sakei*, *Lactobacillus acidophilus*, and *Staphylococcus equorum* as starter cultures in BA accumulation of Dacia sausage (a traditional Romanian sausage). The authors observed that total BA, cadaverine, putrescine, and tyramine contents were reduced after 28 days of ripening by the combination of either *Lactobacillus sakei* with *Staphylococcus equorum* or the three bacteria together.[58] The selection of starter cultures by Mangia *et al.* successfully prevented the production of BAs after the ripening period (21 days at 10–13 °C). According to the authors, the *Staphylococcus xylosus* and *Lactobacillus curvatus* displayed the lowest total BA content (519 mg kg^{-1}) in comparison with other autochthonous starter cultures and non-inoculated sausages (in the range of 626.8–652.1 mg kg^{-1}).[59]

Regarding the use of indigenous starter cultures in fish products, Zhang, Lin, and Nie studied the effect of *Lactobacillus plantarum* on fermentation of Silver carp sausage after 48 h of fermentation. The final content of putrescine and cadaverine were significantly reduced in comparison with spontaneously fermented sausage. However, the inoculated and control sausages had similar tyramine, histamine, spermidine, and spermine contents at the end of the fermenting stage.[61] Udomsil *et al.* reported the use of *Tetragenococcus halophilus* to prevent the formation of BAs. The content of histamine was reduced from 10.24 (control) to 4.50 mg per 100 mL (selected starter cultures). However, no effects or BA forming activity were observed for tryptamine, putrescine, and cadaverine. In addition, the authors also argued that differences among strains influenced the final BA content.[62]

A similar preventive effect was reported by Renes *et al.* who studied the use of *Lactococcus lactis* (2 strains), *Lactobacillus plantarum* (1 strain), and *Enterococcus raffinosus* (1 strain) on ewe's milk cheese. The lowest total and individual BA (phenylethylamine, spermine, and tryptamine) contents were obtained with starter cultures composed by indigenous *Lactococcus lactis* subsp. *lactis* with *Lactococcus lactis* subsp. *cremoris* and the combination of these two strains with *Lactobacillus plantarum* after 210 days. On the other hand, the highest total BA content was observed in cheese inoculated with *Lactococcus lactis* subsp. *lactis*, *Lactococcus lactis* subsp. *cremoris*, and *Enterococcus raffinosus*, followed by cheese added of commercial starter culture. The authors argued that this outcome could be attributed to intense proteolytic activity of these starter cultures. In such conditions, an increase in the concentration of free amino acids (precursor of BAs) was observed.[63]

The use of selected autochthonous starter cultures can also contribute in the processing of traditionally fermented food of vegetable origin. Wu *et al.* produced pickled cabbage and radish by inoculation of *Lactobacillus plantarum* and *Zygosaccharomyces rouxii*. During fermentation, the accumulation of selected BAs (cadaverine, putrescine, and tyramine) was lower than observed in spontaneously fermented products. Moreover, the authors also observed that the concentration of BAs reduced over the fermentation period.[65]

3.4 Microorganisms that Have the Ability to Degrade BAs and Their Technological Relevance in Food

In this section, the technological relevance of microorganism with amine oxidase activity is discussed. The relevance of this class of microorganism is the objective: decompose BAs.[66] This approach has the same potential advantages as indicated for using decarboxylase-negative microorganisms and the additional amine oxidase activity. The reaction of BA oxidation is based in the deamination by amine oxidase that produces hydrogen peroxide, ammonia, and the respective aldehyde.[67] The potential use of isolated BA-degrading strains was explored by several authors in both *in vitro* and food samples (Table 3.4).

In vitro studies indicate the potential use of amine degrading bacteria as potential starter cultures on food. Fadda, Vignolo and Oliver characterized the *in vitro* BA degradation activity of *Lactobacillus casei* and *Kocuria varians* to degrade tyramine. The reduction was species-dependent: reduction of 98% for *Lactobacillus casei* and 39% for *Kocuria varians*.[68] A similar outcome was obtained by Zaman *et al.*, who studied the *in vitro* activity of several selected bacteria, *Bacillus amyloliquefaciens*, *Bacillus subtilis*, and *Staphylococcus intermedius*, on histamine, putrescine, and cadaverine.[69]

In accordance with *in vitro* results, several studies reported the BA degradation of starter cultures in fermented food. Capozzi *et al.* isolated 26 strains of *Lactobacillus plantarum* from wine samples and evaluated their potential use as BA-degrading strains for wine producing. The capacity to degrade histamine, tyramine, cadaverine, and putrescine was mainly observed in five strains. The maximum reduction levels *in vitro* were 31.1, 22.1, 8.6, and 6.7% for putrescine, tyramine, histamine, and cadaverine,

Table 3.4 Selected amine oxidase-positive microorganisms applied to degrade biogenic amine (BA) in food.[a]

Microorganism	Matrix	Target BA	Reduction	Ref.
Lactobacillus casei and *Kocuria varians*	*In vitro*	TYR	From 39% (*Kocuria varians*) to 98% (*L. casei*)	68
Bacillus amyloliquefaciens, *Bacillus subtilis*, and *Staphylococcus intermedius*	*In vitro*	HIS, PUT, and CAD	Up to 59% for HIS (*B. amyloliquefaciens*), 30% for PUT (*S. intermedius*), and 29% for CAD (*B. subtilis*)	69
Lactobacillus plantarum	*In vitro*	HIS, TYR, CAD, and PUT	Up to 31.1% for PUT, 22.1% for TYR, 8.6% for HIS, and 6.7% for CAD	70
Lactobacillus casei, *Lactobacillus hilgardii*, *Lactobacillus plantarum*, *Oenococcus oeni*, *Pediococcus parvulus*, and *Pediococcus pentosaceus*	*In vitro* and wine	HIS, TYR, and PUT	Up to 54% for HYS, 55% for TYR, and 69% for PUT on *in vitro* assay; around 28% for HYS, 11% for TYR, and 35% for PUT on wine sample for all strains	71
Lactobacillus sakei	*In vitro* and ensilage fish slurry	HIS	Up to 56 on *in vitro* assay and 53 g per 100 g for ensilage fish slurry	72
Lactobacillus farciminis, *Lactobacillus plantarum*, and *Pediococcus acidilactici*	Wine	HIS, TYR, and PUT	Up to 27% for HIS, 28% for TYR, and 44% for PUT	73
Natrinema gari	Fish sauce	HIS	Up to 24%	74
Lactobacillus curvatus and *Kocuriu variuns*	Fermented sausage	TYR	Up to 60%	75
Staphylococcus epidermidis	Fermented sausage	HIS	Up to 60%	76
Brevibacterium linens	Munster cheese	HIS and TYR	From 0.9–1 to 0.3–0.5 g kg^{-1}	77

[a]Abbreviations: CAD, cadaverine; HIS, histamine; PUT, putrescine; SPM, spermine; TYR, tyramine.

respectively.[70] Similarly, a study revealed that isolated bacterial from wine raw materials and commercial strains used in wine production revealed similar outcome. The results of the *in vitro* assay indicated that histamine content was reduced by 54% (*Lactobacillus casei*), tyramine by 55% (*Lactobacillus casei*), and putrescine by 69% (*Pediococcus pentosaceus*) *in vitro*. In wine samples, the degradation of BAs by selected bacteria was limited to 23–28 (histamine), 10–12 (tyramine), and 34–36% (putrescine) for all samples. According to the authors, wine components such as ethanol and polyphenolic compounds may explain the drastic reduction in the BA-degrading activity of selected strains.[71]

In a similar way, the *in vitro* assessment of *Lactobacillus sakei* strains capacity to degrade histamine was carried out by Dapkevicius *et al.* The authors observed that all four selected strains (isolated from naturally fermented mackerel paste) degraded histamine (19.6–56.2 g per 100 g) after 30 h of incubation. In addition, the authors also evaluated the performance of the *Lactobacillus sakei* strain to degrade histamine in ensilage fish slurry and noted the activity observed *in vitro* was preserved (degradation in the range of 49.9–53.6 g per 100 g). Another study revealed that bacterial strains isolated from traditionally fermented sausages can also degrade BA *in vitro* (2001). In this study, *Lactobacillus casei* (two strains), *Lactobacillus plantarum* (two strains), *Kocuria varians* (four strains), and *Micrococcus varians* (one strain) displayed different potential to degrade tyramine that ranged from 39 to 98% of degradation.[72]

A recent study carried out by Callejón *et al.* tested the potential use of isolated lactic acid bacteria to reduce the content of histamine, tyramine, and putrescine in wine samples. The authors selected *Lactobacillus farciminis* (one strain), *Lactobacillus plantarum* (four strains), and *Pediococcus acidilactici* (one strain) among 76 strains. The highest level of degradation for the three BAs was obtained for *Lactobacillus plantarum* strains. Interestingly, *Lactobacillus farciminis* did not degrade histamine but had the highest potential to degrade putrescine (44%) among all selected bacteria.[73] Chaikaew *et al.* evaluated the use of a bioreactor to reduce histamine content in fermented fish sauce. In this study, *Natrinema gari* cells (10%) were immobilized in diatomite which degraded up to 24% of initial histamine content (600 mg L^{-1}; sum of natural and added histamine) after 2 h of incubation in the presence of riboflavin (10 mg per 100 mL) under static conditions. Flow rate had a significant effect on histamine degradation efficiency. The content of histamine was decreased by 60% after 100 mL (first cycle) of fish sauce was injected (0.1, 0.5, and 2.0 mL min^{-1}) in the bioreactor. However, the authors observed that the efficiency of histamine degradation was reduced as the number of cycles increased using the same immobilized cells (flows of 0.1 and 0.5 mL min^{-1}) from 60% to 25%. The capacity to degrade histamine in the bioreactor was drastically reduced when 2.0 mL min^{-1} was tested. According to the authors, only 10% of histamine was degraded.[74]

The study carried out by Leuschner and Hammes with *Lactobacillus curvatus* and *Kocuriu variuns* in fermented sausages revealed the protective effect of such bacteria against BA accumulation. The authors observed a sharp decrease in tyramine concentration in the first days from an initial 103 mg kg^{-1} to around 60 mg kg^{-1}. Moreover, the tyramine content remained stable until the end of ripening, which indicated a consumption of initial and forming tyramine during ripening.[75]

More recently, Wang *et al.* evaluated the BA degrading capacity of *Staphylococcus epidermidis* during the ripening of pork sausage (24 days at 15–18 °C). This study revealed that *S. epidermidis* displays amine oxidase activity throughout the processing period (fermenting for 3 days followed by 24 days of ripening). The intense production of histamine observed on non-inoculated sausage on the first 15 days was also observed in inoculated sausage. Interestingly, the histamine maximum concentration in non-inoculated sausage was observed after 15 days (35–40 mg kg^{-1} dry basis) while the histamine concentration in *S. epidermidis* sausage drastically reduced over time. According to the authors, 60% of histamine accumulation was obtained by *S. epidermidis* inoculation.[76]

Particularly for cheese, Leuschner and Hammes evaluated the use of *Brevibacterium linens* as surface inocula to reduce the accumulation of BAs. The authors observed that three selected strains (LTH 456, LTH 3686, and LTH3813) actively degrade tyramine during 27 days of ripening of cheese added with histamine. The initial values were reduced from 0.7–0.8 to 0.2–0.3 g kg^{-1}. The same effect was observed for cheese inoculated with LTH 456 and LTH 3686 strains for histamine content (from 0.9–1 to 0.3–0.5 g kg^{-1}). Only strain LTH3813 had no effect on histamine content.[77]

Although the isolation and characterization of BA degrading activity on selected bacteria is a necessary step for further application in food as a decontamination technology, more studies using these isolated strains as starter cultures in fermented food are necessary. It is crucial to determine the effect of selected stains on ultimate concentration of BAs in final products. Selected strains must prevent the excessive accumulation of BAs beyond safety levels.

3.5 Conclusions

The current use of thermal processing and cold chain management in food processing and storage can reduce the risk of BA poisoning. It is important to highlight that a balance between the inhibition of BA formation and accumulation with adequate processing is a crucial point in selecting appropriate technologies.

Therefore, it is possible to increase the control on the production and accumulation of BAs in food. The control of microbial load and composition is the main strategy to reduce BA accumulation and food poisoning. The selection of the most suitable technique is still dependent on food properties, since important changes in food characteristics may occur. In a broad point of view, most of the technologies (thermal processing, HPP, ozone, and irradiation) and strategies (addition of antimicrobial ingredients) applied to reduce microbial load have a major impact in BA formation. For fermented food in particular, selecting both decarboxylase-negative and amine oxidase-positive microorganisms seems the most suitable strategy for fermented products. It is worth mentioning that irradiation is also a very interesting technology to prevent both formation and accumulation of BAs in food, but such technology is limited by its complex operational conditions.

Acknowledgements

The authors José Manuel Lorenzo and Paulo Eduardo Sichetti Munekata are members of the MARCARNE network, funded by CYTED (ref. 116RT0503). The present chapter has been edited during the visiting period of Prof. José M. Lorenzo to the Department of Veterinary Medicine of Bari, granted by the University A. Moro of Bari (DR 3681 del 22/11/2017). Paulo E. S. Munekata acknowledges postdoctoral fellowship support from Ministry of Economy and Competitiveness (MINECO, Spain) "Juan de la Cierva" program (FJCI-2016-29486).

References

1. M. S. Santos, *Int. J. Food Microbiol.*, 1996, **29**(2-3), 213.
2. G. Suzzi and F. Gardini, *Int. J. Food Microbiol.*, 2003, **88**(1), 41.
3. M. A. Rabie, S. Elsaidy, A. A. el-Badawy, H. Siliha and F. X. Malcata, *J. Food Prot.*, 2011, **74**(4), 681.
4. Y. Wu, Y. Chen, L. Li, X. Yang, S. Yang, W. Lin, Y. Zhao and J. Deng, *J. Ocean Univ. China*, 2016, **15**(4), 681.
5. H. Zhai, X. Yang, L. Li, G. Xia, J. Cen, H. Huang and S. Hao, *Food Control*, 2012, **25**(1), 303.
6. A. Czajkowska-Mysłek and J. Leszczyńska, *Food Chem. Toxicol.*, 2017, **105**, 82.
7. P. Visciano, M. Schirone, R. Tofalo and G. Suzzi, *Front. Microbiol.*, 2014, **5**, 1.
8. A. Bulushi, S. Poole, H. C. Deeth and G. A. Dykes, *Crit. Rev. Food Sci. Nutr.*, 2009, **49**(4), 369.
9. EFSA Panel on Biological Hazards (BIOHAZ), *EFSA J.*, 2011, **9**(10), 1.
10. S. P. Nordt and D. Pomeranz, *Am. J. Emerg. Med.*, 2016, **34**(2), 339e1.
11. B. J. Wilson, R. J. Musto and W. A. Ghali, *J. Gen. Intern. Med.*, 2012, **27**(7), 878.
12. Y. Yu, P. Wang, L. Bian and S. Hong, *J. Forensic Sci.*, 2018, **63**(3), 980.
13. B. ten Brink, C. Damink and H. M. L. J. Joosten, *Int. J. Food Microbiol.*, 1990, **11**(1), 73.
14. A. Halász, A. Barath, L. Simon-Sarkadi and W. Holzapfel, *Trends Food Sci. Technol.*, 1994, **5**(2), 42.
15. R. Shalaby, *Food Res. Int.*, 1996, **29**(7), 675.

16. E. Stratton, R. W. Hutkins and S. L. Taylor, *J. Food Prot.*, 1991, **54**(6), 460.
17. J. Emborg and P. Dalgaard, *J. Food Prot.*, 2006, **69**(4), 897.
18. Y. H. Tsai, H. F. Kung, T. M. Lee, H. C. Chen, S. S. Chou, C. I. Wei and D. F. Hwang, *Food Control*, 2005, **16**(7), 579.
19. S. C. Chang, H. F. Kung, H. C. Chen, C. S. Lin and Y. H. Tsai, *Food Control*, 2008, **19**(1), 16.
20. S. Novella-Rodríguez, M. T. Veciana-Nogues, A. X. Roig-Sagues, A. J. Trujillo-Mesa and M. C. Vidal-Carou, *J. Dairy Res.*, 2004, **71**(2), 245.
21. T. Veciana-Nogués, A. Mariné-Font and M. C. Vidal-Carou, *J. Agric. Food Chem.*, 1997, **45**(11), 4324.
22. S. Bover-Cid, M. Izquierdo-Pulido and M. C. Vidal-Carou, *J. Food Prot.*, 2000, **63**(11), 1544.
23. T. Komprda, D. Smělá, P. Pechová, L. Kalhotka, J. Štencl and B. Klejdus, *Meat Sci.*, 2004, **67**(4), 607.
24. Ş. Kurt and Ö. Zorba, *J. Sci. Food Agric.*, 2010, **90**(15), 2669.
25. W. Fan, Y. Yi, Y. Zhang and P. Diao, *Food Sci. Biotechnol.*, 2015, **24**(2), 421.
26. S. Lu, H. Ji, Q. Wang, B. Li, K. Li, C. Xu and C. Jiang, *Food Control*, 2015, **50**, 869.
27. Y. Wang, F. Li, H. Zhuang, L. Li, X. Chen and J. Zhang, *J. Food Sci.*, 2015, **80**(3), C547.
28. M. Yaldagard, S. A. Mortazavi and F. Tabatabaie, *Afr. J. Biotechnol.*, 2008, 7(16), 2739.
29. E. Georget, R. Sevenich, K. Reineke, A. Mathys, V. Heinz, M. Callanan, C. Rauh and D. Knorr, *Innovative Food Sci. Emerging Technol.*, 2015, **27**, 1.
30. J. Calzada, A. del Olmo, A. Picón, P. Gaya and M. Nuñez, *Appl. Environ. Microbiol.*, 2012, **79**(4), 1277.
31. J. Calzada, A. del Olmo, A. Picón, P. Gaya and M. Nuñez, *Food Bioprocess Technol.*, 2015, **8**(2), 319.
32. S. Novella-Rodríguez, M. T. Veciana-Nogués, J. Saldo and M. C. Vidal-Carou, *J. Agric. Food Chem.*, 2002, **50**(25), 7288.
33. S. Novella-Rodríguez, M. T. Veciana-Nogués, A. J. Trujillo-Mesa and M. C. Vidal-Carou, *J. Food Sci.*, 2002, **67**(8), 2940.
34. J. Gou, H. Xu, G. P. Choi, H. Y. Lee and J. Ahn, *Food Sci. Biotechnol.*, 2010, **19**(4), 923.
35. J. Gou, K. P. Choi, X. He and J. Ahn, *J. Food Sci.*, 2010, **75**(7), M489.
36. M. Křížek, K. Matějková, F. Vácha and E. Dadáková, *Food Chem.*, 2014, **151**, 466.
37. K. Matějková, M. Křížek, F. Vácha and E. Dadáková, *Food Chem.*, 2013, **137**(1-4), 31.
38. C. Ruiz-Capillas, J. Carballo and F. Jiménez-Colmenero, *Food Chem.*, 2007, **104**(1), 202.
39. M. L. Latorre-Moratalla, S. Bover-Cid, T. Aymerich, B. Marcos, M. C. Vidal-Carou and M. Garriga, *Meat Sci.*, 2007, **75**(3), 460.

40. C. Ruiz-Capillas, J. Carballo and F. J. Colmenero, *Meat Sci.*, 2007, **75**(3), 397.
41. L. Simon-Sarkadi, K. Pásztor-Huszár, I. Dalmadi and G. Kiskó, *Food Res. Int.*, 2012, **47**(2), 380.
42. R. Lanciotti, F. Patrignani, L. Iucci, M. E. Guerzoni, G. Suzzi, N. Belletti and F. Gardini, *Food Chem.*, 2007, **104**(2), 693.
43. M. A. Khadre, A. E. Yousef and J. G. Kim, *J. Food Sci.*, 2001, **66**(9), 1242.
44. R. Mercogliano, A. De Felice, N. Murru, S. Santonicola and M. L. Cortesi, *J. Food Process. Technol.*, 2014, **5**, 1.
45. J. Farkas, *Trends Food Sci. Technol.*, 2006, **17**(4), 148.
46. J. Farkas and C. Mohácsi-Farkas, *Trends Food Sci. Technol.*, 2011, **22**(2-3), 121.
47. J. H. Kim, H. J. Ahn, C. Jo, H. J. Park, Y. J. Chung and M. W. Byun, *Food Control*, 2004, **15**(5), 405.
48. J. H. Kim, H. J. Ahn, D. H. Kim, C. Jo, H. S. Yook, H. J. Park and M. W. Byun, *J. Food Sci.*, 2003, **68**(1), 80.
49. R. Mendes, H. A. Silva, M. L. Nunes and J. M. A. Empis, *Eur. Food Res. Technol.*, 2005, **221**(3-4), 329.
50. J. H. Kim, H. J. Ahn, J. W. Lee, H. J. Park, G. H. Ryu, I. J. Kang and M. W. Byun, *Food Chem.*, 2005, **89**(2), 199.
51. M. A. Rabie, H. Siliha, S. El-Saidy, A. A. El-Badawy and F. X. Malcata, *Innovative Food Sci. Emerging Technol.*, 2010, **11**(4), 661.
52. J. S. Min, S. O. Lee, A. Jang, C. Jo and M. Lee, *Poult. Sci.*, 2007, **86**(9), 2034.
53. F. Wei, X. Xu, G. Zhou, G. Zhao, C. Li, Y. Zhang, L. Chen and J. Qi, *Meat Sci.*, 2009, **81**(3), 451.
54. J. M. Lorenzo, P. E. S. Munekata and R. Domínguez, *Curr. Opin. Food Sci.*, 2017, **14**, 61.
55. R. Casquete, M. J. Benito, A. Martín, S. Ruiz-Moyano, E. Aranda and M. G. Córdoba, *J. Food Sci.*, 2012, 77(1), M70.
56. R. Casquete, M. J. Benito, A. Martín, S. Ruiz-Moyano, F. Pérez-Nevado and M. G. Córdoba, *Int. J. Food Sci. Technol.*, 2012, **47**(5), 1011.
57. R. Casquete, M. J. Benito, A. Martín, S. Ruiz-Moyano, A. Hernández and M. G. Córdoba, *LWT--Food Sci. Technol.*, 2011, **44**(7), 1562.
58. M. C. Simion, C. Vizireanu, P. Alexe, I. Franco and J. Carballo, *Food Control*, 2014, **35**(1), 123.
59. P. Mangia, A. Trani, A. Di Luccia, M. Faccia, G. Gambacorta, F. Fancello and P. Deiana, *Eur. Food Res. Technol.*, 2013, **236**(3), 557.
60. C. Xie, H. H. Wang, X. K. Nie, L. Chen, S. L. Deng and X. L. Xu, *CyTA--J. Food*, 2015, **13**(4), 491.
61. Q. Zhang, S. Lin and X. Nie, *Food Control*, 2013, **32**(2), 496.
62. N. Udomsil, S. Rodtong, Y. J. Choi, Y. Hua and J. Yongsawatdigul, *J. Agric. Food Chem.*, 2011, **59**(15), 8401.
63. E. Renes, I. Diezhandino, D. Fernandez, R. E. Ferrazza, M. E. Tornadijo and J. M. Fresno, *Food Microbiol.*, 2014, **44**, 271.
64. Q. Sun, Q. Chen, F. Li, D. Zheng and B. Kong, *Food Control*, 2016, **68**, 358.

65. C. Wu, J. Zheng, J. Huang and R. Zhou, *Ann. Microbiol.*, 2014, **64**(2), 847.
66. X. H. Wang, H. Y. Ren, W. Wang, T. Bai and J. X. Li, *J. Food Saf.*, 2015, **35**(3), 395.
67. M. A. Alvarez and M. V. Moreno-Arribas, *Trends Food Sci. Technol.*, 2014, **39**(2), 146.
68. S. Fadda, G. Vignolo and G. Oliver, *Biotechnol. Lett.*, 2001, **23**(24), 2015.
69. M. Z. Zaman, F. A. Bakar, J. SelaMat and J. A. Bakar, *Czech J. Food Sci.*, 2010, **28**(5), 440.
70. V. Capozzi, P. Russo, V. Ladero, M. Fernández, D. Fiocco, M. A. Alvarez, F. Grieco and G. Spano, *Front. Microbiol.*, 2012, **3**, 1.
71. A. García-Ruiz, E. M. González-Rompinelli, B. Bartolomé and M. V. Moreno-Arribas, *Int. J. Food Microbiol.*, 2011, **148**(2), 115.
72. M. L. E. Dapkevicius, M. R. Nout, F. M. Rombouts, J. H. Houben and W. Wymenga, *Int. J. Food Microbiol.*, 2000, **57**(1-2), 107.
73. S. Callejón, R. Sendra, S. Ferrer and I. Pardo, *Appl. Microbiol. Biotechnol.*, 2014, **98**(1), 185.
74. S. Chaikaew, P. Tepkasikul, G. M. Young, K. Osako, S. Benjakul and W. Visessanguan, *Fish. Sci.*, 2015, **81**(5), 971.
75. R. G. K. Leuschner and W. P. Hammes, *Meat Sci.*, 1998, **49**(3), 289.
76. X. H. Wang, H. Y. Ren, W. Wang, T. Bai and J. X. Li, *J. Food Saf.*, 2015, **35**(3), 395.
77. R. G. K. Leuschner and W. P. Hammes, *J. Food Prot.*, 1998, **61**(7), 874.

CHAPTER 4

Biogenic Amines in Fermented Fish

LIVIA SIMON SARKADI

Department of Food Chemistry and Nutrition, Szent István University, Budapest, Hungary
Email: sarkadi@mail.bme.hu

4.1 Characterisation of Biogenic Amines Important in Fish and Fish Products

Biogenic amines are generally derived from corresponding amino acids by microbial decarboxylation in food. The basic requirements for biogenic amines formation in fish and fish products are the availability of free amino acids and the presence of decarboxylase-positive microorganisms such as lactobacilli, enterococci and micrococci, and many strains of Enterobacteriaceae.[1]

The main representatives of biogenic amines occurring in fish and fish products are the following: among aliphatic biogenic amines are putrescine (Put; 1,4-diaminobutane), cadaverine (Cad; 1,5-diaminopentane) and agmatine (Agm; 1-(4-aminobutyl)guanidine); aromatic amines are tyramine (Tym; 4-(2-aminoethyl)phenol) and phenylethylamine (Phem; 2-phenylethylamine); and heterocyclic amines are histamine (Him; 2-(1*H*-imidazol-4-yl)ethanamine) and tryptamine (Trm; 2-(1*H*-indol-3-yl)ethanamine). Polyamines include spermidine (Spd; [*N*-(3-aminopropyl)-1,4-diaminobutane]) and spermine (Spm; [*N*,*N*′-bis(3-aminopropyl)-1,4-diaminobutane]).[2]

Food Chemistry, Function and Analysis No. 20
Biogenic Amines in Food: Analysis, Occurrence and Toxicity
Edited by Bahruddin Saad and Rosanna Tofalo

Published by the Royal Society of Chemistry, www.rsc.org

The biogenic amine content of foods has been widely studied because of their potential toxicity. Histamine is physiologically the most important biogenic amine. In sensitive individuals, the most common allergy-like symptoms caused by histamine poisoning are nausea, vomiting, headache, hypotension, diarrhoea and skin rash.[3]

Determination of the exact toxicity threshold of biogenic amines in individuals is extremely difficult. The toxic dose is strongly dependent on the efficiency of the detoxification mechanisms of different individuals.[4]

Histamine toxicity is also known as scombroid poisoning. Scombroid poisoning is caused by certain spoiled Scombroid fish such as tuna (*Thunnus* spp.), mackerel (*Scomber* spp.) and bonito (*Sarda* spp.) that contain a high amount of free histidine in their muscle tissue, which is precursor of histamine. Many bacteria species, both Gram-positive and Gram-negative bacteria, associated with seafoods are known to have histidine decarboxylase activity such as *Clostridium perfringens*, *Enterobacter aerogenes*, *Escherichia coli*, *Klebsiella pneumoniae*, *Morganella morganii*, *Raoultella planticola*, *Proteus mirabilis*, *Proteus morganii* and *Proteus rettgeri.*[5] Non-*scombroid* fish, such as mahi-mahi (*Coryphaena* spp.), bluefish (*Pomatomus* spp.), amberjack (*Seriola* spp.), herring (*Clupea* spp.), sardines (*Sardinella* spp.) and anchovies (Engraulidae) have also been implicated in scombroid poisoning.[6]

The occurrence of biogenic amines is not only a risk factor for intoxications but is also an indicator of food quality. Diamines (Cad, Put) and polyamines (Spm, Spd) are considered as indicators of the freshness and quality of foods. Mietz and Karmas[7] were the first who proposed a quality index (BAI: Biogenic Amine Index) based on the increases in putrescine, cadaverine and histamine, and decreases in spermine and spermidine during fish storage.

BAI = (histamine + putrescine + cadaverine)/(1 + spermidine + spermine), where the amine concentrations are expressed in mg kg^{-1}. Fish or meat with a BAI value below 1 is considered to be of first quality, whereas BAI values above 10 indicate a very poor microbial quality. There is a strong correlation between the BAI value and organoleptic acceptability and microbiological quality.

Several studies have investigated the biogenic amine content of different food. Information on developments of this topic is given in some valuable papers.[2,8–12]

4.2 Occurrences of Biogenic Amines in Fish and Fish Products

Fermentation has traditionally been used to preserve fresh fish and enhance nutritive value. In addition to the beneficial properties of fermented products, it must be taken into account that fermentation may also provide the conditions required for production of harmful components such as histamine or other biogenic amines. The factors involved in biogenic amine formation in fermented foods are well documented.[13,14]

Fish and fish products are common sources of biogenic amines. A wide range of microorganisms naturally present in fish and fish products are capable of producing biogenic amines. Fresh fish usually contain only low amounts of biogenic amines. Silva *et al.*[15] investigated the histamine content of fresh and canned tuna in Brazil. The mean levels found in fresh tuna from the wholesale market were 0.19 mg kg^{-1}, fresh tuna from the retail market contained 1.30 mg kg^{-1} and 4.41 mg kg^{-1} histamine was detected in canned tuna from the retail market. Each sample was compliant with the requirements. An Italian study on fresh fish and fish products showed that among 216 fresh fish samples, 6% had histamine level higher than 100 mg kg^{-1}. The highest percentage of non-compliant samples (14%) was detected in anchovies.[16]

Tao *et al.*[17] investigated the histamine content of different seafood samples (scombroid fish, non-scombroid fish, fish sauce, dried fish) from ten countries around the world. Histamine content was higher than 50 mg kg^{-1} in 9% of the fish samples and it was also found in some samples to exceed 500 and 1000 mg kg^{-1} histamine concentration. They stated that regular monitoring of histamine content in seafood is required to avoid safety problems of these products.

White-muscle fish is usually considered as a low-risk food with regard to histamine content. In histidine-poor white-muscle fish such as hake, putrescine and cadaverine are the main biogenic amines and they are useful indicator of the quality of fish.[18,19]

Storage conditions (temperature and time) greatly influence amine formation. Krizek *et al.*[20] investigated the effect of storage temperature on biogenic amine formation in fish. They found considerably increased accumulation of histamine and other biogenic amines at higher temperatures. However, some other studies have demonstrated that biogenic amines can accumulate in fish stored at low temperatures.[21] Koral and Köse[22] studied the biogenic amine formation in anchovy (*Engraulis encrasicolus*) at three different chilled storage conditions. They have found that chilled freshwater with ice application increased the shelf life for 8 days and slightly delayed biogenic amine formation.

Bunka *et al.*[23] made a survey on raw freshwater and sea fish offered in restaurants in the Czech Republic: 15% of samples had high concentrations of biogenic amines (over 100 mg kg^{-1}), and in 8% of fish samples, the histamine content was higher than the limit (200 mg kg^{-1}) established in Commission Regulation (EC) No 2073/2005.[24] The results point out the necessity of keeping the right storage conditions during the distribution chain of fish products.

Biogenic amine formation in fish and in other seafoods was investigated during storage by Hu *et al.*[25] Cadaverine, putrescine, histamine and tyramine were the main biogenic amines in the seafood samples. The result showed that freezing effectively prevented the formation of biogenic mines, while a significant increase in biogenic amine concentration during storage at 4 and 25 °C was found. Total mesophilic viable counts and total psychrophilic viable counts were positively correlated with the formation of biogenic amines.

Fermented fish products contain high amounts of amino acids that are easily broken down to hazardous biogenic amines, including histamine.[26–28] Different methods are applied for fish fermentation according to regions including fermentation with salting and drying, fermentation and drying without salting, and fermentation with salting but without drying.[29] In Europe, fish fermentation lasts from a few days to 1–2 years, and in some cases, additives, such as sugars and spices, are added to the brine. The fermented products are commercialized as a whole, or as a fish sauce. In Africa, the period for fish fermentation is short compared with Asia and Europe (from a few days to several weeks) and the product is not transformed into a paste or sauce. In Asia, fish fermentation lasts for several months up to 1 year. The final product is usually fish sauce or several other types of products.

The most popular fermented product is fish sauce which is usually produced from salted small fish or shellfish. Fish sauces are good sources of biogenic amines.[30] Zaman *et al.*[31] identified histamine, putrescine, cadaverine and tyramine as the main amines found in fish sauce. Tryptamine and phenylethylamine were occasionally present at a low level, while spermine, spermidine and agmatine were trace amines in fish sauce. Tsai *et al.*[32] detected an average histamine content of 394 mg kg^{-1} in fish sauce, 263 mg kg^{-1} fish paste and 382 mg kg^{-1} in shrimp paste, and about 26% of samples contained 4500 mg kg^{-1} histamine. In shrimp sauces, putrescine (331 mg kg^{-1}) and tyramine (449 mg kg^{-1}) were also found at high mean concentration.[33] Relatively high histamine contents (63–393 mg kg^{-1}) were measured in anchovy sauces.[33,34]

Besides fish sauces, the concentration of biogenic amines in dried and smoked fish products could also be high.[32] Huang *et al.*[35] detected more than 50 mg kg^{-1} histamine in about 30% of dried fish products in Taiwan. Hsu *et al.*[36] found toxic histamine levels (above 500 mg kg^{-1}) in 44% of dried milkfish samples.

Formation of biogenic amines was also reported in vacuum packaged and cold smoked fermented fish products.[37]

Canned fish products have wide variation in histamine contents depending on the condition of the production. Tsai *et al.*[6] reported that 4.2% of canned fish samples were unacceptable in Taiwan. Park *et al.*[38] detected high histamine content (500 mg kg^{-1}) in canned fish samples in Korea and similar results were found by Silva *et al.*[15] with fish samples in Brazil.

There are several reports about the incidents of histamine poisoning following ingestion of fermented fish products.[39,40]

4.3 Biogenic Amines as Quality Factors for Fish

Histamine has been traditionally used as an indicator of the quality of histidine-rich fish (dark-muscle fish), while putrescine and cadaverine are the most useful indicators of the quality of histidine-poor fish (white-muscle fish), shellfish and fermented seafood products.[9]

In the case of freshwater fish, the sum of putrescine and cadaverine of 20 mg kg^{-1} or putrescine of 510 mg kg^{-1} were proposed as good criteria of carp (*Cyprinus carpio*) meat quality.[41] Chytiri *et al.*[42] recommended putrescine, tyramine, spermidine and spermine as a marker for rainbow trout (*Oncorhynchus mykiss*). Other authors recommend using putrescine, cadaverine and tyramine due to the low content of spermidine and spermine content of rainbow trout.[43] Putrescine and tyramine were proposed to be good quality factors for crucian carp (*Carassius auratus*).[44]

4.4 Reductions of Biogenic Amines in Fermented Fish

From the point of view of food safety and health, the reduction of the biogenic amine content of food has recently been the focus. There are many possibilities to reduce biogenic amines in foods: handling and processing under sanitary conditions, utilizing of amine-negative starter cultures, adding some probiotic bacterial strains, applying high pressure processing or low dose gamma irradiation.[1]

The use of selected starter cultures with negative decarboxylase activity is one of the successful strategies to reduce biogenic amine formation in fermented food.[45–47] Mah and Hwang[48] achieved lower biogenic amine accumulation in a salted and fermented anchovy by using starter cultures of *Staphylococcus xylosus* during ripening. Application of *Bacillus amyloliquefaciens* and *Staphylococcus carnosus* reduced histamine up to 59.9% and 29.1%, respectively, as well as application of *Staphylococcus intermedius* and *Bacillus subtilis* degraded putrescine and cadaverine up to 30.4% and 28.9% in Malaysian fish sauce.[34] It was also found that *B. amyloliquefaciens* and *S. carnosus* isolates were able to reduce the accumulation of histamine in laboratory fish sauce fermentation.[49] *Lactococcus lactis* subsp. *lactis* was found as an excellent starter among the isolated microbiota from air-dried and fermented fish from China.[50]

The presence of mixed microbial stains is reported to reduce biogenic amines in fish products.[51] Hu *et al.*[52] have successfully used mixed starter cultures of *Lactobacillus plantarum*-15, *Staphylococcus xylosus*-12, *Pediococcus pentosaceus*-ATCC33316 and *Lactobacillus casei* subsp. casei-1001 for fermentation of silver carp sausages and found lower accumulation of biogenic amines in the final products. Nie *et al.*[53] have also demonstrated that a mixed starter culture (*L. plantarum* ZY40 plus *S. cerevisiae* JM19) had great effects on biogenic amine contents in silver carp sausages. The addition of the mixed starter culture was able to inhibit the growth of undesirable bacteria such as *Pseudomonas* and Enterobacteriaceae, and consequently, reduces the accumulation of putrescine and cadaverine, whereas tyramine accumulation was enhanced because of the slow pH decline.

Biogenic amines could also be eliminated by using yeasts. Wang *et al.*[54] reported that the use of the yeast strain Omer Kodak M8 in fermented fish sauce was effective at degrading histamine and tyramine and also improved the smell and taste of the final product.

Recently, the application of bacteria possessing histamine-degrading enzymes has become an emerging method for reducing histamine concentration in fermented foods.[55] Mah and Hwang[48] used a histamine-degrading isolate as starter culture (*S. xylosus* No. 0538) during the ripening of salted and fermented anchovy (*Engraulis japonicus*) and found decreased level of histamine (38%) and tyramine (4%) in the product. Lee *et al.*[56] investigated the amine degrading activity of *Bacillus polymyxa* D05-1 during salted fish fermentation. They established that the used starter culture was effective to inhibit biogenic amines accumulation. The inoculated samples had 34% lower histamine contents than control fish samples at the end of fermentation.

Irradiation has also been considered as a useful method to reduce biogenic amine formation in fish. Atlantic horse mackerel (*Trachurus trachurus*) was investigated by Mendes *et al.*[57] In control samples, which were stored in ice for 23 days, histamine content of 100 mg kg^{-1} was detected, while fish irradiated at 1 kGy contained no histamine within the same time. Another study reported that irradiation used to reduce histamine content in bonito (*Sarda sarda*) was dose dependent.[58]

Krizek *et al.*[59] investigated the effect of vacuum packaging followed by irradiation (0.25–2.0 kGy) on the shelf life of fillets of rainbow trout (*Oncorhynchus mykiss*). The control and irradiated packs were stored at 3.5 °C from 28 to 98 days. Putrescine, cadaverine and tyramine showed very good correlation with the irradiation dose and organoleptic properties. Irradiation prevented formation of histamine in the samples. It was also studied the effect of low dose irradiation (0.25 and 0.50 kGy, 70 days) and smoking (181 days) on biogenic amine formation in vacuum-packed fillets of grass carp (*Ctenopharyngodon idella*) and bighead carp (*Hypophthalmichthys molitrix*). Both treatments were effective in prolonging the shelf life of samples and reduced the formation of biogenic amines.[60]

The other emerging method for reducing biogenic amines in foods is high hydrostatic pressure processing. High hydrostatic pressure (HHP) is a non-thermal technology used to extend the shelf life of perishable foods while preserving their original properties, depending on the type of food and pressure applied.[61]

Matejková *et al.*[43] examined the effects of vacuum packaging followed by high pressure processing on the shelf life of fillets of rainbow trout (*Oncorhynchus mykiss*). Samples were pressure treated at 300 and 500 MPa and were stored at 3.5 and 12 °C for up to 28 days (control–0 MPa) and 42 or 70 days (pressure treated; 12 and 3.5 °C, respectively). Biogenic amine content was below 10 mg kg^{-1}. Putrescine, cadaverine and tyramine content showed good correlation with the applied pressure and organoleptic properties. Tryptamine, phenylethylamine and histamine were not detected in pressure-treated samples kept at 3.5 °C. Similar results were found in biogenic amine content of pressure treated (300 and 500 MPa) common carp (*Cyprinus carpio*). During storage at 3.5 and 12 °C for 28 days (control, 0 MPa) and 70 days (pressure treated) accumulation of putrescine and cadaverine were influenced by all factors (temperature, pressurization level and time of storage).[62]

Another emerging non-thermal technology is ultraviolet radiation type C (UV-C, wavelengths of 200–280 nm).[63] This method has a direct impact on microbial DNA and reduces the bacterial load in foods. Krizek *et al.*[64] investigated the effect of essential oils (thyme, oregano) and UV-C irradiation (121 and 243 mJ cm^{-2}) on fillets of carp (*Cyprinus carpio*) during storage at 3.5 °C for 49 days. The result showed that the application of thyme and oregano oils prolonged the shelf life of samples by five and six times, respectively, and they were effective in suppressing the formation of putrescine, cadaverine, tyramine and phenylethylamine. Histamine was not detected in any of the carp fillets. UV irradiation were less effective in reducing biogenic amines.

Nile tilapia (*Oreochromis niloticus*) fillets were also treated with high hydrostatic pressure (220 MPa for 10 min at 25 °C) and ultraviolet radiation (at dose of 0.103 ± 0.002 J cm^{-2}) and stored at 4 °C for 14 days. The samples were analysed among other parameters for biogenic amines. Although UV-C accelerated the formation of cadaverine, both UV-C and HHP retarded bacterial growth and delayed the increase of biogenic amines during refrigerated storage, as well as extend the shelf life of refrigerated tilapia fillets at least 2.5 times.[65]

There are some studies reporting successful inhibition of biogenic amines using vacuum or modified atmosphere packaging in fish and fish products. A mixture of $CO_2 : O_2 : N_2$ (40 : 40 : 20) was found to prevent histamine production in bigeye tuna (*Thunnus obesus*).[66]

Oxygen removal by oxygen scavenger from packs significantly reduced histamine formation in seer fish and enhanced shelf life of the product.[67]

Reduction of water activity using organic acids and salt combined with high carbon dioxide concentration can prevent histamine formation by inhibiting the growth of histamine-producing bacteria.[68]

4.5 Health Related Risk of Biogenic Amines

Recently, the European Food Safety Authority (EFSA) published a report on risk assessment based on the monitoring of biogenic amine formation in fermented foods. The report states that histamine and tyramine are the most toxic amines in foods. Based on the Rapid Alert System for Food and Feed (RASFF), there were over 100 cases of histamine intoxication outbreaks between 2005 and 2010.[4]

The average European consumption of fish is 23.1 kg per capita per year, ranging from 5.3 kg in Hungary to 56.7 kg in Portugal. For comparison, the worldwide average is 18.9 kg per capita, with figures of over 30 kg per capita in China, over 50 kg in Norway and Japan, and over 90 kg in Iceland. European consumption comprises approximately 75% wild caught and 25% farmed species. The top 13 species consumed by European consumers are tuna, cod, salmon, pollack, herring, mussels, hake, mackerel, pangasius, shrimps, sardines, squid and scallops. Most of these fish are associated with the formation of biogenic amines.[69]

According to Leuschner *et al.*,[70] the current histamine situation continues to be of public health importance considering the number of human cases and the concentrations reported by RASFF, and that the animal origin products concerned with high histamine concentrations are mainly fish products and cheese.

The FAO (Food and Agriculture Organization) and WHO (World Health Organization) recently published a summary report on the public health risks of histamine and other biogenic amines from fish and fishery products.[71] It was established that tyramine, putrescine and cadaverine produced from amino acid decarboxylation during bacterial growth in fish may potentiate histamine's effect, but the contribution of these biogenic amines to scombroid fish poisoning is not clear.

4.6 Regulation of Histamine Content in Fish

Despite the health risk caused by elevated concentrations of biogenic amines in food, there is no general regulation on their content in food. The European Union (European Council Directive, 1991)[72] established limits only for histamine in fish, as histamine is implicated in causing the most frequent food-borne intoxications. Some countries have regulated the maximum amounts of histamine in different foods at a national level. Generally, upper limits of 100 mg kg^{-1} in food and 2 mg L^{-1} in beverages have been suggested. There are recommendations for tyramine (100–800 mg kg^{-1}) and for 2-phenylethylamine (30 mg kg^{-1}) in food.[73]

The recent European regulation on microbiological criteria for foodstuffs lays down standards for fishery products which are associated with high levels of histamine. In the case of fishery products from fish species associated with relatively high concentrations of histidine (families Scombridae, Clupeidae, Engraulidae, Coryphaenidae, Pomatomidae, Scomberesocidae), the sampling plan comprises nine units. In two of the units, the histamine concentration may be between 100 and 200 mg kg^{-1}; in none may it be above the limit of 200 mg kg^{-1}, and the average should be less than 100 mg kg^{-1}. Similarly, in two of the units, the histamine concentration may be between 200 and 400 mg kg^{-1}; in none of the units may it be above the limit of 400 mg kg^{-1} in the case of fishery products that have undergone enzyme maturation treatment in brine.[24,74]

Different maximum levels of histamine are regulated in different countries. The US Food and Drug Administration (FDA) recommends a 50 mg kg^{-1} limit for histamine in fish and seafood products.[75] Canada, Finland, Switzerland and South Africa regulate maximum amounts of biogenic amines in fish and seafood products at 100 mg kg^{-1}.[76,77] In Turkey and Australia, the limit for histamine in processed seafood products is 200 mg kg^{-1}.[27,76]

4.7 New Methods for Determination of Biogenic Amines in Fish and Fish Products

For qualitative and quantitative determination of biogenic amines in foods, various analytical methods are used including spectrophotometry, fluorimetry, electrophoresis and most commonly chromatography, *e.g.* thin-layer chromatography (TLC), high performance liquid chromatography (HPLC), ion exchange chromatography (IEC) and ultra-performance liquid chromatography (UPLC). Relatively new tools for biogenic amine detection are immunoassays, including enzyme-linked immunosorbent assay (ELISA) and biosensors.[1]

Although several methods are available to determine biogenic amines in fish and fish products, the continuous development of new complex analytical techniques offers more and more opportunities. Some examples are listed below.

A new method based on liquid chromatography (LC) and tandem mass spectrometry (MS/MS) using a triple quadrupole (QqQ) analyser was developed for the analysis of eight biogenic amines (cadaverine, histamine, phenylethylamine, putrescine, spermine, spermidine, tyramine and tryptamine) in fish obtained from commercials of Valencia.[78] Sample preparation was performed by extraction with trichloroacetic acid 5% and solid phase extraction clean up with the STRATA-X cartridge. The MS/MS method was validated and compared with a method based on the analysis of dansyl derivatives by LC and fluorescence detector (FD). MS/MS achieved higher sensitivity when compared with FD (from 0.02 mg kg^{-1} for spermidine and phenylethylamine to 0.2 mg kg^{-1} for spermine, from 1 mg kg^{-1} for putrescine and tyramine to 4 mg kg^{-1} for histamine).

Evangelista *et al.*[79] compared the Codex and the EU mandated methods for the analysis of histamine and observed that they underestimated and overestimated the results, respectively. To solve this problem, a simple and efficient procedure for the extraction and quantification of histamine by ion-pair HPLC method with post-column derivatization and fluorimetric detection is proposed. The method was successfully used in the analyses of freshwater fish and fresh and canned tuna.

A brand new method was applied to determine the most commonly occurred biogenic amines in swordfish by HPLC-FLD after extraction with an acid and subsequent derivatization with dansyl chloride.[80] The developed procedure reaches detection limits down to 2.5 μg L^{-1} for cadaverine or 497 μg L^{-1} for histamine in solvent and 0.07 mg kg^{-1} and 14.81 mg kg^{-1} in fish, respectively.

Parchami *et al.*[81] developed an ion mobility spectrometry (IMS) method for biogenic amine determination. Head-space solid phase microextraction (HS-SPME) was applied to extraction and determination of histamine, putrescine, cadaverine and tyramine in canned fish samples. Limits of detection for analytes were in the range of 0.6–1 ng g^{-1}. HS-SPME-MIR-IMS results indicate that the proposed method can be successfully used for biogenic amines analysis food samples.

An easy and fast quantitative method for the determination of biogenic amines in fish and fish products with liquid chromatography coupled to orbitrap tandem mass spectrometry was developed by Kaufmann and Maden.[82] The chromatography is based on a reversed-phase column and an eluent containing an ion-pairing agent (heptafluorobutyric acid). The method was validated for three fish-based matrixes.

Mattsson *et al.*[83] developed a competitive fluorescent molecularly imprinted polymer (MIP) assay to detect biogenic amines in fish samples. The MIP had the highest affinity towards histamine, but also recognized the structurally similar biogenic amines tyramine and tryptamine, as well as spermine and spermidine, providing simultaneous analysis and assessment of the total amount of biogenic amines.

4.8 Conclusions

The recognition of the close relationship between nutrition and health is becoming more and more common among consumers. There is growing demand for good quality and healthier food products. In all foods containing high amounts of amino acids and favourable conditions for the amino acid decarboxylase enzyme activity, biogenic amines are formed. Determining safe levels of biogenic amines in fish and fish products is a key issue in avoiding health risks for sensitive consumers. To ensure food safety and nutritional quality throughout the entire food supply chain will require further research and better cooperation between industry and academia.

References

1. L. Simon-Sarkadi, in *Process-Induced Food Toxicants. Occurrence, Formation, Mitigation, and Health Risks*, ed. R. H. Stadler and R. David, Wiley, Hoboken, NJ, 2009, vol. 3.2, pp. 321–361.
2. L. Simon-Sarkadi, in *Fermented Foods in Health and Disease Prevention*, ed. J. Frias, C. Martinez-Villaluenga and E. Penas, Elsevier Academic Press, London, UK, 2017, vol. 27, pp. 625–651.
3. V. Ladero, M. Calles-Enriquez, M. Fernandez and M. A. Alvarez, *Curr. Nutr. Food Sci.*, 2010, **6**, 145.
4. European Food Safety Authority (EFSA), *EFSA J.*, 2011, **9**, 2393.
5. K. Björnsdóttir-Butler, G. E. Bolton, L. A. Jaykus, P. D. McClellan-Green and D. P. Green, *Int. J. Food Microbiol.*, 2010, **139**, 161.

6. Y. H. Tsai, H. F. Kung, T. M. Lee, H. C. Chen, S. S. Chou, C. I. Wei and D. F. Hwang, *Food Control*, 2005, **16**, 579.
7. J. L. Mietz and E. Karmas, *J. Food Sci.*, 1977, **42**, 155.
8. G. Spano, P. Russo, A. Lonvaud-Funel, P. Lucas, H. Alexandre, C. Grandvalet, E. Coton, M. Coton, L. Barnavon, B. Bach, F. Rattray, A. Bunte, C. Magni, V. Ladero, M. Alvarez, M. Fernández, P. Lopez, P. F. de Palencia, A. Corbi, H. Trip and J. S. Lolkema, *Eur. J. Clin. Nutr.*, 2010, **64**, S95.
9. L. Prester, *Food Addit. Contam., Part A*, 2011, **28**, 1547.
10. P. Kalac, *Food Chem.*, 2014, **161**, 27.
11. M. A. Alvarez and M. V. Moreno-Arribas, *Trends Food Sci. Technol.*, 2014, **39**, 146.
12. K. B. Biji, C. N. Ravishankar, R. Venkateswarlu, C. O. Mohan and T. K. Srinivasa Gopal, *J. Food Sci. Technol.*, 2016, **53**, 2210.
13. C. Ancín-Azpilicueta, A. González-Marco and N. Jiménez-Moreno, *Crit. Rev. Food Sci. Nutr.*, 2008, **48**, 257.
14. D. M. Linares, B. Del Río, V. Ladero, N. Martínez, M. Fernández, M. C. Martín and M. A. Alvarez, *Front. Microbiol.*, 2012, **3**, 180.
15. T. M. Silva, P. S. Sabaini, W. P. Evangelista and M. B. A. Gloria, *Food Control*, 2011, **22**, 323.
16. M. Muscarella, S. Lo Magro, M. Campaniello, A. Armentano and P. Stacchini, *Food Control*, 2013, **31**, 211.
17. Z. Tao, M. Sato, H. Zhang, T. Yamaguchi and T. Nakano, *Food Control*, 2011, **22**, 430.
18. S. Baixas-Nogueras, S. Bover-Cid, M. T. Veciana-Nogues, A. Marine-Font and M. C. Vidal-Carou, *J. Food Prot.*, 2005, **68**, 2433.
19. S. Baixas-Nogueras, S. Bover-Cid, M. T. Veciana-Nogues and M. C. Vidal-Carou, *Eur. Food Res. Technol.*, 2007, **226**, 287.
20. M. Krizek, F. Vácha, L. Vorlová, J. Lukásová and S. Cupáková, *Food Chem.*, 2004, **88**, 185.
21. M. M. Hernández-Herrero, G. Duflos, P. Malle and S. Bouquelet, *J. Food Prot.*, 2002, **65**, 1152.
22. S. Koral and S. Köse, *J. Food Health Sci.*, 2015, **1**, 150.
23. F. Bunka, P. Budinsky, B. Zimáková, M. Merhaut, R. Flasarová, V. Pachlová, V. Kubán and L. Bunková, *Food Control*, 2013, **31**, 49.
24. European Commission Regulation (EC) No 2073/2005, OJ L, 2005, **338**, 1.
25. Y. Hu, Z. Huang, J. Li and H. Yang, *Food Chem.*, 2012, **135**, 2604.
26. M. Rabie, L. Simon Sarkadi, H. Siliha, S. El Seedy and A. A. El Badawy, *Food Chem.*, 2009, **115**, 635.
27. S. Köse, S. Koral, B. Tufan, M. Pompe, A. Scavnicar and D. Kocar, *Eur. Food Res. Technol.*, 2012, **235**, 669.
28. G. Jairath, P. K. Singh, R. S. Dabur, M. Rani and M. Chaudhari, *J. Food Sci. Technol.*, 2015, **52**, 6835.
29. V. B. Anihouvi, J. M. Kindossi and J. D. Hounhouigan, *Int. Res. J. Biol. Sci.*, 2012, **1**, 72.

30. J. Yongsawatdigul, Y. J. Choi and S. Udompom, *J. Food Sci.*, 2004, **69**, 312.
31. M. Z. Zaman, A. S. Abdulamir, F. A. Bakar, J. Selamat and J. Bakar, *Am. J. Appl. Sci.*, 2009, **6**, 1199.
32. Y. H. Tsai, C. Y. Lin, L. T. Chien, T. M. Lee, C. I. Wei and D. F. Hwang, *Food Chem.*, 2006, **98**, 64.
33. M. Saaid, B. Saad, N. H. Hashim, A. S. M. Ali and M. I. Saleh, *Food Chem.*, 2009, **113**, 1356.
34. M. Z. Zaman, F. A. Bakar, J. Selamat and J. Bakar, *Czech J. Food Sci.*, 2010, **28**, 440.
35. Y. R. Huang, K. J. Liu, H. S. Hsieh, C. H. Hsieh, D. F. Hwang and Y. H. Tsai, *Food Control*, 2010, **21**, 1234.
36. H. H. Hsu, T. C. Chuang, H. C. Lin, Y. R. Huang, C. M. Lin, H. F. Kung and Y. H. Tsai, *Food Chem.*, 2009, **114**, 933.
37. E. Tomé, V. L. Pereira, C. I. Lopes, P. A. Gibbs and P. C. Teixeira, *Food Control*, 2008, **19**, 535.
38. P. J. Park, C. H. Lee, E. Y. Kwon, H. J. Lee, J. Y. Kim and S. H. Kim, *Food Control*, 2010, **21**, 1219.
39. M. Kanki, T. Yoda, M. Ishibashi and T. Tsukamoto, *Int. J. Food Microbiol.*, 2004, **92**, 79.
40. Y. H. Tsai, H. S. Hsieh, H. C. Chen, S. H. Cheng, T. Chai and D. F. Hwang, *Food Chem.*, 2007, **104**, 1366.
41. M. Krizek, T. Pavlicek and F. Vácha, *J. Sci. Food Agric.*, 2002, **82**, 1088.
42. S. Chytiri, E. Paleologos, I. Savvaidis and M. G. Kontominas, *J. Food Prot.*, 2004, **67**, 960.
43. K. Matejková, M. Krizek, F. Vácha and E. Dadáková, *Food Chem.*, 2013, **137**, 31.
44. K. F. Li, Y. L. Bao, Y. K. Luo, H. X. Shen and C. Shi, *J. Food Prot.*, 2012, **75**, 2228.
45. M. Fernández, D. M. Linares, A. Rodríguez and M. A. Alvarez, *Appl. Microbiol. Biotechnol.*, 2007, **73**, 1400.
46. J. M. Landete, B. de las Rivas, A. Marcobal and R. Muńoz, *Int. J. Food Microbiol.*, 2007, **117**, 258.
47. R. Casquete, M. J. Benito, A. Martín, S. Ruiz-Moyano, J. J. Córdoba and M. G. Córdoba, *Food Microbiol.*, 2011, **28**, 1432.
48. J. H. Mah and H. J. Hwang, *Food Control*, 2009, **20**, 796.
49. M. Z. Zaman, F. A. Bakar, S. Jinap and J. Bakar, *Int. J. Food Microbiol.*, 2011, **145**, 84.
50. J. R. Zhao, S. M. Cheng, W. D. Hong, R. L. Wang and C. H. Li, *Mod. Food Sci. Technol.*, 2014, **30**, 100.
51. L. Zhong-Yi, L. Zhong-Hai, Z. Miao-Ling and D. Xiao-Ping, *Int. J. Food Sci. Technol.*, 2010, **45**, 930.
52. Y. Hu, W. Xia and X. Liu, *Food Chem.*, 2007, **104**, 188.
53. X. Nie, Q. Zhang and S. Lin, *Food Chem.*, 2014, **153**, 432.
54. H. Wang, X. Fu, W. Wu, Y. Wu, J. Ren, Q. Lin and Z. Li, *J. Chin. Inst. Food Sci. Technol.*, 2014, **14**, 137.

55. A. Naila, S. Flint, G. Fletcher, P. Bremer and G. Meerdink, *J. Food Sci.*, 2010, **75**, R139.
56. Y. C. Lee, H. F. Kung, C. Y. Huang, T. C. Huang and Y. H. Tsai, *J. Food Drug Anal.*, 2016, **24**, 157.
57. R. Mendes, H. A. Silva, M. L. Nunes and J. M. A. Empis, *Eur. Food Res. Technol.*, 2005, **221**, 329.
58. R. Mbarki, S. Sadok and I. Barkallah, *Food Sci. Technol. Int.*, 2008, **14**, 367.
59. M. Krizek, K. Matejková, F. Vácha and E. Dadáková, *Food Chem.*, 2012, **132**, 367.
60. M. Krizek, E. Dadáková, F. Vácha and T. Pelikánová, *Eur. Food Res. Technol.*, 2017, **243**, 1989.
61. Q. Syed, M. Buffa, B. Guamis and J. Saldo, *Crit. Rev. Food Sci. Nutr.*, 2016, **56**, 474.
62. M. Krizek, K. Matejková, E. Dadáková, J. Spicka, F. Vácha and N. Vrchotová, *J. Food Prot.*, 2015, **78**, 1592.
63. T. N. Koutchma, L. J. Forney and C. I. Moraru, *Ultraviolet Light in Food Technology*, CRC Press, Boca Raton, London, New York, 2009, pp. 1–32.
64. M. Krizek, E. Dadáková, F. Vácha, T. Pelikánová and K. Matejková, *J. Food Sci. Technol.*, 2018, **95**, 268.
65. M. L. G. Monteiro1, E. T. Mársico, S. B. Mano, T. da Silveira Alvares, A. Rosenthal, M. Lemos, E. Ferrari, C. A. Lázaro and C. A. Conte-Junior, *Sci. Rep.*, 2018, **8**, 9524.
66. C. Ruiz-Capillas and A. Moral, *Food Chem.*, 2005, **89**, 347.
67. C. O. Mohan, C. N. Ravishankar, T. K. Srinivasa Gopal, K. Ashok Kumar and K. V. Lalitha, *Food Res. Int.*, 2009, **42**, 411.
68. J. Emborg and P. Dalgaard, *Int. J. Food Microbiol.*, 2008, **128**, 234.
69. Directorate-General for Maritime Affairs and Fisheries, *Facts and figures on the Common Fisheries Policy–Basic statistical data*, Publications Office of the European Union, Luxembourg, 2014, pp. 1–44.
70. R. G. K. Leuschner, A. Hristova, T. Robinson and M. Hugas, *J. Food Compos. Anal.*, 2013, **29**, 37.
71. FAO/WHO Meeting Report, *Public Health Risks of Histamine and other Biogenic Amines from Fish and Fishery Products*, FAO/WHO (Food and Agriculture Organization of the United Nations/World Health Organization), 2013, pp. 1–111.
72. European Council Directive, 1991. 91/493/EEC the Health Conditions for the Production and the Placing on the Market of Fishery Products.
73. B. ten Brink, C. Damink, H. M. L. J. Joosten and J. H. J. Huis in't Veld, *Int. J. Food Microbiol.*, 1990, **11**, 73.
74. European Commission Regulation (EU) No 1019/2013 (OJ L 282, p. 46, 2013) amending Annex I to Regulation (EC) No 2073/2005 as regards histamine in fishery products.
75. Food and Drug Administration (FDA), Scombrotoxin (histamine) formation, in *Fish and Fishery Products Hazards and Controls Guide*, Department of Health and Human Services, Public Health Service,

Food and Drug Administration, Center for Food Safety and Applied Nutrition, Office of Seafood, Washington, DC, 4th edn, 2011, pp. 113–119.
76. L. Auerswald, C. Morren and A. L. Lopata, *Food Chem.*, 2006, **98**, 231.
77. J. Lange and C. Witmann, *Anal. Bioanal. Chem.*, 2002, **372**, 276.
78. G. Sagratini, M. Fernández-Franzón, F. De Berardinis, G. Font, S. Vittori and J. Mañes, *Food Chem.*, 2012, **132**, 537.
79. W. P. Evangelista, T. M. Silva, L. R. Guidi, P. A. S. Tette, R. M. D. Byrro, P. Santiago-Silva, C. Fernandes and M. B. A. Gloria, *Food Chem.*, 2016, **211**, 100.
80. A. Herrero, S. Sanllorente, C. Reguera, M. C. Ortiz and L. A. Sarabia, *Anal. Chim. Acta*, 2016, **945**, 31.
81. R. Parchami, M. Kamalabadi and N. Alizadeh, *J. Chromatogr. A*, 2017, **1481**, 37.
82. A. Kaufmann and K. Maden, *J. AOAC Int.*, 2018, **101**, 336.
83. L. Mattsson, J. Xu, C. Preininger, B. T. S. Bui and K. Haupt, *Talanta*, 2018, **181**, 190.

CHAPTER 5

Biogenic Amines in Non-fermented Food

J. L. ORDÓÑEZ[a] AND R. CALLEJÓN*[b]

[a] Department of Food Science and Health, IFAPA-Alameda del Obispo, Avda. Menéndez-Pidal, s/n. 14071, Córdoba, Spain; [b] Área de Nutrición y Bromatología, Dpto. Nutrición y Bromatología, Toxicología y Medicina Legal, Facultad de Farmacia, Universidad de Sevilla, C/P. García González n° 2, E-41012, Sevilla, Spain
*Email: rcallejon@us.es

5.1 Introduction

Biogenic amines are organic nitrogenous bases of low molecular weight. They are mainly synthetized by decarboxylation of amino acids, although some amines can be formed by amination and transamination of aldehydes and ketones.[1] These compounds are metabolites that can be found in the three domains of life: archaea, bacteria and eukaryotes.[2] These biomolecules are synthetized in normal metabolism and are essential in natural metabolic and physiological functions in animals, plants and microorganisms.[1,3] Thus, these amines are formed and degraded by normal metabolic processes in the cells and they are responsible for certain functions such as growth regulation (cadaverine, spermine and spermidine), neural transmission (catecholamines and serotonin) or acting as mediators of inflammation (histamine and tyramine).[4] For example, histamine is produced and stored in humans mainly by mast cells, basophiles and neurons.[5] Polyamines (putrescine, spermidine and spermine) are present

Food Chemistry, Function and Analysis No. 20
Biogenic Amines in Food: Analysis, Occurrence and Toxicity
Edited by Bahruddin Saad and Rosanna Tofalo

Published by the Royal Society of Chemistry, www.rsc.org

in edible plants, including vegetables and fruits. These amines have different functions, from development and growth processes to acting against biotic and abiotic stress.[6,7] On the other hand, polyamines in humans and animals are involved in different functions during cell proliferation, physiological activity and immunity.[8] In these cases, some authors have detected higher concentrations of polyamines in tumour tissues compared with normal cells.[9]

In general, biogenic amines are present in foods in a wide and variable range of concentrations. They can be classified according to their endogenous/natural or biogenic origin. Thus, these compounds are formed by decarboxylase enzymes of metabolic processes or by microbial metabolic action. However, it is sometimes difficult to differentiate what the origin of these compounds is, especially when they are in low concentrations in foods.[10] The concentration of biogenic amines could increase significantly in fermented products, particularly in those derived from yeast or bacteria activities, especially lactic acid bacteria.[11] Higher concentrations of biogenic amines are found in dairy products, fish and fish products, meat and meat products, fermented vegetables and soy products, and fermented beverages or contaminated products.[12] Regarding non-fermented foods, such as fruits, vegetables, meat, fish and milk, among others, biogenic amines are usually present in low concentrations of endogenous origin.[13] However, the presence of biogenic amines in high concentrations in these products is generally indicative of undesired microbial activity.[14] Therefore, any product containing free amino acids or proteins that are subjected to biochemical or microbial activities could contain biogenic amines (see Figure 5.1).[15] Bacterial and thermal activities are two important factors in protein degradation and amino acid release in foods. Thus, the determination of the concentration and profile of biogenic amines depends on several factors, such as the nature of the food, the environmental conditions and the microbial activity.[13]

Despite the low levels of biogenic amines in non-fermented products, the quantification of these compounds is very important, as they could cause toxicological effects, particularly in consumers with a high sensitivity to them.[1,15] Therefore, the identification and quantification of biogenic amines in food are important in terms of food quality and safety, and they could also be used as an indicator of freshness.[10,16,17]

Food poisoning in relation to the consumption of biogenic amines has been widely studied. In general, the toxicity of biogenic amines ingested is related to the consumer's ability to detoxify them with amine oxidase enzymes of the intestinal lumen.[10] This detoxification system is formed by several enzymes, such as monoamine oxidase (MAO), diamine oxidase (DAO), polyamine oxidase (PAO) or histamine methyltransferase (HMT). Their functions may be affected by different factors, including alcohol, genetic alteration, allergic, gastrointestinal diseases or pharmacological agents, such as MAO inhibitors.[18,19] Some consumers can be more sensitive to the intake of some biogenic amines, even in low amounts: for instance, people with respiratory and coronary heart diseases, hypertension problems or

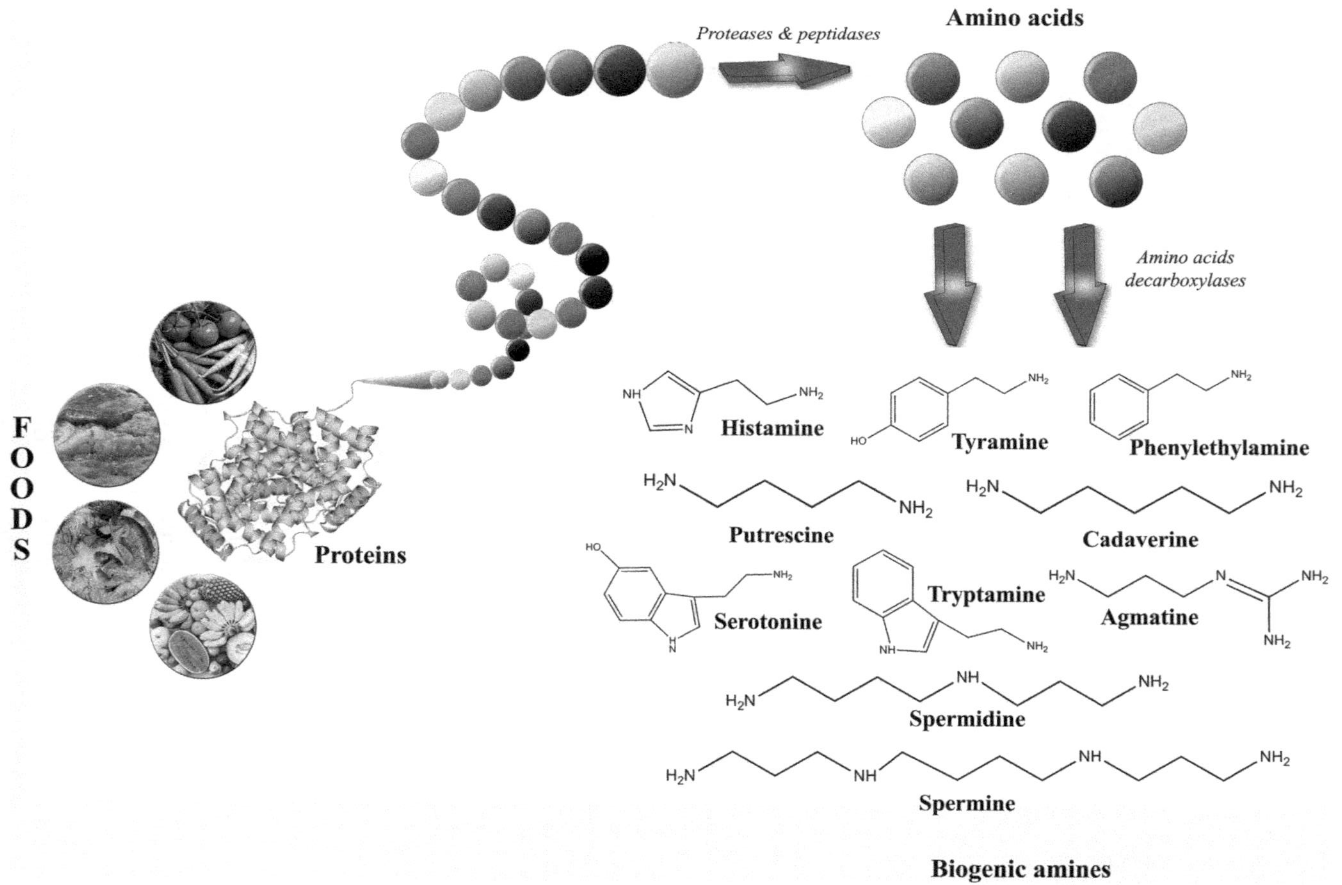

Figure 5.1 Formation of biogenic amines from non-fermented food and their chemical structures. Data from ref. 13, 16, 25 and 32.

vitamin B12 deficiency.[15,20] Similarly, the activity of amine oxidase enzymes may be impaired in some people due to gastrointestinal problems, such as gastritis, irritable bowel syndrome, Crohn's disease, gastric and colon ulcers, *etc.* Besides, some drugs used to treat stress, depression, Alzheimer's and Parkinson's, pulmonary tuberculosis, malaria, panic disorder and social phobia inhibit the activity of enzymes like MAO, DAO and PAO, which also affects the activity of decarboxylating enzymes.[11,15,20,21] Furthermore, the toxicological effects of some amines, such as histamine or tyramine, may be potentiated by other biogenic amines such as putrescine, cadaverine and agmatine.[22] Therefore, the effects of the consumption of food with biogenic amines are variable due to several factors, such as the presence of enhancers or the efficiency of the amine detoxification system.[21] In this context, the consumption of food with histamine, even at low concentrations, can cause food intolerance.[23]

In addition, biogenic amines could affect the organoleptic properties of foods. Thus, putrescine and cadaverine are associated with the smell of putrefaction in foods.[24,25] In fruits and vegetables, these compounds shape the typical and characteristic taste of mature foods and they are precursors for certain aromatic compounds.[26]

On the other hand, the increase of biogenic amines in foods can produce new compounds such as nitrosamines, which are carcinogens.[27,28] Nitrosamines are formed due to reactions of secondary amines in the presence of nitrites, which are usually employed as colouring, flavouring or preservation of fish.[28,29]

Therefore, it is important to evaluate the action of biogenic amines and their levels of toxicity in specific food matrices. In general, there is no legal regulation regarding the levels of biogenic amines in food and beverages, with the exception of fish and fish products. This is due to the differences among the individual toxicological thresholds. These vary greatly and can oscillate from a few mg kg^{-1} in a sensitive person to high levels in a healthy person.[20] Thus, some countries and international organizations have established legal limits in fish and fish-derived products, whilst acceptable concentrations have been recommended in other foods, mainly fermented food.[30] In this context, a qualitative risk assessment concerning biogenic amines in fermented foods was elaborated and published in the European Union.[22]

5.1.1 Biogenic Amines as a Quality Index

As aforementioned in the previous section, the presence of biogenic amines could pose a real risk to human health. For this reason, it is important to control and monitor the formation of biogenic amines in food. The number of biogenic amines in food can be a useful indicator of spoilage, freshness or hygienic quality.[10,16,31] In this context, biogenic amines are more useful as a quality indicator in fresh and heat-treated products than in fermented foods.[32]

Regarding the indicator of food freshness, several biogenic amines indices have been developed to evaluate the food quality using the amine level as an indicator of microbial spoilage. One of these biogenic amines indices is calculated as the sum of the concentrations of biogenic amines, mainly histamine, tyramine, putrescine and cadaverine. In this case, values lower than 5 mg kg^{-1} indicate fresh food and good conditions. On the contrary, values higher than 20–50 mg kg^{-1} show spoilage or putrefaction relating to poor hygiene or quality. On the other hand, another type of index is calculated as the spermidine to spermine ratio. Thus, values lower than 0.5 are related to good conditions, whilst values higher than 0.7 correspond to degradation of food.[25,33] Biogenic amines are thermostable, and the treatments used in food processing and preparation do not inactivate them.[22]

In general, the concentration of each biogenic amine is different and it is related to the food matrix.[30] Particularly, some polyamines, such as spermine and spermidine, can decrease during storage time, so these amines are not useful as a quality index.[31] The variability of these compounds in foods, as well as changes in their concentrations during storage time, makes it difficult to develop a universal biogenic amine index that can be used for all foods. Thus, biogenic amine indices should be evaluated and adapted to specific food products.

5.1.2 Processing and Storage Conditions

Biogenic amines production in foods is influenced by many factors, including food physico-chemical parameters, such as NaCl, pH, ripening temperature, atmosphere, preservatives and food ingredients storage, as well as distribution conditions, manufacturing processes and practices, presence of decarboxylase-positive microorganisms, raw material quality and availability of free amino acids.[13,34] For all these reasons, the concentration and profile of biogenic amines in foods are determined by a large number of variables that may interact among them, therefore making it difficult to evaluate the effects of the different technological steps on the formation of biogenic amines during ripening and storage of food.[13] The synthesis of biogenic amines in food can be controlled with good hygiene in raw materials and production processes, thus avoiding the growth of microorganisms.[14] Both time and temperature are the most important factors in the formation of biogenic amines in fresh food, as well as cooked or ripened products. The control of storage temperature is essential to avoid the production of biogenic amines. Thus, it is necessary to employ production techniques that avoid or limit the increase of biogenic amines in foods.[13] Besides, some kinds of packages and a prolonged shelf life have been reported to increase biogenic amines in commercial products.[35]

Particularly, decarboxylase bacterial enzymes can continue producing biogenic amines, even if the bacteria are eliminated from food by technological treatment, such as heat or high pressures.[36]

In general, most researchers have observed that biogenic amines are stable with a high temperature treatment.[35] Hence, they cannot be eliminated by cooking, smoking, freezing or other types of heat treatment.[37] The formation of biogenic amines by bacteria is limited by decreasing the temperature. Thus, maintaining the cold chain during food storage is a significant factor in avoiding the decarboxylation activity and the accumulation of biogenic amines after manufacturing, especially in non-fermented foods.[36] For instance, in fishery products, a direct correlation between storage temperature and the formation of biogenic amines, mainly histamine, has been observed.[24,36]

Apart from temperature, other factors may have a significant influence in the content of biogenic amines, such as the presence or absence of oxygen. Thus, the storage of food in modified atmosphere packaging or vacuum packaging avoids the formation of biogenic amines, as well as bacterial growth.[38–40]

The addition of salt to food can contribute to avoiding the growth of decarboxylating bacteria and the increase of the content of biogenic amines. In this context, Gram-negative bacteria are more affected by a high concentration of salt than Gram-positive bacteria. However, the addition of a high amount of salt to reduce biogenic amines in foods may conflict with the tendency to decrease it as part of a healthy diet.[36] The addition of other preservatives can also affect the formation of biogenic amines by inhibition of decarboxylase enzymes, such as essential oils, chitosan or trisodium phosphate, among others.[15,41–44]

On the other hand, research on the relationship between pH and biogenic amine concentration in food has been conducted. Thus, bacteria can perform the decarboxylation activity as a mechanism to counteract the acidic stress. The effect of pH is variable according to where it is focused, either on the decarboxylase activity of the bacteria or on the enzyme activity.[36]

The presence of biogenic amines in raw materials is mainly a consequence of poor food handling. In this sense, food quality and safety management related to hazard analysis and critical control point programs (HACCP) are an important tool to avoid their formation.[22] Besides, it is necessary to implement proper cleaning and disinfection methods, from primary production to consumers.[22] Thus, microorganisms producing biogenic amines could be spread from animals, surfaces or equipment during the handling of raw food, and consequently, these compounds could be accumulated during processing and storage. The optimization of handling, processing and storage conditions is the best method to avoid the growth of bacteria producing biogenic amines.[22]

On the other hand, different methods to eliminate biogenic amines have been applied, such as the use of amine oxidase positive organisms[45,46] or by gamma irradiation.[47] However, there is some controversy in the use of techniques to eliminate biogenic amines, because it could cover up improper hygiene and manufacturing practices.[22]

5.2 Biogenic Amines in Animal Derived Products

5.2.1 Fish

In general, fish is more perishable than meat products and it is more likely to accumulate a high concentration of biogenic amines.[39,48] The profile and concentrations of biogenic amines are very variable depending on the fish species or derived product (Table 5.1). As seen earlier, several factors are important in the increase of these compounds, such as availability of substrate, activity of decarboxylases enzymes and environmental conditions.[30]

The main biogenic amine in fish and fish products is histamine, followed by tyramine, tryptamine, putrescine and cadaverine.[49] The consumption of fish with a high concentration of histamine can cause a foodborne intoxication called histamine or scombroid poisoning, which has similar symptoms and treatments as seafood allergies. This intoxication was originally attributed to the consumption of fish from the Scombridae and Scomberesocidae families, and this is where its denomination comes from. Fish belonging to these families are tuna, bluefish and mackerel, among others. These kinds of fish have high concentrations of free histidine in their muscle, which can form histamine by decarboxylation.[21] The concentration of histamine in freshly caught fish is usually low, around 1 mg kg^{-1}.[49,50] However, the histidine present in the muscle of fish stored in high temperatures and for a long period of time can form histamine.[51] Decomposed fish can present high concentrations of histamine and their consumption may cause a wide range of symptoms, such as a peppery or metallic taste sensation, oral numbness, headache, dizziness, palpitations, low blood pressure, difficulty in swallowing, hives and rash.[51,52] Thus, there is much research focused on the toxicological effects of biogenic amines, mainly histamine, after the consumption of fish in poor conditions.[52,53] However, it is difficult to determine the dose-response relationship in humans, as they may react differently, even if they have eaten the same fish. This may be because histamine is not uniformly distributed throughout the parts of the fish.[21,54]

In general, the decomposition of the fish is very complex, therefore the formation of biogenic amines is variable and it depends on the fish species.[52,55] In this case, the muscle of fish is the main source of amino acids, the precursors of biogenic amines. High levels of histamine are more common in histidine-rich or dark muscle fish than in histidine-poor or white muscle fish.[50,56] Histamine may be used as the main quality and safety marker for dark muscle fish. However, the levels of histamine in white muscle fish are significantly lower than the legal limits established for scombroid-like fish species. Thus, there is not much information available regarding histamine poisoning with consumption of white muscle fish.[57,58] In the case of this type of fish, the use of histamine as a quality indicator is not appropriate. Other biogenic amines determined in

Table 5.1 Content of main biogenic amines in animal derived products (mg kg^{-1} or L^{-1}).[a]

Food	Spermidine	Spermine	Histamine	Tyramine	Putrescine	Cadaverine	Ref.
Fish							
Canned fish	n.d.–19.50	n.d.–35.2	n.d.–657.1	n.d.–41.2	n.d.–2.2	n.d.–12.0	13
Amberjack	n.d.–85.1	n.d.–20.1	n.d.–12.5	41.4–222	n.d.–126	n.d.–38.5	27
Anchovy	—	—	0.5–750	2.8	3.1	5.3	54
Anchovy	49.3–55.2	6.2–27.1	14.0–2007	46.9–273	n.d.–260	15.5–863	24
Salted anchovies	—	—	n.d.–2.0	n.d.–22.3	0.1–8.0	0.1–12.0	128
Barracuda	—	—	30	—	—	—	54
Bonito	n.d.–3.0	7.5–20.3	15–30	33.1–99	n.d.–2.5	n.d.–6.2	27
Canned tuna	—	—	n.d.–110.3	n.d.–48.6	n.d.–116.5	n.d.–103.3	128
Carp	—	—	n.d.	0.2–0.4	8.9–11.8	8.1–12.4	54
Caspian kutum	—	—	n.d.	n.d.	5.3	4.2	54
Chub mackerel	—	—	1500	—	80	200	54
Cod	n.d.–4.4	—	2.9–43.7	6.5–688	12.5–664	24.0–513	51
Eel	n.d.	n.d.–6.5	0.5–8.8	8.7–35.7	n.d.	n.d.	27
Emperor bream	—	—	15				54
Fresh mackerel	—	—	9.3–12.4	n.d.–0.4	0.2–2.0	1.2–6.6	128
Hake	—	—	3.1–32.3	2.2–12	3.9–12	3.4–121.5	54
Herring	—	—	236	17	10	147	54
Herring	n.d.–4.5	n.d.–3.4	n.d.–271	n.d.–4.2	n.d.–39.7	8.5–237	24
Mackerel	—	—	40	47	56	n.d.	54
Mackerel	9.3–22.4	—	n.d.–22 326	n.d.–5164	1.5–7050	0.1–9692	51
Mackerel	n.d.–23.8	n.d.–49.7	n.d.–45.5	n.d.–189	n.d.–105.7	n.d.–314.3	27
Mahi-mahi	—	—	2500	—	—	—	54
Mediterranean hake	3.5–10.7	4.6–10.7	0.1–2.2	0.6–2.4	1.7–12.2	0.8–20.3	24
Mediterranean sea bass	n.d.–8.8	n.d.–14.4	n.d.	n.d.–4.3	n.d.–3.1	n.d.–6.5	24
Rainbow trout	—	—	0.4	n.d.	6.6	3.3	54
Rainbow trout	4.1–13.6	0.3–5.1	n.d.–1.6	0.2–2.9	7.5–23.1	n.d.–2.7	24
Sailfish	0.4	2.7	4.6	0.1	0.3	2.1	24
Sardine	—	—	67–620	16–100	114–420	n.d.–100	54

(continued)

Table 5.1 (*Continued*)

Food	Spermidine	Spermine	Histamine	Tyramine	Putrescine	Cadaverine	Ref.
Sardine	1.2–7.6	n.d.–2.9	19.5–203	n.d.–16.3	13.4–114	3.9–100	24
Silver pomfret	n.d.–5.5	n.d.	n.d.	n.d.–5.6	n.d.	n.d.	27
Skipjack tuna	—	—	1533	—	65	649	54
Tuna	4.4–13.3	—	n.d.–48 236	n.d.–170	0.7–811	0.4–491	51
Tuna	6.8	22.4	0.2	n.d.	0.3	0.7	24
Tuna steak	—	—	30	—	5	18	54
Yellowfin tuna	—	—	25.6–111	—	1.7	n.d.	54
Shellfish							
Cuttlefish	n.d.	n.d.–12	n.d.	n.d.–5.6	n.d.–39.9	n.d.–45.1	27
Humpback shrimp	n.d.	n.d.–19.5	n.d.–12.5	n.d.–6.2	n.d.–4.5	n.d.–43.1	27
Japanese octopus	n.d.–5.1	n.d.–8.9	n.d.	n.d.	n.d.–6.1	n.d.–6.1	27
Octopus	—	—	1.3–9.1	n.d.–14.5	2.8–94	0.1–164	128
Baby octopus	n.d.–2.8	0.2–13.1	n.d.	n.d.–1.4	17.0–190	n.d.–161	27
Swimming crab	n.d.	n.d.–33.2	n.d.–8.3	n.d.–2.6	12.0–27.1	6.9–20.1	27
Venus clam	n.d.–64.6	41.7–149	n.d.–9.1	n.d.–20.2	1.9–144	n.d.–20.5	27
Animal meat							
Beef meat	—	—	0.4–7.4	n.d.–17.4	n.d.–202.5	n.d.–221.4	128
Beef meat	1.9–4.2	28.7–44.6	n.d.–1.1	n.d.	n.d.–1.75	n.d.	16
Chicken meat	—	—	0.2	n.d.–221.6	n.d.–45.9	n.d.–33.5	128
Pork meat	—	—	0.9–2.3	n.d.–29.0	n.d.–66.5	1.0–145.4	128
Pork meat	2.2–7.0	25.5–67.1	n.d.–4.7	n.d.–3.5	n.d.–7.8	n.d.–13.3	16
Others							
Milk	—	—	n.d.–0.7	n.d.	n.d.	n.d.–0.1	128
Egg (albumens)	—	—	—	—	1.1–1.8	1.3–4.2	95
Egg (yolk)	n.d.–4.4	n.d.–5.6	—	—	0.8–3.5	n.d.–6.4	95
Honey	—	—	n.d.–3.8	n.d.	n.d.	n.d.	128

[a]Abbreviations: Ref., references; n.d., non-detected; —, non-determined.

decomposed fish, such as tyramine, putrescine and cadaverine, seem to be more suitable as spoilage indicators in white muscle fish, such as hake.[54,59–61]

Ishimaru *et al.*[51] observed that mackerel contained spermidine (22.38 mg kg^{-1}) and cod presented cadaverine (24.01 mg kg^{-1}) and putrescine (12.47 mg kg^{-1}) as the main biogenic amines. In addition, the researchers observed that in several kinds of fish that were stored for 2 days at 30 °C, the concentration of biogenic amines increased significantly. Thus, the total levels of biogenic amines reached by mackerel, tuna and cod were 44.33 g kg^{-1}, 9.73 g kg^{-1} and 1.91 g kg^{-1}, respectively. The increase of histamine was particularly relevant in mackerel and tuna, whilst tyramine and putrescine were higher in cod after 48 hours. In the case of spermidine, it was the only biogenic amine that decreased its concentration after the treatment. On the other hand, tryptamine was not determined in the fish studied. These results have been observed in several researches on fish, where the levels of histamine, putrescine and cadaverine increased during fish decomposition, whilst spermine and spermidine concentrations exhibited a tendency to decrease.[15,61] Conversely, an evaluation of biogenic amines in frozen anchovies during their storage showed that histamine formation decreased markedly.[62]

Therefore, maintaining fish and fish products at a low temperature is the best way to avoid the formation of biogenic amines during processing and storage. Thus, low temperatures (near to 0 °C) avoid the microbial growth of decarboxylating bacteria, and consequently, decarboxylate activity.[61] Therefore, the biogenic amines present in thawed fish products are mainly those in the raw material.[60] As mentioned previously, histamine and other amines are thermostable compounds that are not degraded once they are formed by thermal treatment, such as cooking, freezing or canning.[35,63] Hence, biogenic amines are also a good quality indicator for raw materials in preserved food products.[22] For this reason, it is important to monitor all steps, from catch to consumer. The application of HACCP in the fishery industry is an adequate system to improve the quality and maintain the security of fish products. Regarding biogenic amines, the implementation of HACCP can avoid the formation of these compounds during harvesting, storage and processing.[54] In general, the risk assessment of histamine in fish and fish products makes it possible to evaluate possible microbial contamination, as well as to show a break in the cold chain.[54]

5.2.1.1 *Legal Limits of Biogenic Amines in Fish and Fish Products*

Since the consumption of histamine is a real risk to human health, different institutions have established limits for safe human consumption. Currently, histamine is the only biogenic amine with a maximum legal limit in fish and fish products established by some national and international authorities.[64]

The US Food and Drug Administration (FDA)[65–68] established a limit of 50 mg kg^{-1} of histamine in fish. The level of toxicity for some fish, such as tuna and mahi-mahi, was established at 500 mg kg^{-1}, whilst the defect action level for histamine was 50 mg kg^{-1}. Similarly, a maximum histamine level of 50 mg kg^{-1} in fish was proposed by the Food and Agriculture Organization of the United Nations (FAO) and the World Health Organization (WHO),[69] as well as the European Food Safety Authority (EFSA).[22]

The Codex Alimentarius Standards proposed the histamine level in fish as an indicator for decomposition and hygiene and handling. Thus, in the case of decomposition, the relevant standards state: 'The products shall not contain more than 100 mg kg^{-1} of histamine based on the average of the sample unit tested and it is apply only some species, including Clupeidae, Scombridae, Scomberesocidae, Pomatomidae and Coryphaenidae families'. In addition, for hygiene and handling, the relevant standards state: 'No sample unit shall contain histamine that exceeds 200 mg kg^{-1} and it is applies only to species of the families Scombridae, Clupeidae, Coryphaenidae, Scomberesocidae and Pomatomidae'. These Codex Standards have been established for different fish products, such as quick frozen fish fillets or blocks of fish fillet, minced fish flesh, eviscerated and uneviscerated quick frozen finfish, quick frozen fish sticks (fish fingers), fish portions, fish fillets processed (breaded, battered, canned) and so on.[22]

Regarding the European Union, the European Commission Regulation (EC) No 2073/2005 (including its amendments such as Regulation (EC) No 1441/2007 and Regulation (EU) No 365/2010) sets food safety criteria for histamine in two different fishery products.[70–72] On the one hand, with reference to fish species (including the families Scombridae, Clupeidae, Engraulidae, Coryphaenidae, Pomatomidae and Scomberesocidae) placed on the market during their shelf life, this regulation indicates that nine samples should be taken, of whom only two may be between 100 and 200 mg kg^{-1} of histamine, and none above the limit of 200 mg kg^{-1}. On the other hand, this regulation is also applied to fishery products, especially those that have undergone an enzyme maturation treatment in brine and have been placed on the market during their shelf life. Similar to the previous case, the sampling plan is nine units, of whom only two samples may be between 200 and 400 mg kg^{-1} of histamine and none above the limit of 400 mg kg^{-1}.[22] Besides, the European Community has established a method of HPLC coupled to a fluorescence detector for the detection and quantification of histamine using dansyl chloride as a derivative agent in fish and fish products.[70]

Other countries have also proposed a maximum permissible limit of histamine in fishery products. Thus, a maximum of 100 mg kg^{-1} has been established in Canada, Switzerland, Finland, South Africa and Brazil, whilst fish products in Australia or New Zealand cannot present more than 200 mg kg^{-1} of histamine.[22,50,73]

5.2.2 Shellfish

The deterioration of seafood or shellfish, such as cephalopods, crustaceans or bivalves, is similar to fish (Table 5.1). Thus, temperature and storage time are the main factors involved in the quality of these products. Regarding histamine in shellfish, this amine is formed in low concentrations, so other biogenic amines have been proposed as a freshness indicator. In this case, a possible quality indicator for cuttlefish and squid is agmatine.[54,74] Putrescine has been proposed as an indicator for crustaceans. Thus, the maximum levels of putrescine in shrimp and lobster are 3 mg kg^{-1} and 7 mg kg^{-1}, respectively.[75,76] Similar to putrescine, cadaverine has been determined in shrimp, swimming crab and white prawn. The production of cadaverine followed a pattern similar to putrescine in several shrimp species.[27,75,77]

Regarding bivalve molluscs, the sum of polyamines can reach concentrations higher than 300 mg kg^{-1}.[27] In general, polyamine levels are very variable, mainly due to the amino acid profile and the activity of bacteria in the shellfish. However, further research is essential in order to establish appropriate freshness indicators for the different kinds of seafood.[54]

5.2.3 Animal Meat and Meat Products

Meat and meat products are rich in proteins, so they can contain certain levels of biogenic amines (Table 5.1). Although spermine and spermidine are the main biogenic amines in these products, other related compounds can be formed by microbial activity.[1] The level of biogenic amines is not the same in all meats, and it depends on the conditions of manipulation, treatment, storage and microorganism activity, among others.[16,19,22]

In the case of fresh animal food, biogenic amines are a significant parameter to evaluate conservation and freshness, as these compounds are produced when foods are spoiled and they increase during storage time.[16,17,78] In addition to spermine and spermidine, low concentrations of putrescine can be determined in fresh meat such as pork, beef, poultry and chicken, whilst other amines were almost undetectable.[16,32,79,80] Although the concentrations of spermine and spermidine are variable in meat, the amount of spermine is higher than spermidine, in contrast to vegetable products. Thus, the concentrations of spermine and spermidine were between 20–40 mg kg^{-1} and 1–5 mg kg^{-1}, respectively.[79] The levels of biogenic amines are different among animal species, as well as among parts or organs of each animal. Thus, polyamines are found in higher concentrations in tissues that present a rapid growth and high metabolic activity.[10,81] However, other biogenic amines can appear during processing or storage.[31,82,83] Hence, the profile of biogenic amines is different depending on the kind of meat, and the levels of histamine, tyramine, cadaverine or putrescine are variable.[84]

Thus, Balamatsia *et al.*[31] observed that levels of putrescine, cadaverine and histamine increased significantly in fresh chicken meat stored in different conditions, such as aerobic conditions or modified atmospheric packaging. However, spermine and spermidine decreased during storage time in the same conditions. Therefore, the authors suggested that these polyamines are not suitable as indicators of freshness quality for poultry meat. Similarly, Vinci and Antonelli[78] observed that biogenic amines (mainly cadaverine, tyramine and putrescine) increased in white meat (chicken) and red meat (adult bovine) after 30 days of storage at 4 °C. This trend was similar in pork meat, beef meat, beef liver or frog leg.[85–88] Therefore, the use of spoiled meat during the processing and storage could be a significant factor for the formation of biogenic amines.[30] Černíková *et al.*[89] performed an evaluation of biogenic amines in meat from different parts of crocodile stored for 24 hours. Some biogenic amines, such as histamine, phenylethylamine, putrescine and cadaverine, were not detected in this type of meat. Spermine was the main amine and its concentration oscillated from 7.44 to 9.35 mg kg^{-1}. Other biogenic amines, such as tyramine, tryptamine and spermidine, did not exceed 6 mg kg^{-1}. Researchers did not observe significant changes in the profile of biogenic amines among different body parts.

Biogenic amines indices are a useful tool to evaluate food quality and good hygienic manufacturing practices, providing significant information about quality control processes.[78,86] Some researchers suggested that a certain concentration of cadaverine could be used as an indication of spoilage in chicken and beef meat.[78] On the other hand, some authors assessed the relationship between the growth of some species of Enterobacteriaceae and aerobic mesophilic bacteria and the production of biogenic amines in different meat, including duck, quail and three types of chicken raised in conventional, free range and organic conditions.[17] However, these researches stated that it was difficult to establish a correlation between the formation of biogenic amines in all meats studied, suggesting that it is necessary to select specific biogenic amines to be used as quality indicators for each kind of meat.[17]

In meat products obtained with thermal treatment, such as cooked ham or cooked meat sausages, spermine and spermidine can be determined. Besides, some authors have observed a light reduction of polyamines during thermal treatments, although these compounds are considered as thermal resistant.[90] Regarding other biogenic amines in cooked meat products, low concentrations of tyramine and histamine have usually been detected. On the other hand, cured products do not seem to produce significant concentrations of biogenic amines, as the product is stored with the addition of salt, which can avoid the growing of spoilage bacteria.[91] However, the concentration of biogenic amines in meat products can increase when they are stored under aerobic conditions. Thus, the storage temperature plays an important role in the formation and accumulation of biogenic amines.[80] This way, the formation of biogenic amines has been used as an indicator of microbial proliferation in packaged meat.[92]

As seen earlier, heat treatments and cooking processes do not affect the concentrations of biogenic amine in meat products, although they avoid the growth of microorganisms. Thus, the presence of biogenic amines in these products may be an indicator of spoilage by microorganisms before, during or after the elaboration of the product.

5.2.4 Dairy Products

Regarding raw milk, the main biogenic amines are spermine and spermidine.[48] The levels of biogenic amines in dairy products are influenced by different factors, such as pH and temperature, as well as the presence of microorganisms producing biogenic amines. In addition, milk quality and heat processing can influence the content of biogenic amines.[48,93,94]

Regarding dairy products, the levels of biogenic amines in these products are becoming increasingly important in relation to food safety.[64] Thus, dairy products that have not been processed in good conditions could present high levels of biogenic amines. Generally, fresh, pasteurized or UHT milk have slight contents of biogenic amines, mainly histamine; although their concentrations could increase significantly after the fermentation process.

Regarding cheese, even though it is a fermented product, it is important to highlight that a thermal treatment applied to raw milk before fermentation can eliminate wild decarboxylating bacteria and avoid the production of biogenic amines. Thus, cheeses from pasteurized milk usually present a low concentration of biogenic amines.[36]

5.2.5 Eggs

Eggs are an animal product that may be affected by bacterial contamination, especially if the cold chain is broken during storage.[32] Figueiredo *et al.*[95] performed research to evaluate the relationship between levels of biogenic amines in eggs and their freshness. The researchers analysed the levels of biogenic amines when the temperature during processing and storage was not adequate. The main biogenic amines found in eggs were phenylethylamine in albumens and spermidine in yolks, with a mean concentration of 38 mg kg^{-1} and 1 mg kg^{-1}, respectively. Finally, the authors concluded that it was not possible to establish a quality index for eggs by using these compounds.

5.2.6 Honey

The evaluation of the biogenic amine profile in honey has been poorly studied. One of the few studies was performed by Kelly *et al.*,[96] who analysed the concentration of biogenic amines in different types of honey. The five types of honey studied were oak, chestnut, Mediterranean scrubland, Eyne valley and rhododendron honey, and none of them presented differences in the concentration of ethanolamine, ranging from 0.8 to

1.8 mg kg^{-1}. The concentrations of histamine were more variable. Thus, oak honey presented the highest concentration (3.8 mg kg^{-1}), whilst this amine was not quantified in Eyne valley samples. Besides, several biogenic amines were not detected, such as putrescine, cadaverine, tyramine, isoamylamine and phenylethylamine, in any of the samples. Similarly, only residual levels of biogenic amines were detected in several honeys from different localizations, including the mainland of Portugal, the Madeira Islands and the Canary Islands.[97]

5.3 Biogenic Amines in Vegetables and Plant-derived Products

5.3.1 Vegetables

Polyamines, including putrescine, spermine and spermidine, have been determined in plants and they perform physiological functions such as fruit growth and development (Table 5.2).[98,99] Polyamines are involved in different cell functions in plants such as cell division, cell differentiation, synthesis of nucleic acids and protein, membrane stability, stress responses and delay in senescence.[26,100]

The concentration and profile of biogenic amines in plants are determined by several factors, for example, plant variety, region of cultivation, maturity degree, cultivation practices, water stress, post-harvest manufacturing processes and storage conditions, among others.[101–105] Some biogenic amines, such as cadaverine or putrescine, are important in agriculture, as they are precursors of alkaloids and they act against predators.[7,106] In addition, the high concentration of histamine during germination could be due to a protective effect against predators.[102,105] Furthermore, in some fruits and vegetables, biogenic amines may be part of the taste of mature foods and some aroma compounds.[26]

The main biogenic amines in fruit and green vegetables are putrescine and spermidine, respectively.[20,26] However, polyamines are usually present in low levels in fresh vegetables.[107] In general, foods derived from plants are rich in spermidine, whilst spermine concentration is higher in animal products.[13] Moret *et al.*[26] evaluated several fresh vegetables (Table 5.2). The authors observed that putrescine, followed by spermidine, were the main biogenic amines present in the samples and could be quantified in all types of fresh vegetables. Regarding the concentration of biogenic amines in fresh vegetables, putrescine was determined in higher concentrations in broad bean (65 mg kg^{-1}) and zucchini (49 mg kg^{-1}), whilst broccoli (45 mg kg^{-1}) and cauliflower (38 mg kg^{-1}) presented a higher amount of spermidine. With regard to the rest of the biogenic amines, a high level of histamine was observed in spinach (20 mg kg^{-1}), whilst tyramine showed a high concentration in arugula (12 mg kg^{-1}) and in broad bean (10 mg kg^{-1}). However, these concentrations of biogenic amines are not significant enough to cause poisoning in healthy consumers. After storing these

Table 5.2 Content of main biogenic amines in vegetables and plant-derived products (mg kg^{-1} or L^{-1}).[a]

Food	Spermidine	Spermine	Histamine	Tyramine	Putrescine	Cadaverine	Ref.
Vegetables							
Vegetables	1.1–108.2	n.d.–15.9	n.d.–69.7	n.d.–14.1	n.d.–48.1	n.d.–0.4	13
Legumes	11.0–85.3	5.1–32.4	—	—	0.4–6.4	—	13
Arugula	19	3	—	12	2	—	26
Beans	13–74	2–20	n.d.	n.d.	15–78	n.d.	112
Broad bean	8	3	2	10	65	—	26
Broccoli	45	11	—	1	14	—	26
Cabbage	—	—	1.0	18.6	11.6	7.6	128
Carrot	12	1	tr	3	7	tr	26
Cauliflower	38	7	tr	2	11	tr	26
Cucumber	10	tr	tr	2	25	—	26
Fennel	6	1	—	1	14	tr	26
Lettuce	16	1	—	1	10	—	26
Onion	4	1	—	3	3	—	26
Parsley	26	2	—	tr	11	—	26
Pepper	4	2	1	2	5	—	26
Potato	10	7	—	7	10	—	26
Savoy cabbage	4	1	tr	3	19	—	26
Soybeans/tofu	—	—	n.d.–5.8	n.d.	0.7–5.0	n.d.–3.4	128
Spinach	23	6	20	8	8	1	26
Tomato	6	1	7	2	29	tr	26

(continued)

Table 5.2 (*Continued*)

Food	Spermidine	Spermine	Histamine	Tyramine	Putrescine	Cadaverine	Ref.
Zucchini	19	2	—	tr	40	—	26
Cereals							
Cereals	0.2–327.6	n.d.–84.2	n.d.–0.9	—	n.d.–31.9	n.d.–0.7	13
Barley	—	—	n.d.	8.0–18.9	7.6–21.0	0.5–1.1	128
Fruits and fruit juice							
Fruits	0.8–15.8	n.d.–8.9	n.d.–2.5	n.d.–11.5	n.d.–173.8	—	13
Apples	—	—	n.d.	n.d.	5.1	—	128
Grapes	—	—	n.d.–5.8	n.d.–2.4	n.d.–8.0	n.d.–3.4	128
Orange juice	—	—	n.d.–0.04	n.d.–0.06	0.1–2.2	—	128
Red grape must	—	—	n.d.	0.8–0.9	4.2–15.3	0.5–1.0	129
Nuts							
Nuts	6.2–48.1	2.0–37.81	n.d.–11.9	n.d.–2.6	n.d.–39.5	—	13
Cocoa							
Cocoa beans	—	—	—	1.0–4.8	—	—	122
Chocolate	0.3–4.7	1.2–2.7	0.2–0.6	0.2–5.9	n.d.–2.0	n.d.–2.8	13
Chocolate	—	—	0.3–2.0	3.1–8.1	0.8	0.8	128

[a]Abbreviations: Ref., references; n.d., non-detected; —, non-determined; tr, traces.

vegetables in a refrigerator for 3 weeks, the authors observed a decrease in the content of spermidine in the samples, which did not present significant changes in other amines.

Biogenic amines could be formed in fresh vegetables by different strains of Enterobacteriaceae and *Pseudomonas*. Besides, some researches have isolated and identified several strains of *Klebsiella*, *Pantoea*, *Hafnia*, *Enterobacter*, *Serratia*, *Morganella* and *Aeromonas* with the capacity to synthetize histamine, putrescine or cadaverine during cool storage of vegetables.[108,109]

Biogenic amines have also been determined in different legumes and radish seeds.[102,104,110,111] Thus, cadaverine was the main biogenic amine determined in broccoli seeds, followed by histamine and putrescine. Regarding the total content of biogenic amines in broccoli seeds, the authors observed that they ranged from 24 to 45 mg kg^{-1} of dry matter, according to the cultivars. On the other hand, radish seeds exhibited a higher concentration of cadaverine and histamine, reaching a total concentration from 47.8 to 48.6 mg kg^{-1} of dry matter. These differences in total amounts may be due to the type or variety of seed, water availability, mineral deficiency, chemical-caused damage, osmotic shock and temperature.[104,111] There is some research focused on assessing the different levels of biogenic amines in legume grains in relation to changes in concentrations during germination.[110] Even though biogenic amines have been determined in ungerminated legume seeds, their concentrations may increase significantly after germination, mainly due to the increase of cell proliferation during the first phases.[102,105] In this context, several authors have observed an increase in polyamines during the germination of several types of seeds. Besides, legume sprouts also contained biogenic amines.[110]

On the other hand, the germination of broccoli and radish seeds may cause an impact in the content of biogenic amines. Thus, five days after germination, the concentrations of biogenic amines reached levels of 150 mg kg^{-1} of dry matter in broccoli seeds, whilst radish seeds showed concentrations of 170 mg kg^{-1}.[111] This increase may be attributed to different functions, such as cellular multiplication, increase in cell size, growth during sprouting or their role in the elongation of the root, among others.[102,110,111]

Besides, some authors have observed that cooking processes and heat treatments can affect polyamine concentrations.[99,110] Preti *et al.*[112] observed that some types of beans (green or yellow beans) increased their concentration of biogenic amines after boiling. In addition, the authors proposed some hypotheses to explain the increase of biogenic amines after cooking, such as the possible inactivation of oxidase enzymes due to heat, an increased availability for extraction due to the fragmentation of cell structures or the degradation of more complex compounds (proteins, amino acids).

5.3.2 Fruits and Fruit Juices

In general, there is not a lot of research on the content of biogenic amines in fruits, and available data shows a high variability of amine levels within the same type of fruit (Table 5.2). In general, fruits and fruit juices are rich in putrescine.[20]

Regarding fresh fruits, the main biogenic amines detected are putrescine, tyramine and spermidine. Thus, the highest content of these amines was determined in bananas, with 23.5 mg kg^{-1} of tyramine and 3.23 mg kg^{-1} of putrescine, as well as 36.2 mg kg^{-1} of cadaverine.[113] The fresh fruits with a higher concentration of biogenic amines were bananas (62.9 mg kg^{-1}) and pomelo (48.8 mg kg^{-1}), whilst those with a lower concentration were white grapes (9.1 mg kg^{-1}) and apples (3.9 mg kg^{-1}).[113] Santiago-Silva *et al.*[114] analysed biogenic amines in different tropical fruits. Thus, total biogenic amine levels ranged from 7.7 g kg^{-1} in mango to 75.3 g kg^{-1} in passion fruit. Spermidine prevailed in mango, pineapple, passion fruit and guava (similar concentration to spermine), whilst papaya presented a higher concentration of putrescine. Regarding passion fruit, the total amine levels in the pulp prepared without the seed (79 g kg^{-1}) were significantly higher than in the pulp with the seed (52 g kg^{-1}), mainly due to the high levels of spermidine and spermidine in free form.

Jain *et al.*[115] conducted research to determine biogenic amines in different commercial fruit juices. Thus, mango (692 μg L^{-1}) and grape (508 μg L^{-1}) were the fruit juices with a higher concentration of biogenic amines. Besides, the authors observed that the biogenic amines profiles were very different, and the main amine detected changed according to the type of fruit. Therefore, tryptamine was predominant in grape (351 μg L^{-1}) and mango (305 μg L^{-1}) juices, whilst methylamine was the highest one in litchi (66 μg L^{-1}) and orange (61 μg L^{-1}) juices. In the same way, apple and pineapple juice samples presented 89 μg L^{-1} of phenylethylamine and 125 μg L^{-1} of putrescine, respectively. On the other hand, Jastrzębska *et al.*[113] evaluated different non-filtered juices, obtaining higher levels of biogenic amines. Thus, the non-filtered juices evaluated include apple and plum (4.1 mg kg^{-1}), apple (4.4 mg kg^{-1}), grapefruit (11.6 mg kg^{-1}), red currant (2.2 mg kg^{-1}), black currant (3.4 mg kg^{-1}), cherry (7.4 mg kg^{-1}) and pineapple (9.2 mg kg^{-1}) juices.

Biogenic amines have been evaluated in fruit nectars. Fruit nectars are prepared by adding water, sometimes with the addition of sugars, to fruit juices or concentrated fruit purée, reaching a fruit content in the range of 25–99%.[116] Thus, three types of fruit nectars with different concentrations were analysed. Apricot nectar, with 50% of fruit, exhibited the highest concentration of biogenic amines (28.7 mg L^{-1}), mainly cadaverine and putrescine. In addition, the authors observed an increase in the total amount of biogenic amines in fruit nectars with 70% of fruit compared with those with 50%. Therefore, the total mean concentrations of biogenic amines in peach nectar were 11.5 and 12.7 mg L^{-1} for 50% and 70% nectar, respectively. Similarly, the total concentrations of biogenic amines in pear nectar were 10.1 mg L^{-1} for nectar with 50% and 12.2 mg L^{-1} for nectar with 70%. In both cases, the biogenic amine that significantly increased its concentration was spermidine.[116] Biogenic amines in juices made with fruit concentrates were also evaluated. In general, the total mean levels of biogenic amines ranged from 5.4 mg L^{-1} in apple juices to 84 mg L^{-1} in orange juices. Putrescine was the main biogenic amine in juice concentrates of orange, grapefruit and apples, whilst spermidine predominated in pineapple juice concentrate.

Regarding the fermentation industry, low levels of putrescine were determined in apple musts before being fermented to obtain cider (2.70–6.65 mg L^{-1}).[117] Besides, the authors observed that neither the type of press nor the processing step affected the biogenic amines content. These results are similar to those found in grapes.[118]

5.3.3 Nuts

Nuts, including almond, pine nuts and pistachios, presented mainly spermidine (6.2–48.1 mg kg^{-1}) and spermine (2.0–37.8 mg kg^{-1}). As for the concentrations of histamine and putrescine, these were highly variable, ranging from non-detected to 11.9 mg kg^{-1} and 39.5 mg kg^{-1}, respectively. Tyramine, phenylethylamine, tryptamine and agmatine showed a mean concentration below the limits of detection.[13] On the other hand, serotonin has been determined in high concentrations in some raw nuts, such as walnut (155 mg kg^{-1}), pecan (13.6 mg kg^{-1}) and cashew (1.3 mg kg^{-1}). In addition, roasted Brazil nut showed a content of 3.8 mg kg^{-1} of serotonin.[119]

5.3.4 Cocoa

Several biogenic amines have been determined in cocoa and chocolate, including phenylethylamine, tyramine, tryptamine, serotonin, dopamine and histamine.[120,121]

Some authors have evaluated the effects of thermal processes in the biogenic amine profile in cocoa beans.[122] Thus, the concentration of biogenic amines in cocoa beans ranged from 2.66 to 11.43 mg kg^{-1} of dry weight. Tyramine was the most abundant biogenic amine in cocoa beans, reaching around 32% of the total, followed by tryptamine and phenylethylamine.

According to Oracz and Nebesny,[122] the concentration of biogenic amines increased during the roasting process. This may be due to the formation of biogenic amines during heat conditions caused by chemical decarboxylation of amino acids when lipid peroxidation products are present.[122,123] The content of biogenic amines in cocoa beans increased significantly after roasting. Thus, the temperature and relative humidity of the air had an important effect in the content of biogenic amines. Regarding the biogenic amines profile, phenylethylamine was the most abundant in the cocoa samples after treatment.[122]

5.4 Analytical Methods in the Determination of Biogenic Amines in Food

Biogenic amine analysis is usually employed to monitor foods, including controlling raw, intermediate and final products, fermentation processes, storage changes and the application of storage conservation.[11]

The analytical determination of biogenic amines in food is not simple due to the variety of their chemical structures and because they are usually present at low concentrations in complex matrices.[22,124] Besides, these compounds have strong polar characters, which makes it difficult to separate and evaluate them.[125] Biogenic amines are kinds of molecules that do not have good absorption properties in the visible, ultraviolet or fluorescence wavelength ranges. Thus, a chemical derivatization prior to HPLC is necessary for the determination of these compounds.[126]

Another important factor is the presence of other related molecules in the matrix, such as amino acids. These are usually one or two orders of magnitude higher, so they can interfere in the detection of biogenic amines.[127,128]

The most employed technique is HPLC using C18 reverse phase columns, as this is a methodology with a high resolution and sensitivity.[129] Analytical methods often use extraction and derivatization processes before the analysis by HPLC in order to improve the sensitivity and to avoid interferences. Different types of extraction are used to concentrate the analytes and clean up samples. As aforementioned, these compounds do not have good properties in the visible, ultraviolet or fluorescence wavelength ranges, so a chemical derivatization is performed to improve their sensitivity and selectivity.[126] The most widely used derivatization reagents are dansyl chloride, *o*-phthalaldehyde, benzoyl chloride, 6-aminoquinolyl-*N*-hydroxysuccinimidyl carbamate and diethyl ethoxymethylenemalonate, among others.[4]

On the other hand, other techniques can be used to evaluate biogenic amines in food, such as thin-layer chromatography, gas chromatography, capillary electrophoresis and flow-injection analysis, among others.[4]

5.5 Conclusions

This chapter showed an overview of biogenic amines in non-fermented products. These compounds are naturally present in non-fermented foods, especially polyamines (spermine and spermidine). However, their concentrations may be affected by different technological factors, mainly food processing or during their storage. Thus, biogenic amines can increase significantly due to the growth of aminogenic microorganisms, mainly histamine, tyramine and putrescine.

The variability of biogenic amines in foods makes it difficult to elaborate a universal biogenic amine index that can be used for all foods. Therefore, different indices should be evaluated and adapted to specific food products.

Currently, there is no legal regulation regarding the levels of biogenic amines in non-fermented products, except for fish and fish products. Even though the natural levels of these compounds are not considered to be a hazard to human health, incorrect food handling may cause external contamination by microorganisms that could produce high levels of these compounds. Hence, it is necessary to have analytical tools with an adequate sensitivity and selectivity, as well as establishing a maximum threshold to guarantee the protection of consumer health. In this sense, it is necessary to improve the knowledge about the level of biogenic amines in different foods, as well as their changes during processing and storage, in order to elaborate future regulations.

References

1. B. ten Brink, C. Damink, H. Joosten and J. H. J. Huis In't Veld, *Int. J. Food Microbiol.*, 1990, **11**, 73.
2. A. J. Michael, *J. Biol. Chem.*, 2016, **291**, 14896.
3. J. Kirschbaum, A. Meier and H. Brückner, *Chromatographia*, 1999, **49**, 117.
4. G. I. Mohammed, A. S. Bashammakh, A. A. Alsibaai, H. Alwael and M. S. El-Shahawi, *TrAC, Trends Anal. Chem.*, 2016, **78**, 84.
5. V. Ladero, M. Calles-Enriquez, M. Fernandez and M. A. Alvarez, *Curr. Nutr. Food Sci.*, 2010, **6**, 145.
6. M. D. Groppa and M. P. Benavides, *Amino Acids*, 2008, **34**, 35.
7. V. Sridevi, P. Giridhar and G. A. Ravishankar, *Acta Physiol. Plant.*, 2009, **31**, 757.
8. M. Atiya Ali, E. Poortvliet, R. Strömberg and A. Yngve, *Food Nutr. Res.*, 2011, **55**, 5572.
9. E. Agostinelli and U. Bachrach, *Amino Acids*, 2007, **33**, 173.
10. S. Bardócz, *Trends Food Sci. Technol.*, 1995, **6**, 341.
11. A. Önal, *Food Chem.*, 2007, **103**, 1475.
12. J. M. Landete, S. Ferrer and I. Pardo, *Food Control*, 2007, **18**, 1569.
13. S. Bover-Cid, M. L. Latorre-Moratalla, M. T. Veciana-Nogués and M. C. Vidal-Carou, *Encyclopedia of Food Safety*, Academic Press, Cambridge, 2013, vol. 2, p. 381.
14. E. Chiacchierini, D. Restuccia and G. Vinci, *Talanta*, 2006, **69**, 548.
15. M. H. Silla-Santos, *Int. J. Food Microbiol.*, 1996, **29**, 213.
16. C. Ruiz-Capillas and F. Jiménez-Colmenero, *Crit. Rev. Food Sci. Nutr.*, 2004, **44**, 489.
17. C. A. Lázaro, C. A. Conte-Júnior, A. C. Canto, M. L. G. Monteiro, B. Costa-Lima, A. G. Da Cruz, E. T. Mársico and R. M. Franco, *LWT–Food Sci. Technol.*, 2015, **60**, 15.
18. M. A. Álvarez and M. V. Moreno-Arribas, *Trends Food Sci. Technol.*, 2014, **39**, 146.
19. G. Jairath, P. K. Singh, R. S. Dabur, M. Rani and M. Chaudhari, *J. Food Sci. Technol.*, 2015, **52**, 6835.

20. A. R. Shalaby, *Food Res. Int.*, 1996, **29**, 675.
21. L. Lehane and J. Olley, *Int. J. Food Microbiol.*, 2000, **58**, 1.
22. European Food Safety Authority, Scientific opinion on risk based control of biogenic amine formation in fermented foods, *EFSA J.*, **9**(10), 2393.
23. L. Maintz and N. Novak, *Am. J. Clin. Nutr.*, 2007, **85**, 1185.
24. P. Visciano, M. Schirone, R. Tofalo and G. Suzzi, *Front. Microbiol.*, 2012, **3**, 188.
25. S. Sentellas, Ó Núñez and J. Saurina, *J. Agric. Food Chem.*, 2016, **64**, 7667.
26. S. Moret, D. Smela, T. Populin and L. S. Conte, *Food Chem.*, 2005, **89**, 355.
27. M.-K. Kim, J.-H. Mah and H.-J. Hwang, *Food Chem.*, 2009, **116**, 87.
28. S. Yurchenko and U. Mölder, *Food Chem.*, 2006, **96**, 325.
29. N. G. Hord, Y. Tang and N. S. Bryan, *Am. J. Clin. Nutr.*, 2009, **90**, 1.
30. M. P. Costa, B. L. Rodrigues, B. S. Frasao and C. A. Conte-junior, in *Food Quality: Balancing Health and Disease*, ed. A. M. Grumezescu and A. M. Holban, Academic Press, Cambridge, 13th edn, 2018, pp. 75–108.
31. C. C. Balamatsia, E. K. Paleologos, M. G. Kontominas and I. N. Savvaidis, *Antonie van Leeuwenhoek*, 2006, **89**, 9.
32. C. A. Lázaro de la Torre and C. A. Conte-Junior, in *Food Control and Biosecurity*, ed. A. M. Grumezescu and A. M. Holban, Academic Press, Cambridge, 16th edn, 2018, pp. 225–257.
33. V. Sirocchi, G. Caprioli, C. Cecchini, M. M. Coman, A. Cresci, F. Maggi, F. Papa, M. Ricciutelli, S. Vittori and G. Sagratini, *Int. J. Food Sci. Nutr.*, 2013, **64**, 921.
34. D. M. Linares, B. del Río, V. Ladero, N. Martínez, M. Fernández, M. C. Martín and M. A. Álvarez, *Front. Microbiol.*, 2012, **3**, 1.
35. A. Naila, S. Flint, G. Fletcher, P. Bremer and G. Meerdink, *J. Food Sci.*, 2010, **75**, 139.
36. F. Gardini, Y. Özogul, G. Suzzi, G. Tabanelli and F. Özogul, *Front. Microbiol.*, 2016, 7, 1218.
37. K. Becker, K. Southwick, J. Reardon, R. Berg and J. N. MacCormack, *JAMA, J. Am. Med. Assoc.*, 2001, **285**, 1327.
38. F. Özogul and Y. Özogul, *Food Chem.*, 2006, **99**, 574.
39. B. L. Rodrigues, T. da S. Alvares, G. S. L. Sampaio, C. C. Cabral, J. V. A. Araujo, R. M. Franco, S. B. Mano and C. A. Conte Junior, *Food Control*, 2016, **60**, 596.
40. X. Zhang, H. Wang, N. Li, M. Li and X. Xu, *LWT--Food Sci. Technol.*, 2015, **64**, 1123.
41. K. Bjornsdottir-Butler, D. P. Green, G. E. Bolton and P. D. McClellan-Green, *J. Food Sci.*, 2015, **80**, 1253.
42. L. Cai, A. Cao, Y. Li, Z. Song, L. Leng and J. Li, *Food Control*, 2015, **56**, 1.
43. M. Jouki, F. T. Yazdi, S. A. Mortazavi, A. Koocheki and N. Khazaei, *Int. J. Food Microbiol.*, 2014, **174**, 88.

44. M. Tsiligianni, E. Papavergou, N. Soultos, T. Magra and I. N. Savvaidis, *Int. J. Food Microbiol.*, 2012, **159**, 101.
45. R. G. K. Leuschner and W. P. Hammes, *Meat Sci.*, 1998, **49**, 289.
46. F. Gardini, M. Martuscelli, M. A. Crudele, A. Paparella and G. Suzzi, *Meat Sci.*, 2002, **61**, 275.
47. J. H. Kim, H. J. Ahn, C. Jo, H. J. Park, Y. J. Chung and M. W. Byun, *Food Control*, 2004, **15**, 405.
48. G. Spano, P. Russo, A. Lonvaud-Funel, P. Lucas, H. Alexandre, C. Grandvalet, E. Coton, M. Coton, L. Barnavon, B. Bach, F. Rattray, A. Bunte, C. Magni, V. Ladero, M. Alvarez, M. Fernández, P. Lopez, P. F. de Palencia, A. Corbi, H. Trip and J. S. Lolkema, *Eur. J. Clin. Nutr.*, 2010, **64**, 95.
49. K. B. Biji, C. N. Ravishankar, R. Venkateswarlu, C. O. Mohan and T. K. S. Gopal, *J. Food Sci. Technol.*, 2016, **53**, 2210.
50. L. Auerswald, C. Morren and A. L. Lopata, *Food Chem.*, 2006, **98**, 231.
51. M. Ishimaru, Y. Muto, A. Nakayama, H. Hatate and R. Tanaka, *Food Anal. Methods*, 2019, **12**, 166–175.
52. J. M. Hungerford, *Toxicon*, 2010, **56**, 231.
53. I. Al Bulushi, S. Poole, H. C. Deeth and G. A. Dykes, *Crit. Rev. Food Sci. Nutr.*, 2009, **49**, 369.
54. L. Prester, *Food Addit. Contam., Part A*, 2011, **28**, 1547.
55. C. Ruiz-Capillas and A. Moral, *Amino Acids*, 2004, **26**, 125.
56. L. Prester, J. Macan, V. M. Varnai, T. Orct, J. Vukušić and D. Kipčić, *Food Addit. Contam., Part A*, 2009, **26**, 355.
57. S. Baixas-Nogueras, S. Bover-Cid, M. C. Vidal-Carou and M. T. Veciana-Nogues, *J. Food Sci.*, 2001, **66**, 83.
58. M. Rezaei, N. Montazeri, H. E. Langrudi, B. Mokhayer, M. Parviz and A. Nazarinia, *Food Chem.*, 2007, **103**, 150.
59. S. Baixas-Nogueras, S. Bover-Cid, M. T. Veciana-Nogués, A. Mariné-Font and M. C. Vidal-Carou, *J. Food Prot.*, 2005, **68**, 2433.
60. S. Baixas-Nogueras, S. Bover-Cid, M. T. Veciana-Nogués and M. C. Vidal-Carou, *Eur. Food Res. Technol.*, 2007, **226**, 287.
61. C. Ruiz-Capillas and A. Moral, *J. Food Sci.*, 2001, **66**, 1030.
62. R. Rossano, L. Mastrangelo, N. Ungaro and P. Riccio, *J. Chromatogr. B: Anal. Technol. Biomed. Life Sci.*, 2006, **830**, 161.
63. W. Tapingkae, S. Tanasupawat, K. L. Parkin, S. Benjakul and W. Visessanguan, *Enzyme Microb. Technol.*, 2010, **46**, 92.
64. N. Benkerroum, *Compr. Rev. Food Sci. Food Saf.*, 2016, **15**, 801.
65. Food and Drug Administration, USA, Decomposition and histamine: raw, frozen tuna and mahi-mahi, canned tuna, and related species, revised compliance guide, availability, *Fed. Regist.*, 1995, **60**(149), 39754.
66. Food and Drug Administration, USA, *Guidance for Industry: Bioanalytical Method Validation*, US Department of Health and Human Services, 2001, vol. 2015.

67. Food and Drug Administration, USA, *Fish and Fishery Products Hazards and Controls Guidance*, 4th edn, 2011, ch. 7, p. 113.
68. Food and Drug Administration, USA, *Acceptance Criteria for Confirmation of Identity of Chemical Residues using Exact Mass Data for the FDA Foods and Veterinary Medicine Program*, 2015.
69. Food and Agriculture Organization of the United Nations/World Health Organization, *Public Health Risks of Histamine and Other Biogenic Amines from Fish and Fishery Products*, 2013.
70. Commission Regulation (EC) No. 2073/2005 of 15 November 2005 on microbiological criteria for foodstuffs, 2005, vol. **338**, p. 1.
71. Commission Regulation (EC) No. 1441/2007 of 5 November 2007 amending Regulation (EC) No. 2073/2005 on microbiological criteria for foodstuffs, 2007, vol. **322**, p. 2.
72. Commission Regulation (EU) No 365/2010 of 28 April 2010 amending Regulation (EC) No 2073/2005 on microbiological criteria for foodstuffs, 2010, vol. **107**, p. 9.
73. J. Lange, K. Thomas and C. Wittmann, *J. Chromatogr. B: Anal. Technol. Biomed. Life Sci.*, 2002, **779**, 229.
74. P. Vaz-Pires, P. Seixas, M. Mota, J. Lapa-Guimarães, J. Pickova, A. Lindo and T. Silva, *LWT--Food Sci. Technol.*, 2008, **41**, 1655.
75. R. A. Benner, W. F. Staruszkiewicz, P. L. Rogers and W. S. Otwell, *J. Food Sci.*, 2003, **68**, 2178.
76. L. Prester, T. Orct, J. Macan, J. Vukušić and D. Kipčić, *Arch. Ind. Hyg. Toxicol.*, 2010, **61**, 389.
77. Q. X. Zhao, J. Xu, C. H. Xue, W. J. Sheng, R. C. Gao, Y. Xue and Z. J. Li, *J. Agric. Food Chem.*, 2007, **55**, 3083.
78. G. Vinci and M. L. Antonelli, *Food Control*, 2002, **13**, 519.
79. P. Kalač, *Meat Sci.*, 2006, **73**, 1.
80. T. Hernández-Jover, M. Izquierdo-Pulido, M. T. Veciana-Nogués, A. Mariné-Font and M. C. Vidal-Carou, *J. Agric. Food Chem.*, 1997, **45**, 2098.
81. Y. Ramot, S. Tiede, T. Bíró, M. H. Abu Bakar, K. Sugawara, M. P. Philpott, W. Harrison, M. Pietilä and R. Paus, *PLoS One*, 2011, **6**, e22564.
82. L. Buňková, R. Gál, E. Lorencová, P. Jančová, M. Doležalová, V. Kmeť and F. Buňka, *Eur. J. Wildlife Res.*, 2016, **62**, 341.
83. Z. Hutařová, V. Večerek, I. Steinhauserová, P. Maršálek and G. Bořilová, *Acta Vet. Brno*, 2013, **82**, 61.
84. A. Jastrzębska, *Eur. Food Res. Technol.*, 2012, **235**, 563.
85. M. Li, L. Tian, G. Zhao, Q. Zhang, X. Gao, X. Huang and L. Sun, *Meat Sci.*, 2014, **96**, 843.
86. F. Galgano, F. Favati, M. Bonadio, V. Lorusso and P. Romano, *Food Res. Int.*, 2009, **42**, 1147.
87. F. B. Custódio, K. H. Theodoro and M. B. A. Gloria, *Food Control*, 2016, **60**, 151.

88. M. E. Büyükdeveci, E. K. Boga and G. Ozyurt, *LWT--Food Sci. Technol.*, 2019, **99**, 379.
89. M. Černíková, R. Gál, Z. Polášek, M. Janíček, V. Pachlová and F. Buňka, *J. Food Compos. Anal.*, 2015, **43**, 82.
90. P. Paulsen and F. Bauer, *Eur. Food Res. Technol.*, 2007, **225**, 921.
91. J. M. Lorenzo, S. Martínez, I. Franco and J. Carballo, *Meat Sci.*, 2007, **77**, 287.
92. B. Schumann and M. Schmid, *Innovative Food Sci. Emerging Technol.*, 2018, **47**, 88.
93. M. P. Costa, C. F. Balthazar, B. L. Rodrigues, C. A. Lazaro, A. C. O. Silva, A. G. Cruz and C. A. Conte Junior, *Food Sci. Nutr.*, 2015, **3**, 172.
94. M. Schirone, R. Tofalo, P. Visciano, A. Corsetti and G. Suzzi, *Front. Microbiol.*, 2012, **3**, 1.
95. T. C. Figueiredo, D. C. S. Assis, L. D. M. Menezes, D. D. Oliveira, A. L. Lima, M. R. Souza, L. G. D. Heneine and S. V. Cançado, *Poult. Sci.*, 2014, **93**, 3171.
96. M. T. Kelly, A. Blaise and M. Larroque, *J. Chromatogr. A*, 2010, **1217**, 7385.
97. V. Pereira, M. Pontes, J. S. Câmara and J. C. Marques, *J. Chromatogr. A*, 2008, **1189**, 435.
98. M. Esti, G. Volpe, L. Massignan, D. Compagnone, E. La Notte and G. Palleschi, *J. Agric. Food Chem.*, 1998, **46**, 4233.
99. M. P. G. Cirilo, A. F. S. Coelho, C. M. Araújo, F. R. B. Gonçalves, F. D. Nogueira and M. B. A. Glória, *Food Chem.*, 2003, **82**, 397.
100. A. Bouchereau, P. Guénot and F. Larher, *J. Chromatogr. B: Biomed. Sci. Appl.*, 2000, **747**, 49.
101. S. Casal, E. Mendes, M. R. Alves, R. C. Alves, M. Beatriz, P. P. Oliveira and M. A. Ferreira, *J. Agric. Food Chem.*, 2004, **52**, 6188.
102. M. B. A. Glória, J. Tavares-Neto, R. A. Labanca and M. S. Carvalho, *J. Agric. Food Chem.*, 2005, **53**, 7480.
103. S. D. Oliveira, A. S. Franca, M. B. A. Glória and M. L. A. Borges, *Food Chem.*, 2005, **90**, 287.
104. J. Frías, C. Martínez-Villaluenga, P. Gulewicz, A. Perez-Romero, R. Pilarski, K. Gulewicz and C. Vidal-Valverde, *Food Chem.*, 2007, **105**, 959.
105. C. M. Bandeira, W. P. Evangelista and M. B. A. Gloria, *Food Chem.*, 2012, **131**, 1355.
106. A. Bouchereau, A. Aziz, F. Larher and J. Martin-Tanguy, *Plant Sci.*, 1999, **140**, 103.
107. P. Kalač, S. Švecová and T. Pelikánová, *Food Chem.*, 2002, **77**, 349.
108. A. Halász, Á Baráth and W. H. Holzapfel, *Z. Lebensm.-Unters. -Forsch. A*, 1999, **208**, 418.
109. T. Lavizzari, M. Breccia, S. Bover-Cid, M. C. Vidal-Carou and M. T. Veciana-Nogués, *J. Food Prot.*, 2010, **73**, 385.
110. A. R. Shalaby, *Nahrung*, 2000, **44**, 23.

111. C. Martínez-Villaluenga, J. Frías, P. Gulewicz, K. Gulewicz and C. Vidal-Valverde, *Food Chem. Toxicol.*, 2008, **46**, 1635.
112. R. Preti, M. Rapa and G. Vinci, *J. Food Qual.*, 2017, **2017**, 1.
113. A. Jastrzębska, A. Piasta and E. Szłyk, *J. Anal. Chem.*, 2015, **70**, 1131.
114. P. Santiago-Silva, R. A. Labanca and M. B. A. Gloria, *Food Res. Int.*, 2011, **44**, 1264.
115. A. Jain, M. Gupta and K. K. Verma, *J. Chromatogr. A*, 2015, **1422**, 60.
116. R. Preti, R. Bernacchia and G. Vinci, *Eur. Food Res. Technol.*, 2016, **242**, 2031.
117. G. Garai-Ibabe, A. Irastorza, M. T. Dueñas, P. J. Martín-Álvarez and V. M. Moreno-Arribas, *Int. J. Food Sci. Technol.*, 2013, **48**, 375.
118. V. Del Prete, A. Costantini, F. Cecchini, M. Morassut and E. Garcia-Moruno, *Food Chem.*, 2009, **112**, 474.
119. C. Yılmaz, N. G. Taş, T. Kocadağlı and V. Gökmen, *Food Chem.*, 2019, **272**, 347.
120. V. Guillén-Casla, N. Rosales-Conrado, M. E. León-González, L. V. Pérez-Arribas and L. M. Polo-Díez, *J. Chromatogr. A*, 2012, **1232**, 158.
121. P. Pastore, G. Favaro, D. Badocco, A. Tapparo, S. Cavalli and G. Saccani, *J. Chromatogr. A*, 2005, **1098**, 111.
122. J. Oracz and E. Nebesny, *Food Res. Int.*, 2014, **55**, 1.
123. R. Zamora, R. M. Delgado and F. J. Hidalgo, *Food Res. Int.*, 2012, **46**, 321.
124. J. L. Ordóñez, R. M. Callejón, M. L. Morales and M. C. García-Parrilla, *Food Chem.*, 2013, **141**, 2713.
125. F. Bedia Erim, *TrAC, Trends Anal. Chem.*, 2013, **52**, 239.
126. A. Peña-Gallego, P. Hernández-Orte, J. Cacho and V. Ferreira, *Food Rev. Int.*, 2012, **28**, 71.
127. S. Hernández-Cassou and J. Saurina, *J. Chromatogr. B: Anal. Technol. Biomed. Life Sci.*, 2011, **879**, 1270.
128. J. L. Ordóñez, A. M. Troncoso, M. D. C. García-Parrilla and R. M. Callejón, *Anal. Chim. Acta*, 2016, **939**, 10.
129. P. Hernández-Orte, A. Peña-Gallego, M. J. Ibarz, J. Cacho and V. Ferreira, *J. Chromatogr. A*, 2006, **1129**, 160.

CHAPTER 6

Fermented Sausages: A Potential Source of Biogenic Amines

A. PAPARELLA AND R. TOFALO*

Faculty of Bioscience and Technology for Food, Agriculture and Environment, University of Teramo, via Balzarini 1, 64100 Teramo, TE, Italy
*Email: rtofalo@unite.it

6.1 Introduction

Fermented sausages are highly appreciated all around the world for their quality in terms of sliceability, firmness, colour, and taste. In Mediterranean countries, they have a long tradition that traces back to Roman times. Nowadays, they are still frequently manufactured in small-scale units according to traditional recipes, which arise from the use of local ingredients and specific production techniques without the addition of selected starter cultures.[1–3] However, most of the products sold for modern distribution are manufactured with starter cultures to speed up ripening, ensure colour development, enhance flavour, and improve product safety.[2]

In general, dry fermented sausages are made by chopping or mincing meat, fat, salt, spices, and in some cases, sugar, herbs, and/or other ingredients are added. The mix is stuffed into casings, which can be natural or artificial such as collagen, cellulose, or plastic.[4] After casing and during fermentation and ripening, several modifications occur in the sausage.

Food Chemistry, Function and Analysis No. 20
Biogenic Amines in Food: Analysis, Occurrence and Toxicity
Edited by Bahruddin Saad and Rosanna Tofalo

Published by the Royal Society of Chemistry, www.rsc.org

Regarding preservatives, nitrates (sodium nitrate, E251 and potassium nitrate, E252) and nitrites (sodium nitrite, E250 and potassium nitrite, E249) are usually added according to Reg. EC 1333/2008 (and subsequent modifications) and the maximum amount is 150 mg kg^{-1}.[5] However, due to the well-known impact of nitrates and nitrites on food safety, some countries, *e.g.* Denmark, are decreasing the legal limits for these additives, and manufactures are facing the challenge of producing industrial sausages at low content of added nitrates/nitrites or without the addition of these additives. After all, industrial manufacturing of sausages without added nitrates/nitrites is common in specific geographical regions for small calibre sausages, *e.g.* kaminwurst in Tyrol.

In Europe, on the basis of production process, dry fermented sausages can be divided into northern and Mediterranean types.[6] In Mediterranean products, fermentation occurs at 18–24 °C for 1–2 days or at lower temperatures (10–12 °C) for 1 week.[7,8] Drying/ripening is carried out at 10–14 °C during a period ranging from 4–12 weeks.

Traditionally, northern European sausages used to have a pH below 5.0, while Mediterranean products had a pH of 5.3–6.2 and were highly desiccated (for a review see Oliveira *et al.*, 2018).[9] As a matter of fact, to improve product acceptability, final pH is often increased by the addition of specific combinations of starters, and for this reason, northern and Mediterranean sausages can now have similar pH values, even close to 7.0. American fermented sausages are completely different. They are often semi cured, fermented rapidly at relatively high temperatures, with a short drying period.[10] In some Asian products, *e.g.* Mam in Thailand, drying is minimal, and sausage stability and safety entirely depend on low pH.

Throughout the world, sausages are manufactured with different meats, recipes, and methods. In Germany alone, over 350 types of sausages are produced. Even the word “sausage” can be a matter of debate, because some traditional fermented meats are not stuffed into a casing, *e.g.* salami spreads as nduja or ventricina teramana in Italy. On the other hand, some non-fermented meats can be processed and sold in a casing, *e.g.* bacon-like products and Italian Lonza.

A common criterion for classification is final water content. On this basis, fermented sausages can be classified as semi-dry or dry. The discrimination between semi-dry and dry sausages is very important in the USA and in other countries, because semi-dry products, which generally have higher water content, should be heated to meet the time-temperature requirements designed to inactivate *Trichinella* spp., *Salmonella* spp. and *Escherichia coli*.

In dry fermented sausages, drying removes 20–50% of the moisture, and final moisture to protein ration (M/P) is not higher that 2.3 to 1.0.[11] Traditionally, final a_w level used to be below 0.89, which would make them shelf-stable even though the pH is not lowered. However, to decrease product cost and increase its acceptability, dry fermented sausages are also sold at a_w above 0.89, and therefore they are commonly displayed at refrigeration temperatures.

Semi-dry fermented sausages lose 15–20% of moisture during processing, thus have a softer texture and a different flavour profile than dry fermented sausages. Final a_w is usually 0.92 or higher, and M/P is 2.0–3.7 to 1.0. Heating, smoking, and cooking are commonly applied to increase product safety and stability. However, semi-dry sausages can also be considered shelf-stable if pH is 5.0 or less and M/P is 3.0 to 1.0 or less, or have a combination of a pH of 5.2 and a_w of 0.95 or less.[11]

It needs to be pointed out that M/P ratio is much more reliable than final a_w when processing length has to be evaluated, in particular to discriminate long curing (*e.g.* Italian bresaola) from short curing (*e.g.* Italian carpaccio). In fact, final a_w is strongly affected by fat content, and therefore a final a_w below 0.92 can be reached in a few days if fat content is high. In this respect, other parameters have been proposed for legal applications in different countries to be used in combination with M/P and final a_w: HPD in France (Humidité du Produit Dégraissé – Moisture on Defatted Product $= 100 \times$ %Water/100 – % Fat) and UPSD in Italy (Umidità del Prodotto Sgrassato e Deadditivato – Moisture on Defatted Product, after subtraction of additives $=$ [% Moisture/(100 – % Fat – F)] $\times$ 100, where: $F = 100 -$ (% Moisture + % Protein + % Fat + % Ashes).

A different basis for sausage classification can be obtained by combining processing length, final water content, and final a_w. On this basis, fermented sausages can be divided into three groups: spreadable, sliceable and short processed, sliceable and long processed.[12]

Spreadable sausages can be ready for distribution after 3–5 days processing, with a final water content ranging from 34 to 42%, and an a_w around 0.95–0.96 depending on fat content.

Sliceable and short processed sausages include a wide range of products that usually require 1–4 weeks of processing, depending on calibre, casing, starter, composition, and process conditions. Final water content is 30–40%, and final a_w is usually 0.92–0.94.

Sliceable and long processed sausages include the most famous traditional European salamis, *e.g.* Italian salame Milano. Process length extends to 3–4 months, final water content can reach 20–30%, and final a_w is usually below 0.89. Therefore, the first two groups of sausages are semi-dry, while the third group is dry.

Finally, a different classification can be obtained when processing conditions and product stability are considered. In this respect, fermented sausages can be grouped into two categories: constant weight sausages and variable weight sausages. Traditionally, all fermented sausages have a variable weight because they lose moisture during storage. However, a constant weight can be achieved by combining a series of technological solutions: careful selection of meats with a constant fat content, high precision in mixing and stuffing, starter selection, artificial casing, low calibre, intense drying up to a final a_w below 0.85, and modified atmosphere packaging. Thus, constant weight salamis are shelf-stable if pH is not lowered. Constant weight sausages are becoming increasingly popular throughout the world, due to product stability and convenience. Their microbiology is completely different in comparison with traditional products due to final drying.

Fermentation and ripening depend on the activities of heterogeneous microbial communities. In particular, meat fermentation is mainly driven by lactic acid bacteria (LAB) that lowers pH through the production of organic acids and favour certain microbial, physico-chemical and sensory changes that are essential to give an appealing colour, taste, flavour, and texture (Figure 6.1).[2]

The quality of the final product mainly relies on the raw materials used, the specific conditions of processing and ripening, and the composition of the microbial community.[13]

The typical flavour and aroma of dry fermented sausages is the result of complex chemical reactions involving proteins, hydrocarbons, and lipids. Moreover, other factors including process conditions, raw materials, and starter cultures, among others, can favour the production of volatile compounds (alcohol, ketones, aldehydes, and furans) and non-volatile compounds (amino acids, peptides, sugars, and nucleotides) shaping the final sensory quality of the product.[10,14]

In this context, proteolysis is one of the main phenomena that takes place during ripening. Free amino acids and small oligopeptides released can be further metabolized by microorganisms resulting in energy production (*i.e.* arginine metabolism), development of flavour compounds (for branched chain or aromatic amino acids), and accumulation of biogenic amines (BAs). Several studies report the accumulation of BAs in dry fermented sausages of different origin.[15–17] Therefore, there is an evident interest in developing new strategies to control aminogenesis during meat fermentation.[18]

6.2 Occurrence of Biogenic Amines in Fermented Sausages

The most common BAs found in fermented sausages are tyramine (TYR), tryptamine (TRY), phenylethylamine (PHE), putrescine (PUT), and cadaverine (CAD) (Table 6.1). Their presence is often associated to the decarboxylating activity of LAB. However, the high levels of CAD and HIS detected in some fermented sausages have also been associated to the low quality of raw materials. In general, different BA levels can be found in final products. These differences rely on the occurrence of decarboxylating microorganisms in raw materials, environmental contamination, and starter cultures, and on favourable growth conditions for BA-producing bacteria (Figure 6.2). For this reason, it is important to apply proper measures to guarantee the hygienic quality of both raw material and processing units to avoid the development of aminogenic contaminant bacteria and in turn to reduce BAs content. The improvement of hygienic conditions during sausage manufacture processes, and the use of non-aminogenic starter cultures, is important to reduce the BA levels in fermented sausages. However, proper

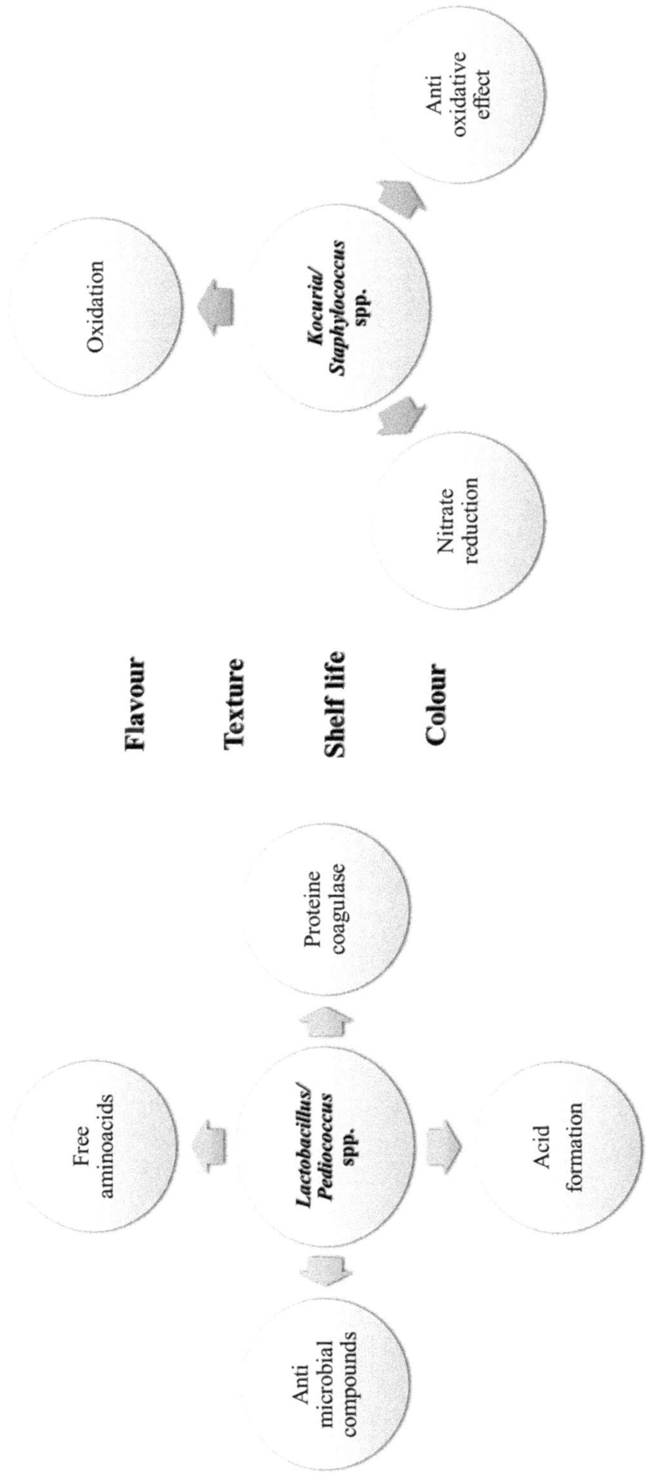

Figure 6.1 Starter cultures and indigenous microbial groups impart desired effects during fermentation of sausages through their metabolic activity.

Table 6.1 BAs content detected in fermented sausages at retail in different countries. Results are expressed as mg kg^{-1} fresh matter. Adapted from ref. 17 with permission from John Wiley and Sons, Copyright © 2015 John Wiley & Sons, Ltd.[a]

Country	Products	TYR	HIS	PHE	TRY	CAD	PUT	SPD	SPM
Italy	Soppressata (9[b])	0–557	0–101	0–20	—	0–270	0–416	0–91	0–98
	Salsiccia (10)	0–339	nd	nd	—	0–39	0–78	0–57	0–28
	Italian salamini (10)	60–372	8–165	nd–53	nd–69	—	—	—	—
	Ciauscolo	41–207	0–48	0–19	0–14	3–259	19–182	1–9	3–36
Spain	Chorizo (20)	30–627	0–314	0–52	0–88	0–658	3–416	2–10	14–44
	Salchichón (22)	53–513	0–151	0–35	0–65	0–342	6–400	1–14	7–43
	Salami (5)	3–12	2–102	—	—	—	—	—	—
	Sobrasada (7)	58–501	3–143	0–39	0–65	3–42	2–501	2–7	10–18
	Fuet (11)	32–743	0–358	0–34	0–68	5–51	2–222	1–11	9–30
	Thin fuet (15)	1–128	0–5	0–29	0–39	1–115	1–513	1–21	2–44
	Dry sausage (industrial) (33)	6–675	0–186	0–186	0–194	2–466	0–410	1–20	8–44
	Dry sausage (traditional) (67)	1–433	0–205	0–61	0–52	0–156	0–537	1–29	2–78
France	Saucisson (industrial) (5)	172–268	16–151	0–8	0–9	31–192	195–410	4–6	59–119
	Saucisson (traditional) (3)	84–217	15–16	0–4	nd	39–110	61–317	2–6	82–86
Finland	Finnish sausage (11)	4–200	0–180	2–248	0–43	0–270	0–230	2–7	19–46
	Russian sausage (4)	6–240	0–200	1–33	0–43	3–18	3–310	2–8	23–40
	Danish sausage (8)	5–110	1–56	0–4	0–91	0–790	0–450	3–9	23–47
	Meatwurst (12)	5–320	0–170	0–5	0–54	0–16	2–580	3–11	22–38
	Lubeck (9)	9–150	0–40	0–7	0–20	0–8	0–220	3–7	20–40
	Salami (13)	3–200	0–9	0–8	0–51	0–71	0–210	1–8	19–45
	Pepperoni (11)	5–190	0–200	0–48	0–42	0–390	0–230	3–11	21–48
Greece	Greek sausage (50)	0–510	0–515	0–28	0–50	0–690	0–505	2–10	15–37
Egypt	Egyptian sausage (50)	10–53	7–41	2–81	3–34	6–39	12–100	5–12	2–5
Turkey	Sucuk (factory made) (19)	1–189	4–255	0–87	0–47	—	0–383	—	0–50
	Sucuk (butcher made) (31)	2–316	2–478	0–32	0–7	0–7	0–919	—	0–42
Thailand	Som-fug (7)	19–228	55–291	—	19–86	20–328	17–275	—	—
Netherlands	Dutch sausage (14)	40–310	1–63	5–45	—	1–150	1–190	—	—
Austria	Austrian sausages (24)	24	—	—	—	0–529	0–554	—	—

[a]Abbreviations: TYR, tyramine; HIS, histamine; PHE, phenylethylamine; TRY, tryptamine; CAD, cadaverine; PUT, putrescine; SPD, spermidine; SPM, spermine.
[b]n. of samples studied.

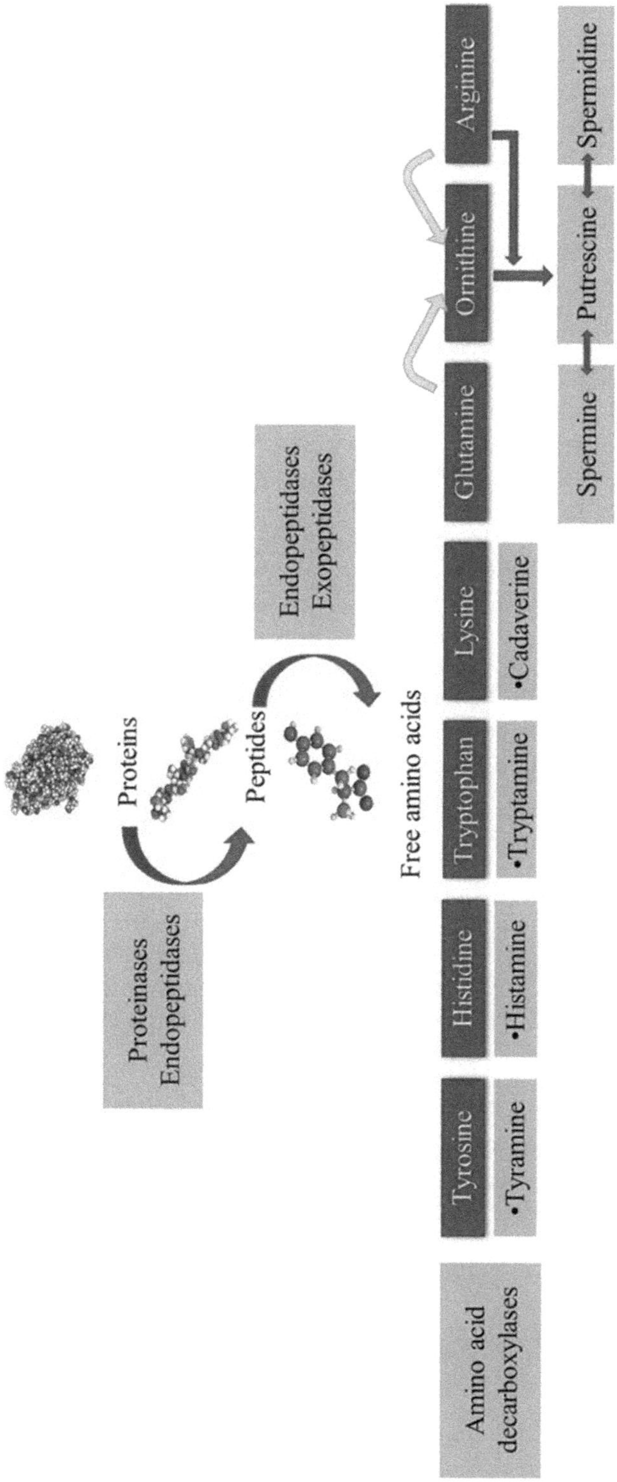

Figure 6.2 Formation of biogenic amines. Adapted from ref. 26 with permission from Taylor & Francis Ltd (http://www.tandfonline.com), Copyright 2005.

hygiene may not be enough to avoid some BAs formation, and other technological measures must be applied.[17,19–22] In fact, microorganisms already present in the product or added as starter cultures use free amino acids as nutrients for energetic purposes producing BAs.[23]

A way to reduce BAs occurrence is the selection of decarboxylase negative starter cultures. *Lactobacillus sakei*, *Lactobacillus plantarum*, and *Staphylococcus xylosus* are considered as weak or non-aminogenic microorganisms.[19,24] However, this feature is strain dependent, therefore, a case-by-case evaluation of the aminogenic activity of the strains is required. For this purpose, several procedures have been reported.[25] However, these microorganisms could not be well adapted to the stressing conditions of meat fermentation or more specifically to traditional fermenting conditions. To face this challenge, several authors recommended the use of autochthonous starter cultures consisting of selected strains originating from each specific fermented meat product.[18] Another alternative to reduce BAs accumulation during the fermentation process is the use of mono- and di-aminoxidase positive bacteria. Several studies highlighted the presence of this metabolic activity in LAB involved in sausages fermentation (for a review see Latorre-Moratalla *et al.*, 2012).[18] However, scientific studies are controversial, in fact, in some cases, amine-oxidizing microorganisms have shown a limited effect on TYR and HIS levels, probably because of low oxygen availability inside the sausage and/or an insufficient number of amine-oxidizing bacteria.[18] The use of tailored starter cultures, high quality raw materials, and optimal technological conditions are essential to reduce BAs accumulation in fermented sausages.

As reported in Table 6.1, BAs levels in fermented sausages in retail are high. In Mediterranean countries, such as Spain and Italy, CAD, PUT, TYR, and HIS are the most detected, followed by PHE, TRY, spermine (SPM), and spermidine (SPD). Different Spanish (*e.g.* chorizo, fuet, sobrasada, salsichón), Italian (*e.g.* soppressata, salsiccia, salamini, ciauscolo), and French sausages showed that TYR and CAD could reach 600 mg kg^{-1}, whereas PUT could be present up to 500 mg kg^{-1}, and PHE and TRY up to 50 mg kg^{-1}, with the only exception of Spanish industrial dry sausages (190 mg kg^{-1}). HIS levels could be above 300 mg kg^{-1} in chorizo and fuet, which are thought to be hazardous to human health. On the other hand, TYR, CAD, and PHE are the most abundant BAs in dry Finnish sausages, and high levels of CAD have been detected in Danish sausages. High levels of TYR and HIS have been reported in Russian sausages, but usually not above 200 mg kg^{-1}. High amounts of BAs are present in Turkish sausages as amounts of 300 mg kg^{-1} of TYR and 400 mg kg^{-1} of HIS were found in sucuks (Table 6.1).[17,25]

6.3 Aminogenic Microorganisms in Fermented Sausages

The first requirements for BAs production are the presence of precursors (*i.e.* amino acids), the presence of decarboxylase positive microorganisms, and favourable conditions for their growth and decarboxylating activity.[27] The decarboxylation of amino acids is essential for energetic purposes, especially in a nutritionally poor environment. When the decarboxylase system is active, a translocation of charge occurs and consequently, the membrane potential is modified.[28] Fermented sausages present high protein levels, and the high proteolytic activity during ripening provides the precursors for decarboxylase activity of microorganisms, permitting accumulation of BAs. Nonetheless, sausages that present a similar microbiological profile may differ for BAs content, indicating that the presence of microorganisms is not the only factor that allows the production of BAs. Moreover, decarboxylating activity is strain specific.[29] Fermentation and ripening involve the activity of the microorganisms present in raw meat or added as starters.

Decarboxylase positive microorganisms are pathogens belonging to the *Bacillus*, *Pseudomonas*, *Escherichia*, and *Salmonella* genera, but they can also be foodborne starters such as *Lactobacillus*, *Enterococcus*, *Lactococcus*, and *Leuconostoc* spp. Halász *et al.*[30] showed that Enterobacteriaceae are high BAs producers, especially for CAD and PUT synthesis. Other authors found that *Enterobacter cloacae*, *Enterobacter aerogenes*, *Klebsiella oxytoca*, *E. coli*, and *Morganella* (*Proteus*) *morganii* were high HIS producers.[24,30–32] These microorganisms are normally present at low numbers in fermented sausages, but a wrong storage of raw materials or an incorrect fermentation can induce their proliferation and consequently a higher production of decarboxylases in the first stages of sausages manufacturing.[33]

Some LAB desirably present in sausages can produce BAs. Lactococci, pediococci and streptococci, such as *Streptococcus thermophilus* and *Oenococcus oeni* are not decarboxylase positive. On the contrary, some strains of *Lactococcus* and *Lactobacillus* (*i.e. L. buchneri*, *L. alimentarius*, *L. plantarum*, *L. curvatus*, *L. farciminis*, *L. bavaricus*, *L. homohiochii*, *L. reuteri*, and *L. sakei*) are considered BA producers, especially TYR.[20] Other TYR producers are enterococci[24,35] but they are not able to produce relevant amounts of PUT and CAD.[24,29] Masson *et al.*[34,35] found that *Carnobacterium* (*i.e. C. divergens*, *C. piscicola*, and *C. gallinarum*) can be also considered TYR producers.

6.4 Factors Influencing Biogenic Amine Content in Fermented Sausages

The level and the type of BAs in fermented sausages depend on different variables, all of which interact, making it difficult to depict the effects of each singular technological factor on BAs production. To explain when and why aminogenesis occurs, it is necessary to analyse each particular case separately. In the literature, many technological and physico-chemical factors affecting the BA content in fermented sausages are described. Factors such as raw material quality, pH, redox potential, salt concentration, temperature, additives, hygienic conditions of manufacturing practices, and diameter of sausages, can influence the growth and the activity of aminogenic

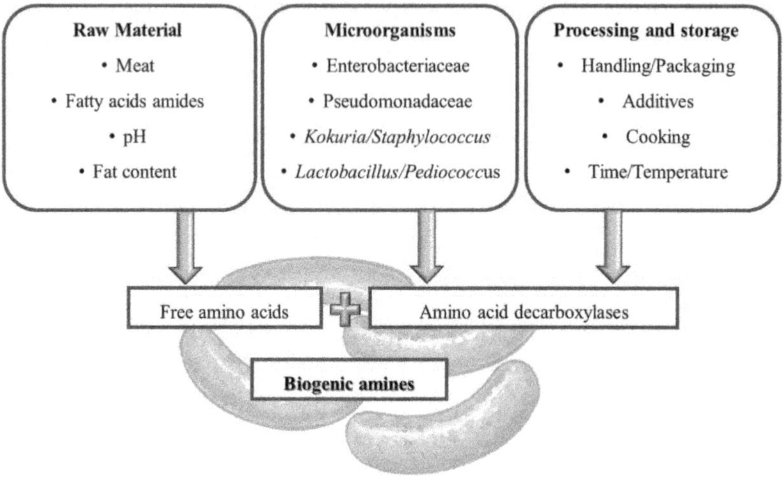

Figure 6.3 Factors influencing the formation of biogenic amines. Adapted from ref. 26 with permission from Taylor & Francis Ltd (http://www.tandfonline.com), Copyright 2005.

microorganisms during the manufacturing processes of fermented sausages.[15,20,22,25,36] The main factors summarized in Figure 6.3, are:

- *pH:* Microbial decarboxylase activity is regulated by pH as a response to stress conditions. The optimum pH for decarboxylase positive microorganisms is acid.[37] As a result, pH decrease in fermented sausages increases decarboxylase activity as a physiological system to neutralize a stressful acidic environment. On the other hand, a rapid drop in pH can inhibit contaminant aminogenic bacteria, leading to a reduction of BA levels.[38,39] Maijala *et al.*[38,40] showed that the acidification of the synthetic media MRS broth with glucono-γ-lactone led to a decrease in levels of amines and Enterococci and Enterobacteriaceae cell counts. The same effects were found when glucono-γ-lactone was added directly to dry sausages, giving lower HIS and TYR concentrations.
- *Redox potential:* A reduced redox potential increases BAs content, stimulating HIS production. However, decarboxylase activity seems to be inactivated or destroyed in the presence of oxygen.[41]
- *Temperature:* This variable has different effects on BAs accumulation, since it is correlated to growth kinetics, cell yields, and enzymatic activity of microorganisms. The optimal temperature for aminogenic microorganisms is between 20 °C and 37 °C. Lower temperatures reduce their growth,[41] while BAs content increases with time and storage temperature.[42] Traditional sausage fermentation generally occurs at low temperatures, controlling or reducing BAs formation from microorganisms.[19]
- *Salt concentration:* High NaCl concentration leads to a decrease in the accumulation of BAs.[20] The presence of NaCl concentrations ranging

from 3.5 to 5.5% stimulate tyrosine decarboxylase activity but obstruct histidine decarboxylase activity.[41] Moreover, the rate of BAs production of a *Lactobacillus bulgaricus* (now *L. delbrueckii* subsp. *bulgaricus*) strain was considerably reduced when salt concentration in the medium increased from 0% to 6%.[43]

- *Calibre and size of sausages:* Since the calibre of sausages influences the environment in which microorganisms grow (different degrees of anaerobiosis, redox potential, salt concentration, and a_w values), a larger calibre is considered to be one of the reasons that leads to a higher production of BAs, especially of TYR and PUT.[44,45] Different authors found a relationship between BAs content and the size of sausages.[20,26,29] BAs content was generally higher in larger sausage calibres than in the smaller ones, especially in the central part, more than in the edge. Treviño *et al.*[46] analysed cervelat sausage in three different sections. They found that the edge and the central part had a lower BAs amount (due to the higher drying level and to the higher NaCl concentration, respectively) in comparison with the central area in which the environmental conditions (higher a_w) allowed a more intense microbial activity.
- *Hygienic conditions of manufacturing practices:* Hygiene of raw materials and during processing helps to avoid the growth of amine positive contaminant microorganisms. Nevertheless, correct hygiene may not be enough to avoid some formation of biogenic amines and other technological measures must be applied.[19]
- *Additives:* The addition of ingredients, such as sugars in fermented meat, can enhance microbial growth and BAs production. Masson *et al.*[34] showed that when sugars are not added to the product, Enterococci and Enterobacteriaceae counts are higher. In this case, the most important BAs produced are TYR and CAD. Moreover, Bover-Cid *et al.*[47] demonstrated that the addition of sodium sulphide to fermented sausages did not reduce TYR levels, instead, it reduced CAD levels. Cantoni *et al.*[48] showed that the addition of sodium nitrite in Italian dry sausages reduced PUT and CAD levels but caused a high increase in HIS production.
- *Packaging:* Preliminary studies showed that BA content is also related to packaging. Sliced sausages packaged under vacuum or modified atmosphere (MAP), and ready-to-eat (RTE) products have a higher BA level than products commercialized as a whole piece.[18] MAP and vacuum packaging create a particular environment that prevents a_w from continuing to decrease during storage, favouring decarboxylase activity of microorganisms.

6.5 Nitrosamines

Plenty of studies on the safety of meat products showed that BAs could react with nitrite to produce volatile *N*-nitrosamines.[49–51] *N*-nitrosamines are classified as carcinogens by International Agency for Research on Cancer (IARC) and United States Environmental Protection Agency (US EPA) and can be found in various foods, including meat products. First evidences of the presence of worrying levels of *N*-nitrosamines were found in fried bacon in the first years of 1970s, when high concentrations of *N*-nitrosodimethylamine (NDMA) and *N*-nitrosopyrrolidine (NPYR) were detected. *N*-nitrosamines are formed by chemical and/or microbial reactions between a nitrosating agent and a substance having an amino group, and their production is faster at pH 3.5 and decreases as pH increases.[52,53] In meat and meat products, like dry fermented sausages, BAs create an additional toxicological risk[26] since they may be transformed to the secondary amines, which react with sodium nitrite, the nitrosating agent added to the meat as a preservative and colouring agent, to form *N*-nitrosamines. PUT, CAD, SPM, and SPD can be the precursors of *N*-nitrosamines. *N*-nitrosopyrrolidine is produced from PUT, SPM, and SPD after deamination and cyclization to secondary amines, before reacting with the nitrosating agent (NO^+); *N*-nitrosopiperidine (NPIP) is produced by the same pathway with CAD as precursor.[54] Gavinelli *et al.*,[55] in 1988, found several *N*-nitrosamine in Italian salami: NDMA (0.59–7.76 μg kg^{-1}), *N*-nitrosodimethylamine (NDEA, nd-4.04 μg kg^{-1}), *N*-nitrosodibutylamine (NDBA) (0.73–50.12 μg kg^{-1}), and NPIP (nd-0.38 μg kg^{-1}). Recently, De Mey *et al.*[56] analysed 101 samples to evaluate volatile *N*-nitrosamine concentrations in dry fermented sausages available on the Belgian market. Only in one product the amounts of CAD and PUT reached unsafe levels. NPIP and *N*-nitrosomorpholine (NMOR) were detected in 22% and 28% of analysed samples, respectively. The authors found no correlation between the presence of *N*-nitrosamines and BAs content, suggesting that the occurrence of NPIP could be attributed to the use of pepper. Other studies on occurrence of *N*-nitrosamine in processed meat products are summarized in Table 6.2,

Table 6.2 Contents of nitrosamine in fermented sausages (μg kg^{-1}). Number of samples are reported in brackets. Data from ref. 58.[a]

Products	Sausages (32)				Salami (4)	Various smoked sausages (15)
NDMA	0.31–1.54	0.04–4.5	0.3–0.6	0.2–3.28	0.04–2.1	0.04–9.3
NDEA	nd	0.02–7.9	0.3–0.6	0.2–1.78	1.5–12	0.91–10.3
NDBA	nd	0.11–1.0	—	0.2–0.60	0.29–2.0	0.11–2.0
NPYR	nd	—	—	—	—	0.16–3.7
NPIP	nd-0.91	0.09–0.90	0.3–1.1	0.2–1.02	0.09–0.58	0.09–2.2
NMOR	nd	—	0.3–0.4	—	—	0.19–7.2
NMEA	nd	—	—	—	—	—
References	58	59	60	61	59	59

[a]Abbreviations: NDMA, *N*-nitrosodimethylamine; NDEA, *N*-nitrosodiethylamine; NDBA, *N*-nitrosodibutylamine; NPYR, *N*-nitrosopyrrolidine; NPIP, *N*-nitrosopiperidine; NMOR, *N*-nitrosomorpholine; NMEA, *N*-Nitrosomethylethylamine.

in which NDMA showed the highest detection rate in meat and meat products. Even if the use of nitrite in fermented sausages has a crucial technological role (*i.e.* development and stability of colour and flavour, and prevention of undesirable and pathogenic bacteria), it is important to reduce its levels and, consequently, *N*-nitrosamine concentration. Plant polyphenols (*i.e.* green tea polyphenols), ascorbic acid, and α-tocopherol ingredients can be used as alternatives to nitrates and nitrites to process meat, improving quality, shelf-stable, and the safety of processed meat products.[57]

6.6 Conclusion

Biogenic amines are found in varying concentrations in fermented sausages. Despite their negative impact on human health, no standards or guidelines exist for the allowable concentrations of BAs in fermented sausages. Tyramine is usually the most frequent and abundant biogenic amine found in fermented sausages. In terms of average values, it has been reported that fermented sausages show the highest TYR content among fermented products (BIOHAZ, 2011).[25] A possible strategy to reduce the content of BAs in these products could be the selection of aminoxidase starter cultures. Moreover, the use of autochthonous strains could represent a useful possibility for the manufacture of traditional fermented sausages. High quality raw materials and optimal technological conditions are also crucial factors to ensure a proper performance of starter cultures for the reduction of BAs accumulation in fermented sausages.

References

1. M. Elias and A. V. Carrascosa, *Food Control*, 2010, **21**, 97.
2. I. Franciosa, V. Alessandria, P. Dolci, K. Rantsiou and L. Cocolin, *Int. J. Food Microbiol.*, 2018, **279**, 26.
3. A. Olivares, J. L. Navarro, A. Salvador and M. Flores, *Meat Sci.*, 2010, **86**, 251.
4. Y. C. Wu and S. P. Chi, Casings, in *Handbook of Fermented Meat and Poultry*, ed. F. Toldrà, John Wiley & Sons, Hoboken, NJ, 2nd edn, 2014, pp. 101–110.
5. Commission Regulation (EU) No 1129/2011 of 11 November 2011 Amending Annex II to Regulation (EC) No 1333/2008 of the European Parliament and of the Council by Establishing a Union List of Food Additives.
6. D. Demeyer, M. Raemaekers, A. Rizzo, A. Holck, A. De Smedt, B. ten Brink, B. Hagen, C. Montel, E. Zanardi, E. Murbrekk, F. Leroy, F. Vandendriessche, K. Lorentsen, K. Venema, L. Sunesen, L. Stahnke, L. De Vuyst, R. Talon, R. Chizzolini and S. Eerola, *Food Res. Int.*, 2000, **33**, 171.

7. I. Lebert, S. Leroy, P. Giammarinaro, A. Lebert, J. P. Chacornac, S. Bover-Cid, M. C. Vidal-Carou and R. Talon, *Meat Sci.*, 2007, **76**, 112.
8. G. Mauriello, A. Casaburi, G. Blaiotta and F. Villani, *Meat Sci.*, 2004, **67**, 149.
9. M. Oliveira, V. Ferreira, R. Magalhães and P. Teixeira, *Food Res. Int.*, 2018, **103**, 438.
10. J. A. Ordóñez and L. de la Hoz, Mediterranean products, in *Handbook of Fermented Meat and Poultry*, ed. F. Toldrà, John Wiley & Sons, Hoboken, NJ, 2nd edn, 2014, pp. 333–348.
11. H. W. Ockerman and L. Basu, Production and consumption of fermented meat products, in *Handbook of Fermented Meat and Poultry*, ed. F. Toldrà, John Wiley & Sons, Hoboken, NJ, 2nd edn, 2014, pp. 9–15.
12. M. Roca and K. Incze, Fermented sausages, *Food Rev. Int.*, 1990, **6**, 91.
13. L. Aquilanti, S. Santarelli, G. Silvestri, A. Osimani, A. Petruzzelli and F. Clementi, *Int. J. Food Microbiol.*, 2007, **120**, 136.
14. J. A. Ordóñez, E. M. Hierro, J. M. Bruna and L. Hoz, *Crit. Rev. Food Sci. Nutr.*, 1999, **39**, 329.
15. M. L. Latorre-Moratalla, T. Veciana-Nogués, S. Bover-Cid, M. Garriga, T. Aymerich, E. Zanardi, A. Ianieri, M. J. Fraqueza, L. Patarata, E. H. Drosinos, A. Lauková, R. Talon and M. C. Vidal-Carou, *Food Chem.*, 2008, **107**, 912.
16. S. Lu, X. Xu, G. Zhou, Z. Zhu, Y. Meng and Y. Sun, *Food Control*, 2010, **21**, 444.
17. M. C. Vidal-Carou, M. T. Veciana-Nogués M, L. Latorre-Moratalla and S. Bover-Cid, Biogenic Amines: Risks and Control, in *Handbook of Fermented Meat and Poultry*, ed. F. Toldrà, John Wiley & Sons, Hoboken, NJ, 2nd edn, 2014, pp. 413–428.
18. M. L. Latorre-Moratalla, S. Bover-Cid, M. T. Veciana-Nogués and M. C. Vidal-Carou, *Front. Microbiol.*, 2012, **3**, 169.
19. M. L. Latorre-Moratalla, S. Bover-Cid, R. Talon, M. Garriga, E. Zanardi, A. Ianieri, M. J. Fraqueza, M. Elias, E. H. Drosinos and M. C. Vidal-Carou, *LWT–Food Sci. Technol.*, 2010, **43**, 20.
20. G. Suzzi and F. Gardini, *Int. J. Food Microbiol.*, 2003, **88**, 41.
21. T. Komprda, J. Neznalova, S. Standara and S. Bover-Cid, *Meat Sci.*, 2001, **59**, 267.
22. T. Komprda, D. Smělá, P. Pechová, L. Kalhotka, J. Štencl and B. Klejdus, *Meat Sci.*, 2004, **67**, 607.
23. W. N. Konings, J. S. Lolkema, H. Bolhuis, H. W. Van Veen, B. Poolman and A. J. M. Driessen, *Antonie van Leeuwenhoek*, 1997, **71**, 117.
24. S. Bover-Cid, M. Hugas, M. Izquierdo-Pulido and M. C. Vidal-Carou, *Int. J. Food Microbiol.*, 2001, **66**, 185.
25. BIOHAZ (EFSA Panel on biological hazards), Scientific opinion on risk based control biogenic amines formation in fermented foods, *EFSA J.*, 2011, **9**, 2393.
26. C. Ruiz-Capillas and F. Jiménez-Colmenero, *Crit. Rev. Food Sci. Nutr.*, 2005, **44**, 489.

27. B. ten Brink, C. Damink, H. M. L. J. Joosten and J. H. In't Veld, *Int. J. Food Microbiol.*, 1990, **11**, 73.
28. W. N. Konings, J. S. Lolkema, H. Bolhuis, H. W. Van Veen, B. Poolman and A. J. M. Driessen, *Antonie van Leeuwenhoek*, 1997, **71**, 117.
29. S. Bover-Cid and W. H. Holzapfel, *Int. J. Food Microbiol.*, 1999, **53**, 33.
30. A. Halász, A. Barath, L. Simon-Sarkadi and W. Holzapfel, *Trends Food Sci. Technol.*, 1994, **5**, 42.
31. A. X. Roig-Sagues, M. Hernandez-Herrero, E. I. Lopez-Sabater, J. J. Rodriguez-Jerez and M. T. Mora-Ventura, *J. Food Prot.*, 1996, **59**, 516.
32. M. H. S. Silla-Santos, *Int. J. Food Microbiol.*, 1998, **39**, 227.
33. S. Bover-Cid, M. Izquierdo-Pulido and M. C. Vidal-Carou, *Meat Sci.*, 2001, **57**, 215.
34. F. Masson, G. Johansson and M. C. Montel, *Meat Sci.*, 1999, **52**, 65.
35. F. Masson, R. Talon and M. C. Montel, *Int. J. Food Microbiol.*, 1996, **32**, 199.
36. S. Eerola, R. Maijala, A. X. R. Sagués, M. Salminen and T. Hirvi, *J. Food Sci.*, 1996, **61**, 1243.
37. E. F. Gale, *Adv. Enzymol. Relat. Areas Mol. Biol.*, 1946, **6**, 1.
38. R. Maijala, E. Nurmi and A. Fischer, *Meat Sci.*, 1995, **39**, 9.
39. S. Bover-Cid, M. Izquierdo-Pulido and M. C. Vidal-Carou, *Int. J. Food Microbiol.*, 2001, **65**, 113.
40. R. Maijala and S. Eerola, *Meat Sci.*, 1993, **35**, 387.
41. J. Karovičová and Z. Kohajdová, *Chem. Pap.*, 2005, **59**, 70.
42. A. Halász, A. Barath, L. Simon-Sarkadi and W. Holzapfel, *Trends Food Sci. Technol.*, 1994, **5**, 42.
43. H. Chander, V. K. Batish, S. Babu and R. S. Singh, *J. Food Sci.*, 1989, **54**, 940.
44. E. Parente, M. Martuscelli, F. Gardini, S. Grieco, M. A. Crudele and G. Suzzi, *J. Appl. Microbiol.*, 2001, **90**, 882.
45. S. Bover-Cid, S. Schoppen, M. Izquierdo-Pulido and M. C. Vidal-Carou, *Meat Sci.*, 1999, **51**, 305.
46. E. Treviño, D. Beil and H. Steinhart, *Food Chem.*, 1997, **60**, 521.
47. S. Bover-Cid, M. Izquierdo-Pulido and M. C. Vidal-Carou, *J. Food Prot.*, 2001, **64**, 367.
48. C. Cantoni, C. Bersani, L. Damenis and G. Comi, *Industrie Alimentari*, 1994, **23**, 1239.
49. F. Wei, X. Xu, G. Zhou, G. Zhao, C. Li, Y. Zhang, L. Chen and J. Qi, *Meat Sci.*, 2009, **81**, 451.
50. M. Cantwell and C. Elliott, *J. Clin. Nutr. Diet.*, 2017, **3**, 27.
51. E. De Mey, H. De Maere, H. Paelinck and I. Fraeye, *Crit. Rev. Food Sci. Nutr.*, 2017, **57**, 2909.
52. A. Ayanaba and M. Alexander, *Appl. Microbiol.*, 1973, **25**, 862.
53. K. Domanska, K. and B and O. G. D. A. N. Kowalski, *Bull. Vet. Inst. Pulawy*, 2002, **46**, 317.

54. T. Hernández-Jover, M. Izquierdo-Pulido, M. T. Veciana-Nogués, A. Mariné-Font and M. C. Vidal-Carou, *J. Agric. Food Chem.*, 1997, **45**, 2098.
55. M. Gavinelli, R. Fanelli, M. Bonfanti, E. Davoli and L. Airoldi, *Bull. Environ. Contam. Toxicol.*, 1988, **40**, 41.
56. E. De Mey, K. De Klerck, H. De Maere, L. Dewulf, G. Derdelinckx, M. C. Peeters, I. Fraeye, Y. Vander Heyden and H. Paelinck, *Meat Sci.*, 2014, **96**, 821.
57. F. Toldrá and M. Reig, *Trends Food Sci. Technol.*, 2011, **22**, 517.
58. J. E. Park, J. E. Seo, J. Y. Lee and H. Kwon, *Toxicol. Res.*, 2015, **31**, 279.
59. T. Mavelle, B. Bouchikhi and G. Debry, *Food Chem.*, 1991, **42**, 321.
60. S. Raoul, E. Gremaud, H. Biaudet and R. J. Turesky, *J. Agric. Food Chem.*, 1997, **45**, 4706.
61. C. H. Jo, H. R. Park, D. S. Kim, K. H. Lee and M. H. Kim, *Food Sci. Biotechnol.*, 2010, **5**, 541.

CHAPTER 7

Occurrence of Biogenic Amines in Cheese: An Overview

L. M. PERIN* AND L. A. NERO

Departamento de Veterinária, Universidade Federal de Viçosa, Campus Universitário, Centro, 36570-900, Viçosa, Minas Gerais, Brazil
*Email: luana.mperin@gmail.com

7.1 Introduction

This chapter will focus on the occurrence of biogenic amines (BAs) in cheese, especially in artisanal cheeses, and will present the main aspects related to BA production by lactic acid bacteria (LAB) during cheese manufacture. It is well known that cheeses produced with raw milk have better sensory characteristics than those prepared with pasteurized milk. The high interest in artisanal cheeses is mainly due to their uniqueness. The absence of heat treatment of the milk allows the development of its rich autochthonous microbiota, which contributes to their organoleptic and qualitative characteristics. Moreover, the increase in consumer demand for natural products and minimally processed dairy products (without chemical preservatives addition or the use of thermal treatments) has generated the strengthening of artisanal cheese production all over the world.

However, despite all the beneficial aspects presented by artisanal cheeses, the main challenge related to this matrix is the guarantee of its microbiological quality, requiring them to be produced under good and controlled hygienic conditions.[1] The main concern in using raw milk is the possible contamination by foodborne pathogens and the occurrence of spoilage

Food Chemistry, Function and Analysis No. 20
Biogenic Amines in Food: Analysis, Occurrence and Toxicity
Edited by Bahruddin Saad and Rosanna Tofalo

Published by the Royal Society of Chemistry, www.rsc.org

microorganisms, which can lead to foodborne intoxications due to contamination with pre-formed toxins of microbial origin or with production of undesirable substances, such as BAs.[2] Cheese is already reported as one of the most prevalent products associated with amine poisoning and also having the highest content of amines.[3]

BAs are basic nitrogen compounds of low molecular weight that possess biological activity; these compounds are formed mainly by decarboxylation of amino acids or by amination and transamination of aldehydes and ketones synthetized.[4–6] The main BAs occurring in foods and their molecular formula and classification are as follows: histamine ($C_5H_9N_3$; monoamine, heterocyclic); putrescine ($C_4H_{12}N_2$; diamine, aliphatic); cadaverine ($C_5H_{14}N_2$; diamine, aliphatic); tyramine ($C_8H_{11}NO$; monoamine, aromatic); tryptamine ($C_{10}H_{12}N_2$; monoamine, heterocyclic); β-phenylethylamine ($C_8H_{11}N$; monoamine, aromatic), spermidine ($C_7H_{19}N_3$; polyamine, aliphatic); and spermine ($C_{10}H_{26}N_4$; polyamine, aliphatic). BA accumulation can vary between types of cheese and within the same type of cheese.

Foods containing low quantities of BA are not considered a health concern because, normally, small amounts of BA are metabolized in the human gut to physiologically less active forms.[3] In cheeses, LAB are the main cause of BA occurrence.

7.2 The Role of Lactic Acid Bacteria in the Production of Biogenic Amines in Cheeses

LAB are described as Gram-positive bacilli or cocci, non-spore forming and catalase negative. According to Zhang *et al.*,[7] LAB are generally classified as belonging to four families and seven genera: Lactobacillaeae (genus *Lactobacillus* and *Pediococcus*); Leuconostocaceae (genus *Oenococcus* and *Leuconostoc*); Enterococcoceae (genus *Enterococcus*) and Streptococcaceae (genus *Lactococcus* and *Streptococcus*).

The term LAB refers mainly to their basal metabolism, which results in lactic acid as the main product of sugar fermentation.[8] Due to the fast production of lactic acid, LAB strains are industrially important and used worldwide in the production of fermented foods, such as cheese. In addition, they are capable of producing other desirable substances which contribute to typical sensory characteristics of the final product by modifying taste, texture and flavor. However, some LAB are capable of converting amino acids present in cheese into BAs by decarboxylation or amination and transamination during the ripening process. For this reason, amino acid catabolism by LAB may affect the quality and safety of fermented foods.[9]

According to Zuljan *et al.*,[10] cheese fermentation can be separated into two stages and LAB is the main group of bacteria involved. The first step is carried out by starter LAB, which is controlled by acid production. The second step or ripening is carried out by non-starter LAB; during ripening most of the starter strains die and cell autolysis occurs, whereas the numbers of non-starter LAB increase and dominate the ripening process.[11] The development of sensory characteristics occurs in this step, mainly by mesophilic lactobacilli and enterococci strains.[10,12]

LAB strains produce amines mainly in response to the stress conditions of fermentation and storage. For example, the acid produced during fermentation can pass through the cell membrane and cause acidification of the cytoplasm acting as an antimicrobial. This causes LAB to have to encode specific genetic mechanisms aiming the expression of regulatory elements for acid neutralization and to regulate the intracellular homeostasis, allowing its survival. One of the strategies to regulate pH homeostasis and respond to acid stress is the expression of decarboxylases, an enzyme that belongs to the pyridoxal-phosphate-dependent enzyme group, whose members use pyridoxal-5′-phosphate (PLP) as a coenzyme.[10,13] It is already demonstrated that these decarboxylases show no substrate specificity, providing a mechanism for 'emergency' pH homeostasis, being independent from the available extracellular amino acids.[14] In general, this pathway consumes protons and converts the amino acid into biogenic amines, ATP, CO_2 and NH_4^+, helping to restore the internal pH. The uptake of amino acids into the cell and the excretion of BA from the cytoplasm are coupled events and carried out by an inner membrane antiporter (Figure 7.1). For microorganisms lacking a respiratory chain to generate ATP, BA formation also displays an energetic contribution; the antiporter is an electrogenic system and the enzyme (decarboxylase) promotes a proton consumption reaction, leading to the generation of the proton motive force.[10,13,15]

According to Silla Santos,[16] the prerequisites for BA production are as follows: presence of decarboxylase-positive microorganisms, such as LAB; availability of free amino acids; and specific conditions that allow bacterial development and decarboxylase synthesis and activity. Concerning LAB capability to produce BA, they may be naturally derived from raw milk or introduced intentionally or by contamination as starter and non-starter LAB. Among them, enterococci are mainly responsible for BA formation,[17,18] but also *Lactococcus*, *Lactobacillus*, *Leuconostoc* and *Streptococcus* have been described as BA producers.[19,20] However, Ladero *et al.*[21] reported that BA production is a species-level trait in LAB. Also, horizontal gene transfer has been proposed as one of the mechanisms by which the BA-producing ability is acquired by LAB.[22,23] Thus, concerning cheese production, the selection of potential starter and non-starter LAB cultures must focus not only on their functional properties but also on the absence of production of undesirable factors, such as BA. However, the quantification of BA production by LAB is essential because only large amounts of BA constitute a health risk to consumers.

The amino acid precursors of the main BA involved in food poisoning are, respectively: histidine–histamine; tyrosine–tyramine; tryptophan–tryptamine; lysine–cadaverine; ornithine–putrescine; arginine–spermine; arginine–spermidine; phenylalanine–β-phenylethylamine.[16] Histamine and

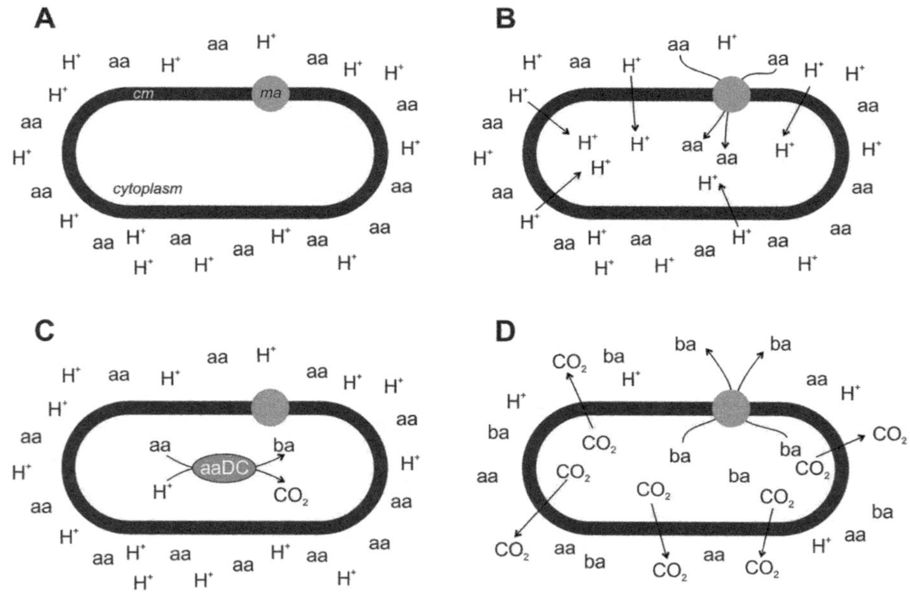

Figure 7.1 Biogenic amine biosynthesis pathway during acid stress. (A) Bacteria in an acid environment (H^+) and with free amino acids (aa); (B) weak acid passes through the cell membrane (cm) and membrane antiporter (ma) imports aa into the cytoplasm; (C) amino acid decarboxylase (aaDC) converts H^+ and aa into CO_2 and the biogenic amine; (D) CO_2 and the biogenic amine are excreted from the cytoplasm and intracellular homeostasis is neutralized (data from EFSA[13]).

tyramine are the most studied BA due to their toxicological effects – vasoactive and psychoactive properties – and along with putrescine are the most important BAs in dairy products.

In normal conditions, histamine and tyramine are metabolized by monoamine oxidase (MOA), an enzyme located in the digestive tract and liver. But in patients treated with antidepressants and antiparkinsonian, drugs that contain MAO inhibitors, the detoxification mechanism become inactive. This occurs because MAO inhibitors maintain high levels of neurotransmitters (serotonin, tryptamine, dopamine and phenylethylamine) due to the prevention of catabolism of amines. Thus, tyramine is not metabolized, causing food migraine and hypertensive crises (leading to brain hemorrhage and heart failure).[24,25] This phenomenon can receive the popular name of a 'cheese reaction'. The reaction depends on the amount of free tyrosine and the presence of microorganisms capable of expressing tyrosine decarboxylase. In cheese, other factors will affect tyramine formation by bacteria, such as growth phase, temperature, salt concentration, ripening conditions and pH.[10] Perin *et al.*[18] investigated the formation of BA by lactococci and enterococci strains isolated from goat milk through high-performance liquid chromatography (HPLC) and identified that some strains were capable of forming 2-phenylethylamine and tyramine. The authors stated that ripening cheese contains the ideal environment for BA production by starter and non-starter LAB.

According to the European Food Safety Agency (EFSA),[13] histamine is a potent vasodilator. However, among Gram-positive bacteria, histidine decarboxylation is an uncommon feature. Histamine is known to be the causative agent of scombroid poisoning, associated with the consumption of some fish[26] and usually causes blood pressure drop, headache, skin rashes, vomiting and diarrhea. Histamine excretion acts as an advantage for bacteria when inside the host; due to the vasodilator effect, essential nutrients are transferred from the host to bacteria. Also, histamine can cause tissue damage, facilitating colonization in the gut. According to Joosten,[27] enterococci are mainly responsible for tyramine formation in cheese and heterofermentative lactobacilli the main histamine producer. However, recent studies have shown this is a strain-specific trait.[28,29]

Cadaverine and putrescine are diamines that can react with nitrite forming carcinogenic nitrosamines. These BAs have special relevance in cured and fermented food, such as cheese and meat. Putrescine is frequently isolated from cheese; however, ornithine decarboxylation is mainly present in Gram-negative bacteria, specially *Enterobacteriaceae*,[30] that means these BAs could be indicators of poor hygiene in milk and cheese manufacturing.

7.3 Occurrence of Biogenic Amines in Artisanal Cheeses

In cheese, the ripening process and proteolysis play an important role in BA formation. Ripening is important in developing the specific flavor and texture responsible for the uniqueness of the cheese variety.[3] During this step, proteolysis results from a breaking down of caseins, which are degraded by proteolytic enzymes. A part of proteolysis is caused by LAB, in which the proteolytic system comprises a number of intracellular proteinases and peptidases, a cell envelope-associated proteinase and a transport system for amino acids and peptidases.[31] As describe earlier, during ripening, most starter LAB suffer cell autolysis, leading to the release of intracellular enzymes such as peptidases and lipases, and that catalyze amino acid conversions. Although starter autolysis is usually beneficial, because it directly influences ripening and the development of flavor, from a toxicological point of view, the free amino acid generated by proteolysis could be transformed into BAs by the decarboxylase activity of other microorganisms present in cheese, mainly other LAB, and often as an adaptation to acid stress.

Besides low pH, BA formation in cheese is strongly influenced by other intrinsic (water activity, nutrient composition, natural microbiota) and extrinsic (time and temperature of ripening and storage) characteristics. The usual temperature for cheese fermentation (25–44 °C) and maturation (10–20 °C) can be favorable to proteolysis and BA formation, especially if under temperature abuse conditions.[2] BA accumulation is minimized at low temperatures due to the control of microbial growth and enzymatic activity.[13] However, a low temperature should be associated with the good microbial quality of the product; otherwise it would allow the maintenance and development of BA-producing psychrophilic and psychrotrophic bacteria (*Pseudomonas* spp. and *Proteus* spp.).

A number of studies have already demonstrated a direct correlation between length of the ripening period and type and amount of BA formed.[32,33] Perin *et al.*[34] described significant increase ($p < 0.05$) of tyramine after 30 days of ripening in artisanal Minas cheese. Martuscelli *et al.*[32] analyzed total BA content in Pecorino Abruzzese cheese (made from raw milk) and found that amounts increased from 34 mg kg^{-1} (in the curd) to 401 mg kg^{-1} after 14 days of ripening, increasing slightly from 14 to 60 days.

pH has an ambiguous influence on cheeses; the low pH could prevent the growth of undesirable microorganisms, but the acid stress could trigger BA formation, mainly by LAB strains (as described earlier). In general, levels of water activity in cheese are around 0.90 to 1.00, representing optimal conditions for the growth of the BA producer's microorganisms. The increase in salt concentrations in cheese can display an inhibitory effect on the growth rate of BA-forming microorganisms.[33]

Regarding milk pasteurization to prevent BA formation, Torracca *et al.*[35] studied the effect of this process and ripening on BA content and the sensory properties of cheeses. The authors described how heat treatment affected the sensory characteristics of the cheese and how it was not sufficient to guarantee low BA amounts, since ripening is also a determining factor. Also, Martuscelli *et al.*[32] reported that pasteurization did not reduce the quantity of tyramine in Pecorino Abruzzese. Thus, it is often not worth sacrificing specific organoleptic characteristics, appreciated by the consumers, using heat treatment. The best alternative is to control the conditions of cheese manufacture, ripening and storage, mainly to avoid the development of carboxylase positive microorganisms.

Benkerroum[2] described the mean values of BAs (tyramine, cadaverine, putrescine and histamine) found in some types of cheese as: 1460 mg kg^{-1} in acid-curd cheese (the highest mean value); 220–388 mg kg^{-1} in washed rind cheese; 188–351 mg kg^{-1} in blue cheese; 167–318 mg kg^{-1} in hard cheese; and 32–172 mg kg^{-1} in fresh cheese. EFSA[13] performed a survey (a public call for data on the presence of BA in food and beverages) and described the results for cheese samples: 15% of 2143 samples presented histamine around 20–62 mg kg^{-1}; 25% of 2377 samples presented tyramine around 68.5–104 mg kg^{-1}; 18% of 2136 samples presented putrescine around 25.4–65 mg kg^{-1}; and 23% of 2149 samples presented cadaverine amounts around 72–109 mg kg^{-1}. Acid-curd cheese presented higher values for all BAs. For phenylethylamine and tryptamine, the results show large numbers of non-quantified values. These data show that fresh cheeses have a lower amount of BAs and higher values are found in acid-curd cheese. Rauscher-Gabernig *et al.*[36] collected data, from surveys conducted from 2000 to 2011, of putrescine and cadaverine concentration in ripened cheese from an Austrian market and in 23 samples found 58 mg kg^{-1} of putrescine and 122 mg kg^{-1} of cadaverine, with maximum values of 725 and 1353 mg kg^{-1}, respectively. Mayer and Fiechter[37] determined BA content in different types of cheese using ultra-high performance liquid chromatography (UHPLC): tyramine was found in 72% of samples with highest values in acid-curd cheeses (160 mg per 100 g); histamine was found in 79% of samples, with maximum amounts in hard cheese (116 mg per 100 g); putrescine was found in 70% of samples (up to 84 mg per 100 g in some acid-curd cheeses); cadaverine was found in 48% of samples, with highest values in some acid-curd cheeses (127 mg per 100 g); and tryptamine, with the lowest occurrence, was found in 15% of samples, with highest values in semi-hard cheese (31 mg per 100 g).

7.4 Intoxication by Biogenic Amines Related to Cheese Consumption

BAs are naturally synthetized by living organism and they are involved in some biological processes (blood pressure control, synaptic transmission, cellular grow control and allergic response). Under normal conditions, BAs, naturally produced and small amounts ingested with foods, are degraded during cellular metabolism activities (through the action of monoamine oxidases – MAO and diamine oxidases – DAO, that detoxify these compounds). The detoxification of BAs is a genetic trait in humans and the ingestion of food with large amounts of BAs may lead to toxicological effects, which is a serious threat to public health. This situation becomes more serious in groups at risk – children, elderly people, pregnant women, those under antidepressant treatment, those with a high alcohol intake, smokers – generally due to weakened BA oxidative activity.[2,16] The toxicological effects of the main BAs involved in outbreaks associated with cheese intake are described in the following sections.

Histamine. Endogenous histamine has important physiological functions related to local immune responses, gastric acid secretion and neuromodulation.[38] Sensitive individuals and/or being on drug therapy (that inhibit DAO or MAO) can suffer with food poisoning after the consumption of histamine-rich foods, with fish and cheese the most commonly foods associated with histamine poisoning. This syndrome is the most notorious foodborne intoxication caused by BAs, inducing hypotension (by dilatation of blood vessels, capillaries and arteries), flushing and headache;[39] the symptoms mimicking an allergic reaction. Histamine intoxication is characterized by an incubation period ranging from a few minutes to hours and symptoms are usually noticeable for a few hours.[13] Doeglas *et al.*[40] reported the first case of histamine poisoning, involving Gouda cheese in 1967 in the Netherlands. According to EFSA,[13] consumption of cheeses with histamine concentration around 850 and 1870 mg kg^{-1} may cause an outbreak. Cheeses made from raw and pasteurized milk, such as Gouda, Swiss, Cheddar, Gruyere and Cheshire, were already associated with histamine poisoning outbreaks.[13,40–42] The toxicological effect of histamine depends on individual characteristics (amino oxidase activity and intestinal physiology), intake concentrations and the presence of other BAs (such as cadaverine and putrescine), hampering the establishment of a maximum limit of safety.[16] Thus, there is limited information about threshold level of adverse health effects in humans, but it is suggested a potential acute reference dose (ARfD) of 50 mg of histamine per healthy person. According to EFSA,[13] consumption of 75 mg may cause some symptoms, such as flushing, diarrhea and headache.

Tyramine. Together with histamine, it is considered a major concern and the most frequently detected in ripened cheeses. Tyramine, tryptamine and β-phenylethylamine belong to the presser amine group, due to their vasoactive effects. In the brain, tyramine can act as indirect sympathomimetic through the release of noradrenaline, causing vasoconstriction and an increase in heart rate and blood pressure.[38,43] Individuals with a MAO deficiency associated with a tyramine-rich diet can suffer from a hypertensive crisis. Other symptoms include pupil dilatation, headache, lacrimation and salivation, increase of respiration and blood sugar, and migraine. According to data collected by EFSA,[13] concentrations around 600–2000 mg of tyramine can cause a minimal systolic blood pressure increase, but in patients treated with MAO inhibitors (MAOI), a smaller amount is needed to induce the same effects. This problem can be mitigated using different therapeutic protocols, mainly using new generation MAOI – a range of 50–150 mg of tyramine can be tolerated by these patients.

Putrescine and cadaverine. Putrescine is a polycationic molecule that stabilizes the structure of DNA and plays an important role on cell membranes, being involved in the regulation of cell growth and division and in apoptosis. It is required to maintain the high rate of cell proliferation in intestinal and colonic mucosa. The role of cadaverine in humans is less known.[13]

According to FAO/WHO,[38] cadaverine and putrescine are less toxic when compared with tyramine and histamine and, in general, can potentiate histamine effects. In normal conditions, histamine is physiologically metabolized by DAO and histamine-*N*-methyltransferase (HMT) and the end product is excreted in urine, but studies have shown that putrescine and cadaverine act as functional inhibitors of DAO and HMT, hindering histamine detoxification and potentiating its toxic effect, since they also are responsible for disruption of the physical barrier function of the small intestine, facilitating histamine passage into blood.[2,44] The level for potentiate histamine toxicity is still unknown and differences in its toxicity in food may be influenced on putrescine and cadaverine amounts.

Another concern related to these BAs is their implication in chronic diseases, such as neurodegenerative disease and cancer. There is limited scientific information about maximum tolerable levels of these BAs, but Rauscher-Gabernig *et al.*[36] suggested a maximum of 180 m kg^{-1} of putrescine and 540 mg kg^{-1} of cadaverine in cheese; and estimated that daily intake via fermented foods corresponds with 19.2 and 23.1 mg per person, respectively.

Spermidine and spermine. Putrescine acts as precursor of spermidine and spermine, and together with putrescine, these BAs belong to the polyamines group and are essential components of all living cells. These BAs are the least toxic and are not frequently involved in food poisoning, but their relevance in food safety is related to potentiate the toxicity of other BAs, mainly tyramine and histamine.

β-Phenylethylamine. Phenylalanine acts as precursor of this BA. β-Phenylethylamine is also a DAO and HMT inhibitor[44] and has a vasoactive effect (similar as tyramine); it causes vasoconstriction and neurological disorders by releases of noradrenaline from sympathetic nervous system. Its threshold toxic level is 30 mg; but its acceptable level in cheese is not available.[2]

Tryptamine. Tryptophan acts as its precursor. Tryptamine can potentiate toxic effects of other BAs by inhibition of DAO and it also causes alteration of adrenergic receptors sensitivity, leading to vasoconstriction and neural and brain cell death.[2] The threshold toxic level and acceptable level in cheese are not available.

In addition, some BAs (cadaverine, putrescine, spermine and spermidine) in the presence of nitrite may form *N*-nitrosamines, which are carcinogenic for animals and a health hazard to humans. Also, some technological processes, such as salting and smoking, may induce nitrosamine formation.[41]

7.5 Biogenic Amine Control in Cheeses

Due to the high toxic effects related to BA ingestion by humans, regulation and determination of BAs in food is extremely important. Some authors have already suggested their use as indicators of freshness or spoilage in cheese, since high levels may indicate poor microbiological quality of raw milk and contamination during cheese manufacturing. However, this information is controversial, because BAs in cheese is produced by starter and non-starter LAB, essential for fermentation and specific flavor.

BA control in dairy products constitutes a challenge since their precursor amino acids are naturally found in milk in a free state or released by the proteolytic process. As described earlier, LAB are mainly responsible for BA formation in cheese; thus, the inability to form BAs is an indispensable condition for strains intended to be used as starters in cheese, being preferable to select strains with 'Qualified Presumption of Safety' (QPS) status. EFSA introduced the concept of QPS. The safety of microorganisms used in food and feed is assessed before their commercial usage and safety strains become part of the QPS list.

A promising alternative to degrading BAs and avoiding their accumulation is the use of food microorganisms capable of synthesizing amino oxidases (responsible for detoxification of dietary BAs) or enzymes with the same activity. Herrero-Fresno *et al.*[45] found two *Lactobacillus casei* strains, isolated from cheese, capable of high degradation rates of histamine and tyramine in a Cabrales-like mini-cheese manufacturing model; the strains reduced both BAs over four months of ripening. The authors suggested the use of BA-degrading LAB particularly for raw milk cheeses as a strategy to avoid BA accumulation. In fact, BA-degrading microorganisms have high technological relevance in fermented food production, mainly in artisanal cheese, on which specific starter and non-starter LAB is essential for organoleptic characteristics. Milk pasteurization should be avoided to not interfere with these specific attributes.

There is still a vacuum of information on toxic doses for all BAs in cheese and limited information about BA interaction. Legal limits are stablished only for histamine in fishery products. Different countries/regions established different limits for histamine, and some do not have any regulation for BAs and follow international legislations. For example, in South America, MERCOSUR Resolution established a maximum level of 10 mg per 100 g of histamine in fish; in the European Union, European Commission Regulation (EC) No. 2073/2005 established the maximum level as 200 and 400 mg kg^{-1} in fresh and cured fisheries, respectively, and Regulation (EC) No. 178/2002 established the legal basis of the Rapid Alert System for Food and Feed (RASFF); in the US, the Food and Drug Administration established a maximum of 50 mg kg^{-1}; and the Australian and New Zealand Food Standards Code states a maximum of 200 mg kg^{-1} of histamine in fish and fish products. For other foodstuffs, including cheese, and for other BAs, there are no legal limits stablished; but amounts between 50 and

150 mg kg^{-1} of tyramine can be a risk for sensitive consumers[13] and some authors suggested other limits: 100 mg kg^{-1} for histamine alone and 900 mg kg^{-1} for a sum of histamine, tyramine, putrescine and cadaverine.[33,41,46]

In fact, the consumption of raw milk cheese should be avoided in risky groups of people, since BA toxicity is influenced by different factors, such as the synergetic effect of combined BAs in food, use of MAOI drugs, alcohol intake and gastrointestinal diseases.

7.6 Analysis of Biogenic Amines

Various methods for qualitative and quantitative analysis of BAs have been published, but there is not a single method for all foods. Cheese represents a complex matrix and can contain high levels of fat, peptides, amino acids and inorganic acids. The presence of these interfering compounds, a combination of different BAs in cheese, cheese manufacture and ripening, microbiota and enzymatic activity can jeopardize the detection and efficiency of the analytical method used, making BA determination in cheese more difficult than other matrices.[24,41]

The extraction of BAs from the cheese matrix is a crucial step and the main techniques for BA determination are thin-layer chromatography (TLC), HPLC, UHPLC, gas chromatography (GC) and capillary electrophoresis (CE).[13,37,47] Şanlı and Şenel[3] summarized in a table the pre-separation procedures and HPLC and UHPLC conditions described to detect BAs in cheese.

Regulation (EC) No. 2073/2005 specified HPLC (after dansyl derivatization) as the reference method for the determination of histamine in fishery products. This technique has a detection limit around 0.1 mg kg^{-1} and according to EFSA,[13] this method should be applied for official control.

Before chromatographic techniques, fluorometric assay (AOAC Official Method – Method AOAC 977.13) was commonly used for detecting histamine, but only in fish, and is the reference method for the US Food and Drug Administration.[48]

Besides the relevance of determining the BA concentrations in cheeses to ensure public health, it is also very important to detect amino acid decarboxylase-positive microorganisms, mainly LAB strains. Bover-Cid and Holzapfel[26] described a qualitative screening, using a differential medium (with initial pH of 5.3) containing the precursor amino acid and bromocresol purple as pH indicator. BA production changes the medium color to purple in response to increased pH. HPLC can also be used for the quantification of BA produced by LAB.[18,26] Moreover, PCR can be used to detect decarboxylase genes in producing strains[29,49] and to quantify BAs in milk and cheese by real-time quantitative PCR.[21] Molecular methods allow rapid and sensitive detection of BAs and BA producers. The characterization of LAB as a decarboxylase-negative strain is very important concerning the selection of starter cultures used in cheese manufacture to avoid BA accumulation.

7.7 Conclusion

The importance is clear of the continuous study of BA-producing strains and BA accumulation in food, mainly in cheese, as a way of avoiding food poisoning and ensuring public health. Some studies have already demonstrated the occurrence of high amounts of BAs in cheeses, especially ripened and artisanal cheese. Due to this, it is very important to follow the recommendations of official control agencies and ensure the good microbiological quality of raw milk, as well as maintaining good hygienic practices during cheese manufacture to reduce risk. Also, use of a carboxylase-negative starter LAB (with QPS status) is a worthwhile and inexpensive practice for cheese factories.

Acknowledgements

Sources of funding: Conselho Nacional de Desenvolvimento Científico e Tecnológico (CNPq), Fundação de Amparo à Pesquisa do Estado de Minas Gerais (FAPEMIG) and Coordenação de Aperfeiçoamento de Pessoal de Nível Superior (CAPES).

References

1. M. Schirone, R. Tofalo, G. Fasoli, G. Perpetuini, A. Corsetti, A. C. Manetta, A. Ciarrocchi and G. Suzzi, *Food Microbiol.*, 2013, **34**, 137–144.
2. N. Benkerroum, *Compr. Rev. Food Sci. Food Saf.*, 2016, **15**, 801–825.
3. T. Şanlı and E. Şenel, *Processing and Impact on Active Components in Food*, Elsevier, Oxford, 2015, pp. 223–230.
4. I. Butor, H. Pištěková, K. Purevdorj, P. Jančová, F. Buňka and L. Buňková, *Potravinarstvo*, 2017, **11**, 302–308.
5. A. Askar and H. Treptow, *Biogene Amine in Lebensmitteln: Vorkommen, Bedeutung und Bestimmung*, Ulmer, 1986.
6. R. Maijala, *Lett. Appl. Microbiol.*, 1993, **17**, 40–43.
7. Z.-G. Zhang, Z.-Q. Ye, L. Yu and P. Shi, *BMC Evol. Biol.*, 2011, **11**, 1.
8. L. Axelsson, in *Lactic Acid Bacteria: Microbiological and Functional Aspects*, ed. S. Salmnen, A. Von Wright and A. Ouwehand, Marcel Dekker Inc., New York, 2004, pp. 1–66.
9. M. Coton, A. Romano, G. Spano, K. Ziegler, C. Vetrana, C. Desmarais, A. Lonvaud-Funel, P. Lucas and E. Coton, *Food Microbiol.*, 2010, **27**, 1078–1085.
10. F. A. Zuljan, P. Mortera, S. H. Alarcón, V. S. Blancato, M. Espariz and C. Magni, *Int. Dairy J.*, 2016, **62**, 53–62.
11. V. Pachlová, L. Buňková, R. Flasarová, R.-N. Salek, A. Dlabajová, I. Butor and F. Buňka, *LWT–Food Sci. Technol.*, 2018, **97**, 730–735.
12. L. Buňková, F. Buňka, G. Mantlová, A. Čablová, I. Sedláček, P. Švec, V. Pachlová and S. Kráčmar, *Food Microbiol.*, 2010, **27**, 880–888.

13. European Food Safety Authority, *EFSA J.*, 2011, **9**, 2393.
14. J. W. Foster and H. K. Hall, *J. Bacteriol.*, 1991, **173**, 5129–5135.
15. J. Lolkema, B. Poolman and W. Konings, *J. Bioenerg. Biomembr.*, 1995, **27**, 467–473.
16. M. H. Silla Santos, *Int. J. Food Microbiol.*, 1996, **29**, 213–231.
17. G. Giraffa, G. Pepe, F. Locci, E. Neviani and D. Carminati, *Ital. J. Food Sci.*, 1995, 7, 341–349.
18. L. M. Perin, S. Belviso, B. dal Bello, L. A. Nero and L. Cocolin, *J. Food Prot.*, 2016, **80**, 151–157.
19. D. Linares, B. Del Rio, V. Ladero, N. Martinez, M. Fernandez, M. Martin and M. Alvarez, *Front. Microbiol.*, 2012, **3**, 180.
20. L. M. Perin, M. L. S. Sardaro, L. A. Nero, E. Neviani and M. Gatti, *Food Microbiol.*, 2017, **65**, 160–169.
21. V. Ladero, D. M. Linares, M. Fernández and M. A. Alvarez, *Food Res. Int.*, 2008, **41**, 1015–1019.
22. V. Ladero, M. Fernández, M. Calles-Enríquez, E. Sánchez-Llana, E. Cañedo, M. C. Martín and M. A. Alvarez, *Food Microbiol.*, 2012, **30**, 132–138.
23. Á Marcobal, B. de las Rivas, M. V. Moreno-Arribas and R. Muñoz, *Appl. Environ. Microbiol.*, 2006, **72**, 7954–7958.
24. M. Nuñez and M. Medina, *Reference Module in Food Science*, Elsevier, Madrid, Spain, 2016.
25. D. Morgan, in *COST 917 – Biogenically Active Amines in Food*, ed. S. Bardócz, A. White and G. Hajós, Office for Official Publications of the European Communities, Luxembourg, 1998, vol. 2, pp. 48–63.
26. S. Bover-Cid and W. H. Holzapfel, *Int. J. Food Microbiol.*, 1999, **53**, 33–41.
27. H. Joosten, *Neth. Milk Dairy J.*, 1987, **41**, 329–357.
28. P. M. Moraes, L. M. Perin, S. D. Todorov, A. Silva, B. Franco and L. A. Nero, *J. Appl. Microbiol.*, 2012, **113**, 318–328.
29. L. M. Perin, R. O. Miranda, S. D. Todorov, B. D. G. D. M. Franco and L. A. Nero, *Int. J. Food Microbiol.*, 2014, **185**, 121–126.
30. D. M. Linares, M. Martín, V. Ladero, M. A. Alvarez and M. Fernández, *Crit. Rev. Food Sci. Nutr.*, 2011, **51**, 691–703.
31. P. Piraino, T. Zotta, A. Ricciardi, P. L. McSweeney and E. Parente, *Int. Dairy J.*, 2008, **18**, 81–92.
32. M. Martuscelli, F. Gardini, S. Torriani, D. Mastrocola, A. Serio, C. Chaves-López, M. Schirone and G. Suzzi, *Int. Dairy J.*, 2005, **15**, 571–578.
33. K. Valsamaki, A. Michaelidou and A. Polychroniadou, *Food Chem.*, 2000, **71**, 259–266.
34. L. M. Perin, B. Dal Bello, S. Belviso, G. Zeppa, A. F. de Carvalho, L. Cocolin and L. A. Nero, *Int. J. Food Microbiol.*, 2015, **214**, 159–167.
35. B. Torracca, F. Pedonese, M. B. López, B. Turchi, F. Fratini and R. Nuvoloni, *Int. Dairy J.*, 2016, **61**, 189–195.
36. E. Rauscher-Gabernig, R. Gabernig, W. Brueller, R. Grossgut, F. Bauer and P. Paulsen, *Eur. Food Res. Technol.*, 2012, **235**, 209–220.

37. H. K. Mayer and G. Fiechter, *Food Control*, 2018, **93**, 9–16.
38. FAO/WHO, *Joint FAO/WHO Expert Meeting on the Public Health Risks of Histamine and Other Biogenic Amines from Fish and Fishery Products*, Meeting report, Food and Agriculture Organization of the United Nations/World Health Organization, Rome, Italy, 2012.
39. L. Sahu, S. K. Panda, S. Paramithiotis, N. Zdolec and R. C. Ray, *Fermented Foods Part I: Biochemistry and Biotechnology*, CRC Press, Boca Raton, 2015, pp. 318–332.
40. H. Doeglas, J. Huisman and J. Nater, *Lancet*, 1967, **290**, 1361–1362.
41. A. R. Shalaby, *Food Res. Int.*, 1996, **29**, 675–690.
42. B. Lund, T. Baird-Parker and G. Gould, *The Microbiological Safety and Quality of Food*, Aspen Publishers, Maryland, USA, 2000.
43. S. C. Jansen, M. van Dusseldorp, K. C. Bottema and A. E. Dubois, *Ann. Allergy, Asthma, Immunol.*, 2003, **91**, 233–241.
44. J. E. Stratton, R. W. Hutkins and S. L. Taylor, *J. Food Prot.*, 1991, **54**, 460–470.
45. A. Herrero-Fresno, N. Martínez, E. Sánchez-Llana, M. Díaz, M. Fernández, M. C. Martin, V. Ladero and M. A. Alvarez, *Int. J. Food Microbiol.* 2012, **157**, 297–304.
46. U. G. Spizzirri, D. Restuccia, M. Curcio, O. I. Parisi, F. Iemma and N. Picci, *J. Food Compos. Anal.*, 2013, **29**, 43–51.
47. M. R. Loizzo, F. Menichini, N. Picci, F. Puoci, U. G. Spizzirri and D. Restuccia, *Trends Food Sci. Technol.*, 2013, **30**, 38–55.
48. Codex Alimentarius Commission, 1999, Available at http://www.fao.org/fileadmin/user_upload/agns/pdf/CXS_234e.pdf.
49. J. M. Landete, B. de las Rivas, A. Marcobal and R. Muñoz, *Int. J. Food Microbiol.*, 2007, **117**, 258–269.

CHAPTER 8

Biogenic Amines in Global Beverages

G. PERPETUINI, F. TITTARELLI, N. BATTISTELLI, G. ARFELLI, G. SUZZI AND R. TOFALO*

Faculty of Bioscience and Technology for Food, Agriculture and Environment, University of Teramo, via Balzarini 1, 64100 Teramo, TE, Italy
*Email: rtofalo@unite.it

Biogenic amines (BAs) are among the major factors determining the quality of fermented beverages and beyond a safety level may be hazardous to human health,[1] causing food intoxication or allergies.[2] The anabolic pathway of BAs is regulated by specific decarboxylases belonging to the pyridoxal-phosphate-dependent enzyme group, whose members use pyridoxal-5′-phosphate (PLP) as a coenzyme, which convert amino acids to BAs. Their formation is linked to the availability of the corresponding free amino acid, the presence of decarboxylase-positive microorganisms, and favourable conditions for their growth and decarboxylation activity. The amount and type of BAs formed in foods is strongly influenced by the intrinsic food characteristics including pH, water activity, composition, microbiota, and by extrinsic parameters such as storage time and temperature, which allows bacterial growth during food processing and storage.[3]

The accumulation of high amounts of BAs can result in cases of food intoxication or intolerance. Therefore, their determination in foods and beverages is an essential prerequisite for food safety. The main methods applied for this aim are thin-layer chromatography, gas chromatography,

Food Chemistry, Function and Analysis No. 20
Biogenic Amines in Food: Analysis, Occurrence and Toxicity
Edited by Bahruddin Saad and Rosanna Tofalo

Published by the Royal Society of Chemistry, www.rsc.org

capillary electrophoretic method, high-performance liquid chromatography, ELISA, and electrochemical biosensors.[3] Moreover, several polymerase chain reaction (PCR) protocols have been implemented to detect BA-producing microorganisms. Generally, these methods detect bacteria harbouring BA-producing pathway genes, since a relationship between the presence of the gene encoding the decarboxylase and the capacity to synthesize BAs has been reported by several authors.[3,4]

8.1 Wine

Winemaking practices have changed radically over centuries. What remains unchanged is that everything begins from grapes. Figure 8.1 summarizes the main steps of the winemaking process. White and red wines follow different pathways because maceration lasts longer for red wines, and for white wines, the grape juice is separated away from the skins and clarified *via* cold settling, filtration, centrifugation, or flotation. Fermentations are typically carried for one to two weeks at temperatures around 10 to 18 °C. For red wines, grape skins are left in the fermentation to retain colour.

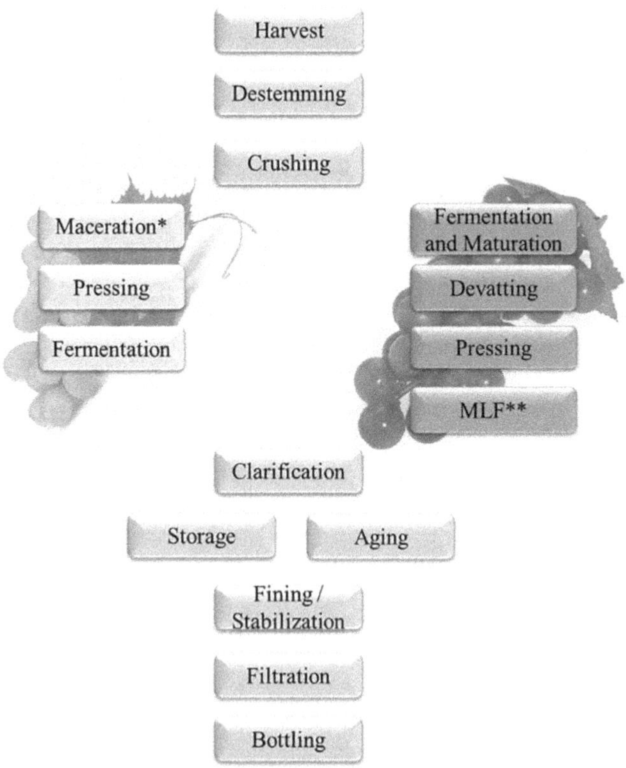

Figure 8.1 Flow sheet of the winemaking process. *Possible, pre-fermentative, low temperature. **If needed.

Yeasts (autochthonous or inoculated) start the alcoholic fermentation and, afterwards, wine can be separated from grape skins, and fermentation is completed in another tank. Malolactic fermentation (MLF) will take place after alcoholic fermentation and it is guided by indigenous or inoculated lactic acid bacteria (LAB). White wines are not stored for long periods of time, while reds are frequently stored in oak barrels for several months or years. Then, the wine is racked to eliminate sediments (yeast and bacterial cells, lees, *etc*) that settle out during clarification and could produce off-odours and spoilage. Bottling is the last stage of winemaking but, prior to this, wines can undergo fining to remove traces of dissolved proteins or excess of tannins. Bottles are usually aged for several months to years before release to harmonize odours and flavours (Figure 8.1).

From a microbiological point of view, vinification is a complex process involving a wide array of microorganisms.

Generally, grapes are characterized by the presence of different non-*Saccharomyces* genera (*e.g. Brettanomyces*, *Candida*, *Kloeckera*/*Hanseniaspora*, *Kluyveromyces*, *Metschnikowia*, *Pichia*, *Saccharomyces*, and *Zygosaccharomyces*).[5] These yeasts generally do not survive long during fermentation because of the high levels of alcohol (over 5% v/v) but can contribute to wine flavour and aroma. The fermentation process is dominated by *Saccharomyces cerevisiae* populations.

Lactic acid bacteria are also associated with grapes and wines and are fundamental to conduct MLF.[6] *Oenococcus oeni*, thanks to its ability to grow at low pH (>2.9), in the presence of high ethanol concentration (up to 16%) and SO_2, with nutrient starvation, and low temperature, is the dominant species used in winemaking.[6] Other LAB belonging to the species of *Lactobacillus* and *Pediococcus* are able to conduct MLF but they are usually associated with wine spoilage.[7]

The presence of BAs in wine has been widely discussed in the last few years because of the potential risks for consumer's health also empowered by the synergic effect of BAs and ethanol. Biogenic amines are considered the most important reason for wine intolerance. However, OIV (*Organization International de la Vigne et du Vin*) have not yet established official limits for BAs in wine. In any case, some countries introduced maximum limits for histamine (2 mg L^{-1} in Germany, 6 mg L^{-1} in Belgium, 10 mg L^{-1} in Switzerland and Austria, 8 mg L^{-1} in France, and 4 mg L^{-1} in The Netherlands).

There are two different sources of BAs in wine: raw materials (spermidine and putrescine are normal constituents of grapes) and fermentation processes. In general, BAs accumulation in wine is influenced by the presence of precursors, grape variety and degree of ripeness, anti-fungal treatment of grapes, the climate and soil of the viticulture area, cultivation practices (*e.g.* degree of irrigation), addition of starter cultures, alcohol content, sulphur dioxide, pH, temperature, microbial contamination in wineries, and contact time of must and skin.[8] For instance, Leitao *et al.*[9] showed an increased level of BAs in Portuguese wines, probably because of climatic conditions, sanitary conditions of grapes, vinification method, and/or dominance of

LAB species in the must. Moreover, red and botrytized wines present higher levels of BAs than white and rosé wines, which can be explained by the different winemaking techniques.[8,10]

Wine is considered a good medium for microbial decarboxylating capacity. Generally, all microbial groups acting in winemaking could be associated with BAs production. There are conflicting opinions regarding the formation of BAs in wine. However, it is generally agreed that BAs concentration is low at the end of alcoholic fermentation and increases during MLF, suggesting that BA production is related to bacterial metabolism.[11–13] Yeasts are the most important microorganisms in wine but their contribution to BAs production is less significant than the contribution on LAB.[14] Caruso *et al.*[15] checked BAs production of 50 different strains isolated from musts. *Saccharomyces cerevisiae* and *Candida stellata* were able to produce ethanolamine and agmatine while *Brettanomyces bruxellensis* produced phenyletylamine and agmatine, pointing out its already known spoilage attitude. Even if yeasts cannot produce high levels of BAs, their interaction with bacterial strains could lead to a level of total BAs in wine which can be hazardous to human health. Landete *et al.*[16] evaluated the decarboxylating potential of 36 wine yeast strains and none of them were able to produce BAs. Tables 8.1 and 8.2 report the content of BAs in spontaneous and controlled fermentations of different wines at the end of alcoholic fermentation.

Malolactic fermentation is the decarboxylation of L-malic acid to L-lactic acid *via* the malolactic enzyme and results in deacidification, enhancing of

Table 8.1 Biogenic amines (BAs) after alcoholic fermentation in different wines obtained with spontaneous fermentation.[a] Data from ref. 18.

	BAs	Total BAs content (mg L^{-1})	Wine	References
Spontaneous AF	HIS; AGM; SPD; TYR; PUT; TRY; CAD; PHE	22.37	Tempranillo	17
	PHE; PUT; HIS; TYR; SPD; SPM	23.7	Arvino	18
		41.9	Gaglioppo	
		44.0	Greco Nero	
		63.1	Magliocco Canino	
		36.6	Magliocco Dolce	
		46.8	Nocera	
	HIS; AGM; SPD; TYR; PUT; TRY; CAD; PHE	6.9	Tempranillo	12
		10.2	Monastel	
		9.6	Maturana Tinta de Navarrete	

[a]Abbreviations: HIS, histamine; MET, methionine; ETH, ethylamine; TYR, tyramine; PHE, phenylamine; PUT, putrescine; CAD, cadaverine; AGM, agmatine; ETA, ethanolamine; TRY, tryptamine; SPD, spermidine; SPM, spermine.

Table 8.2 Biogenic amines (BAs) after alcoholic fermentation in different wines obtained with guided fermentation with starter cultures.[a] Data from ref. 18.

	BAs	Total BAs content (mg L^{-1})	Wine	Starter	References
Fermentation with starter cultures	AGM; ETA; ETH; PUT; TYR	25.87 23.67 14.02 21.31 18.11 20.59 27.53	Merlot Syrah Cabernet Franc. Montepulciano Sangiovese Carmenere Cesanese D'Affile	*S. cerevisiae* (10^6 cells mL^{-1})	19
	CAD; TRY; PHE; TYR; HIS; ETA; ETH; PUT	21–24	Montepulciano d'Abruzzo	*S. cerevisiae* SRS1 (10^6 cells mL^{-1})	20
	ETA; MET; AGM; TRY; PHE; PUT; CAD; HIS	15.01 12.4 6.21 6.73 9.60	Aglianico of Vulture	*Dekkera/B. bruxellensis* (5%) *S. cerevisiae* (5%) *Kloeckera apiculata* (5%) *Candida stellata* (5%) *Metschnikowia pulcherrima* (5%)	15
	PUT; CAD; SPM	0.40	Italian red wine	*Dekkera/B. bruxellensis* (CBS2336 and CBS4601)	21
	HIS; TYR; PHE; PUT; CAD	2.89	Tempranillo	*Kluyveromyces thermotolerans/Schizosaccharomyces pombe* V2	22
	HIS; TYR; PHE; PUT; CAD	1.47 1.55 1.39 1.47 1.50 1.50	Tempranillo	*S. pombe* (Y470) *S. pombe* (CBS1057) *S. pombe* (NCYC3422) *S. pombe* V1 *S. pombe* (CECT12774) *S. pombe* (CECT12774)	23

[a]Abbreviations: HIS, histamine; MET, methionine; ETH, ethylamine; TYR, tyramine; PHE, phenylamine; PUT, putrescine; CAD, cadaverine; AGM, agmatine; ETA, ethanolamine; TRY, tryptamine; SPD, spermidine; SPM, spermine.

organoleptic properties of wine, and improvement of microbiological stability.[24] Lactic acid bacteria are responsible for MLF with *O. oeni* as the main actor.[25] Malolactic fermentation is a crucial step for BAs production: 30 mg L^{-1} of total BAs have been found in wine by Marques *et al.*[26] In particular, putrescine, histamine, and tyramine are the most abundant BAs found[27,28] and their release is influenced by aminogenic capacity of LAB and the inoculum of selected *O. oeni* strains. Tables 8.3 and 8.4 show the results of different studies concerning BAs formation during MLF.

In June 2010, EFSA published a public call for data on the presence of BAs in food and beverages. In 2011, a scientific opinion was published and the occurrence of BAs in wine is shown in Table 8.5. Red wines have a high amount of histamine, while in white wine and sparkling wines, tyramine is the major BA found.[3]

Histamine is considered the most important cause of wine intolerance[30] and LAB are mainly responsible for its production in wine. Histamine production is a high variable among strains and is associated to the presence

Table 8.3 Biogenic amines (BAs) after malolactic fermentation in different wines obtained with spontaneous fermentation.[a] Data from ref. 18.

	BAs	Total BAs content (mg L^{-1})	Wine	References
Spontaneous MLF fermentation	TYR; PUT; HIS	27.6	Periquita	26
	HIS; AGM; SPD; TYR; PUT; TRY; CAD; PHE	1.313 2.236 1.646	Tempranillo Monastel Maturana Tinta de Navarrete	12
	PUT; SPD; SPM; AGM; CAD; SRT; HIS; TYR; TRY; PHE	<0.40 7.94 12.4 12.9 6.88 9.20 6.43	Merlot	29
	PHE; PUT; HIS; TYR; SPD; SPM	30.0 50.3 54.4 74.1 43.3 54.5	Arvino Gaglioppo Greco Nero Magliocco Canino Magliocco Dolce Nocera	18
	HIS; MET; ETH; TYR; PHE; PUT; CAD	22.37	Tempranillo	17

[a]Abbreviations: MLF, malolactic fermentation; HIS, histamine; MET, methionine; ETH, ethylamine; TYR, tyramine; PHE, phenylamine; PUT, putrescine; CAD, cadaverine; AGM, agmatine; ETA, ethanolamine; TRY, tryptamine; SPD, spermidine; SPM, spermine.

Table 8.4 Biogenic amines (BAs) after malolactic fermentation in different wines obtained with guided fermentation with starter cultures.[a] Data from ref. 18.

	BAs	Total BAs content (mg L^{-1})	Wine	Starter	References
MLF with starter cultures	TYR	2.0	Periquita	CMS2 (inducer of MLF)	26
	AGM; ETA; ETH; PUT; TYR	37.55	Merlot	*O. oeni* (5 × 10^{6} cells mL^{-1})	19
		47.59	Syrah		
		31.26	Cabernet Franc.		
		33.85	Montepulciano		
		34.09	Sangiovese		
		29.71	Carmenere		
		37.80	Cesanese d'Affile		
	PUT; SPD; SPM; AGM; CAD; SRT; HIS; TYR; TRY; PHE	1.93	Merlot	*O. oeni* DSM 7008 (6 mg L^{-1})	29
		15.5		*O. oeni* DSM 12 923 (6 mg L^{-1})	
		14.3		*L. plantarum* DSM 4361 (200 mg L^{-1})	
		7.4		*O. oeni* DSM 7008 (6 mg L^{-1})	
		7.7		*O. oeni* DSM 12 923 (6 mg L^{-1})	
		24.1		*L. plantarum* DSM 4361 (200 mg L^{-1})	
		9.08		*O. oeni* DSM 7008 (6 mg L^{-1})	
		6.23		*O. oeni* DSM 12 923 (6 mg L^{-1})	
		14.6		*L. plantarum* DSM 4361 (200 mg L^{-1})	
		6.13		*O. oeni* DSM 7008 (6 mg L^{-1})	
		9.81		*O. oeni* DSM 12 923 (6 mg L^{-1})	
		17.7		*L. plantarum* DSM 4361 (200 mg L^{-1})	
	HIS; MET; ETH; TYR; PHE; PUT; CAD	14.75	Tempranillo	Commercial malolactic bacteria	17

[a]Abbreviations: MLF, malolactic fermentation; HIS, histamine; MET, methionine; ETH, ethylamine; TYR, tyramine; PHE, phenylamine; PUT, putrescine; CAD, cadaverine; AGM, agmatine; ETA, ethanolamine; TRY, tryptamine; SPD, spermidine; SPM, spermine.

Table 8.5 Biogenic amines in wines expressed in mg kg^{-1} (modified from EFSA, 2011[3]).[a]

Wines	HIS	TYR	PUT	CAD
Red wine	12.3–12.4	7.8–8.5	9.5–11.5	0.6–1.6
White wine	2.6	4.3–4.5	3.9–4.3	0.3–0.4
Sparkling white wine	5.2	26.4	15	0.2

[a]Abbreviations: HIS, histamine; TYR, tyramine; PUT, putrescine; CAD, cadaverine.

of a histidine decarboxylase (HDC; EC 4.1.1.22) firstly isolated by Lonvaud-Funel and Joyeux in 1994[31] from *O. oeni* 9204. Recently, Lucas *et al.*[32] demonstrated the plasmidic localization of the *hdc* gene cluster in a strain of *Lactobacillus hilgardii*, giving an explanation to the random distribution of HDC-positive bacteria.

Less is known about the presence of these amines in natural sparkling wines, in which a possible increase could be expected because the starting wine is subjected to a second fermentation by yeasts. Vidal-Carau *et al.*[33] studied the evolution of these amines during the second fermentation process in 38 natural sparkling wines from the Penedés region (Spain). The levels of histamine and tyramine in the natural sparkling wines were very low. The concentrations found in sparkling wines ranged from trace amounts to 3.31 mg L^{-1} for histamine and from 0.36 to 2.36 mg L^{-1} for tyramine. The average value for histamine was 0.48 mg L^{-1} and for tyramine it was 1.03 mg L^{-1}.

Normally, red wines have higher levels of biogenic amines than white wines. This difference is not evident in white and rosé natural sparkling wines. Vidal-Carau *et al.*[33] showed that there are no statistically significant differences between the average levels of both amines in white and rose natural sparkling wines.

Sparkling wines are classified according to their level of reducing sugars: 'brut', ~20 g L^{-1}; 'dry', 20–30 g L^{-1}; 'semi-dry', 30–50 g L^{-1}; 'sweet', >50 g L^{-1}, but no differences were found between the average levels of tyramine in the natural sparkling wines examined with respect to sugar content. On the contrary, statistically significant differences were observed in histamine levels, even if a direct correlation cannot be established between sugar and histamine contents, in view of the facts that the concentrations found were low.[33]

8.2 Beer

The brewing of beer is one of the most ancient practices of somewhere between 6000 and 8000 years.[34] It is produced with water, malted cereal grains (generally barley and wheat), hops, and yeast, and the main process is reported in Figure 8.2. The two main brewing yeasts are *S. cerevisiae* (ale yeast) and *Saccharomyces pastorianus* (lager yeast).[34]

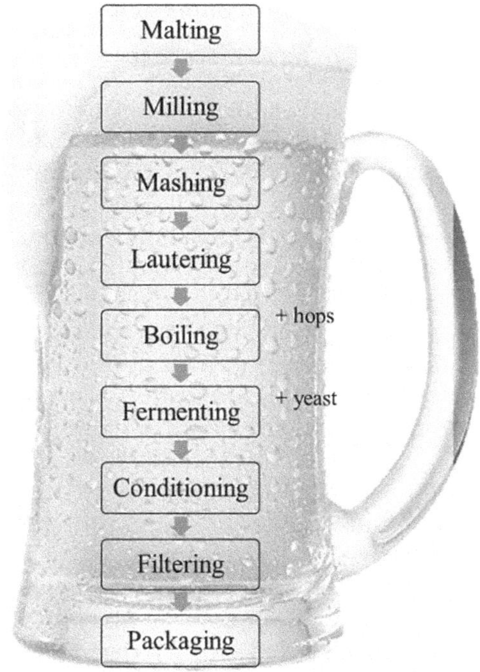

Figure 8.2 Flow sheet showing the main steps involved in beer production.

Recently, the popularity of beers has increased, and the market is progressively shifting from mass-produced lager brands to richer flavours, quirky ingredients, and striking aesthetics of craft beers, which are characterized by a unique aroma and taste.[34]

One of the main concerns about beer consumption is the presence of BAs. These compounds in beer are formed primarily during fermentation but can also be produced by microorganisms in barley, malt, wort, and hops.[35,36] Lactic acid bacteria are commonly implicated in BAs formation, but Enterobacteriaceae and some strains of *Saccharomyces* may also play a role.[35]

The occurrence of BAs in beer is variable and depends on several factors, *e.g.* quality of raw materials, brewing technique, microbial contamination during fermentation (hygiene), and storage conditions.[37,38]

Some of them are present in raw materials (histamine, tyramine, cadaverine, agmatine, spermine, and spermidine) and therefore, these BAs can be considered as 'natural' constituents of beer. Generally, their levels were seen to increase slowly during a five-day barley germination. High levels of putrescine, agmatine, spermine, and spermidine and low levels of histamine, 2-phenylethylamine, tryptamine, and cadaverine are usually found in malt. Tyramine, putrescine, spermine, spermidine, and agmatine have been detected in barley. Relatively high levels of tyramine, 2-phenylethylamine, putrescine, spermine, spermidine, and agmatine have been detected in hops, however, its contribution to BA levels in beer is not significant since the amount used is very small.[39]

They can be formed during mashing and wort boiling. Moreover, during fermentation, wort contamination by amino acid decarboxylase bacteria, such as LAB, might occur leading to the formation of BAs.[40] Other BAs such as histamine, tyramine, and cadaverine are generally produced by LAB during brewing.[41]

The presence of BAs in bottled beer mainly depends on the metabolic activity of LAB. In particular, Kalač *et al.*[42] revealed that tyramine and histamine are mainly formed by lactobacilli, surviving insufficient pasteurization. The same authors proposed that increased levels of BAs can be encountered in beers contaminated by LAB during production. Therefore, high BAs content can be expected in some wheat beers where LAB are dominant components of the microbiota. Finally, contaminated yeasts can be a source of decarboxylating bacteria in bottle-conditioned beers.

The content of BAs in beer is not influenced by the content of alcohol. Izquierdo-Pulido *et al.*[40] and Buiatti *et al.*[43] showed that non-alcoholic beers did not have significantly lower amounts of BAs than alcoholic ones, suggesting that methods used to produce them do not remove amines. Some differences have been detected as a function of a beer's origin (Table 8.7). Among European beers, Belgian beers showed high concentrations of tyramine, histamine, and cadaverine. A reason could be that several beers are produced through a 'spontaneous fermentation'. German beers showed a wide range of tyramine and cadaverine levels, mainly because of 'Weissbier' beers, whereas some particular 'pils', Spanish beers, showed relatively high levels of tyramine.[40]

On the basis of these observations, different types of beers are characterized by different amount of BAs (Tables 8.6 and 8.7). 'Kriek' and 'spontaneous fermentation' beers are characterized by the highest average values of tyramine and histamine. 'Spontaneous fermentation' beers also show the

Table 8.6 Content of biogenic amines (BAs) on the basis of origin.[39–41,44 a]

	mg L^{-1}			
BAs	Europe	Brazil	Cuba	Nigeria
HIS	n.d.–21.6	—	0.2–1.5	0.02–0.1
TYR	0.5–67.5	1.2–9.3	0.3–36.8	0.1–0.35
β-PHE	n.d.–8.3	—	0.2–1.7	—
TRY	n.d.–5.4	—	0.35–10.1	0.02–0.29
CAD	n.d.–39.9	—	0.2–2.6	0.04–0.2
PUT	1.5–15.2	—	—	n.d.–0.4
AGM	0.5–40.9	—	2.1–4.6	—
SPM	n.d.–3.9	—	0.2–2.1	—
SPD	n.d.–6.8	—	0.2–6	—

[a]Abbreviations: HIS, histamine; MET, methionine; ETH, ethylamine; TYR, tyramine; PHE, phenylamine; PUT, putrescine; CAD, cadaverine; AGM, agmatine; ETA, ethanolamine; TRY, tryptamine; SPD, spermidine; SPM, spermine.

Table 8.7 Biogenic amines content (mg L^{-1}) in different type of beers.[a] Data from ref. 40.

Type of beer	AGM	PUT	TYR	CAD	HIS
Top-fermented					
Ale	1.1–15.7	2.6–9.7	1.9–17.4	nd–4.2	0.5–2.0
Stout and Porter	5.2–10.8	2.4–8.2	1.1–9.1	nd–1.9	nd–3.2
Weissbier	3.0–10.7	2.4–6.7	1.3–33.6	0.4–17.7	nd–2.4
Kriek	1.1–3.4	3.5–5.1	6.7–36.4	1.9–15.2	1.6–14.0
Trappsite	3.2–19.4	3.2–9.1	1.8–4.5	nd–1.9	nd–1.6
Bottom-fermented					
Lager	0.9–27.2	1.5–9.7	1.6–20.9	nd–6.2	nd–2.6
Pils	5.1–40.9	2.6–8.8	0.5–46.8	nd–31.4	nd–17.0
Dortmunder	6.1–13.4	3.0–5.6	1.4–7.6	nd–0.5	nd–1.3
Bock	0.4–21.5	2.1–12.4	2.1–10.2	nd–1.5	nd–2.9
Spontaneous fermentation					
Lambic and Gueuze	1.0–18.8	2.8–15.2	0.8–67.6	0.4–39.9	nd–21.6
Non-alcoholic	2.6–16.8	1.6–4.7	2.1–31.5	nd–5.3	nd–3.2

[a]Abbreviations: HIS, histamine; TYR, tyramine; PUT, putrescine; CAD, cadaverine; AGM, agmatine; nd, not detected.

highest average values of cadaverine. 'Weissbier' beers present high levels of tyramine and cadaverine. The high BAs concentration found in these beers is related to the fact that in all cases a lactic fermentation occurs, therefore it could be implicated in amine formation.[40] Bottom-fermented beers show similar levels of amines. Generally, 'Lager' and 'Pils' beers have relatively low average levels of tyramine, cadaverine, and histamine.

Finally, it is important to underline that beers produced in different breweries can differ in terms of BAs, suggesting that the higher amount of some BAs in beers could be an essential indicator of proper hygiene during brewing.[37,38]

8.3 Cider

Cider is obtained from the pressing of fruits (such as apples) and the subsequent fermentation process. Cider can be consumed as a beverage or it can be used as a raw material in vinegar making. The juice colour can vary from a pale yellow to a dark amber rose. It has a fruity flavour and the taste can vary from very sweet (non-alcoholic) to tart. In brief, the cidermaking process involves three steps: apple crushing and pressing out the juice, followed by fermentation. This last process includes classical alcoholic fermentation of sugars into ethanol performed by yeast strains and MLF processed by LAB that can occur during maturation.

Although starter cultures can be added, in general, alcoholic and malolactic fermentations are driven by the indigenous microbiota present in apples, production equipment, and in the cellar. Spontaneous fermentation begins within a few hours if the temperature of the must rises above 10 °C. This process is usually slow, requiring at least 2–3 weeks for the main fermentation and several months for the maturation. Maturation takes place in wooden, polyester, or stainless-steel casks at a controlled temperature of 3 °C–12 °C. The entire process can take from 1 to 6 months depending on the country.[45] Before bottling, ciders are not microbiologically stabilized, therefore, heterofermentative LAB can develop, promoting BAs formation.[46–48]

The presence of BAs in ciders was first reported by Zee *et al.*[49] and Vidal-Carou *et al.*[33]

Table 8.8 Biogenic amine contents in the commercial ciders.[a]

	Biogenic amines (mg L^{-1})			
Cider samples	HIS	TYR	PUT	References
1	3.03	4.81	9.39	46
2	nd	7.31	7.31	
3	3.29	5.03	10.82	
4	nd	20.1	4.89	
5	1.64	nd	3.32	
6	1.08	nd	4.27	
7	1.03	3.95	5.21	
8	5.83	4.25	10.08	
9	0.93	3.03	4.51	
10	nd	4.03	7.31	
11	2.21	nd	6.01	
12	4.63	4.08	12.25	
13	1.42	2.93	nd	33
14	nd	nd	0.2	49

[a]Abbreviations: HIS, histamine; TYR, tyramine; PUT, putrescine; nd, not detected.

Table 8.9 Biogenic amine production by lactic acid bacteria isolated from ciders.[46] [a]

		HIS		TYR		PUT	
Lactic acid bacteria	N[b]	MDA[c]	HPLC[d]	MDA	HPLC	MDA	HPLC
Lactobacillus brevis CECT 216		0	—	1	1	1	—
Lactobacillus 30a		1	1	0	—	1	1
Lactobacillus collinoides	15	1	1	0	—	0	—
Lactobacillus diolivorans	16	3	3	1	1	0	—
Lactobacillus plantarum	2	0	—	0	—	0	—
Lactobacillus suebicus	6	0	—	0	—	0	—
Oenococcus oeni	10	0	—	1	1	0	—
Pediococcus parvulus	8	0	—	0	—	0	—

[a]Abbreviations: HIS, histamine; TYR, tyramine; PUT, putrescine.
[b]N, number of strains analysed.
[c]MDA, number of positive strains in decarboxylase media (MDA).
[d]HPLC, number of positive strains by reversed-phase high-performance liquid chromatography (RP-HPLC).

Table 8.10 Concentration of BAs (mg L^{-1}) produced by LAB strains in the decarboxylase synthetic broth and in cider (mean ± standard deviation).[a]

	Amine production in MDA		HIS production in cider		TYR production in cider	
Bacterial strains	HIS	TYR	pH 4	pH 2	pH 4	pH 2
Lactobacillus 30a	1402.01 ± 3.22	nd	—	—	—	—
L. brevis CECT 216	nd	854.01 ±3.38	—	—	—	—
L. diolivorans 18	1232.4 ± 3.17	—	152.09 ± 0.98	284.32 ± 0.84	nd	nd
L. diolivorans 20	793.2 ± 4.32	—	35.22 ± 0.87	77.33 ± 0.43	nd	nd
L. diolivorans 22	602.23 ± 7.09	—	15.31 ± 0.65	27.55 ± 0.21	nd	nd
L. diolivorans 54	—	746.92 ± 2.13	nd	nd	5.02 ± 0.11	8.54 ±0.44
O. oeni 38	—	543.99 ± 3.11	nd	nd	21.33 ± 0.09	22.11 ± 0.99
L. collinoides 13	1207.3 ± 9.34	—	12.34 ± 0.54	13.21 ± 0.32	nd	nd

[a]Abbreviations: HIS, histamine; TYR, tyramine; nd, not detected.

Studies performed on commercial cider from Spain and France revealed the presence of BAs in almost 90% of the analysed products with a higher prevalence of tyramine, histamine, putrescine, and cadaverine.[48,50] However, some differences were found. In particular, Spanish ciders presented the highest concentrations of cadaverine and putrescine, while tyramine was the most frequently detected BAs in French samples. Histamine was detected at relatively low levels in both French and Spanish samples. These differences are probably related to different elaboration processes and to the microbiota hosted in apples.[50]

A more recent study performed by Garai-Ibabe *et al.*[48] reported that putrescine was the main BA detected in Spanish ciders followed by histamine and tyramine, probably because of the low concentration of amino acids. In particular, amino acids decreased during cidermaking, while after bottling, an increase was detected, probably related to yeast and bacterial autolysis. However, BAs were present in low concentrations and therefore did not affect cider quality.

Biogenic amine concentration in Spanish ciders is much higher than those previously reported in Canadian ciders,[49] in which concentrations of putrescine of 0.2 mg L^{-1} were reported, while histamine and tyramine were not detected. The low amounts of BAs detected in Canadian ciders could be explained by the fact that apple juice is normally pasteurized before fermentation, thus eliminating all microbiota responsible for decarboxylase activities, Canadian ciders are rarely subjected to MLF; thus, little or no decarboxylation of amino acids occurs through bacterial metabolism.[48,49]

In Table 8.8, the amount of histamine, tyramine, and putrescine found in various natural ciders is reported. Table 8.9 reports the decarboxylating strains, while the quantitative production of BAs by these LAB strains is shown in Table 8.10.

8.4 Vinegar

Vinegar is a widespread and common acetic acid-diluted solution product with a long history (Figure 8.3).[51]

Vinegar is commonly used as a food flavouring and preservative, however, recent investigations showed its positive effects on human health. For instance, it presents antibacterial activity, blood pressure reduction, antioxidant activity, reduction in the effects of diabetes, prevention of cardiovascular disease, and improves blood glucose response which would be of benefit to diabetic patients (for a review see Budak *et al.*[52]).

According to Solieri and Giudici,[51] vinegars can be classified into three categories:

- vegetable vinegars (rice vinegar, onion vinegar, and tomato vinegar);
- fruit vinegars (cider vinegar, mango vinegar, and pineapple vinegar); and
- animal vinegars (honey vinegar and whey vinegar).

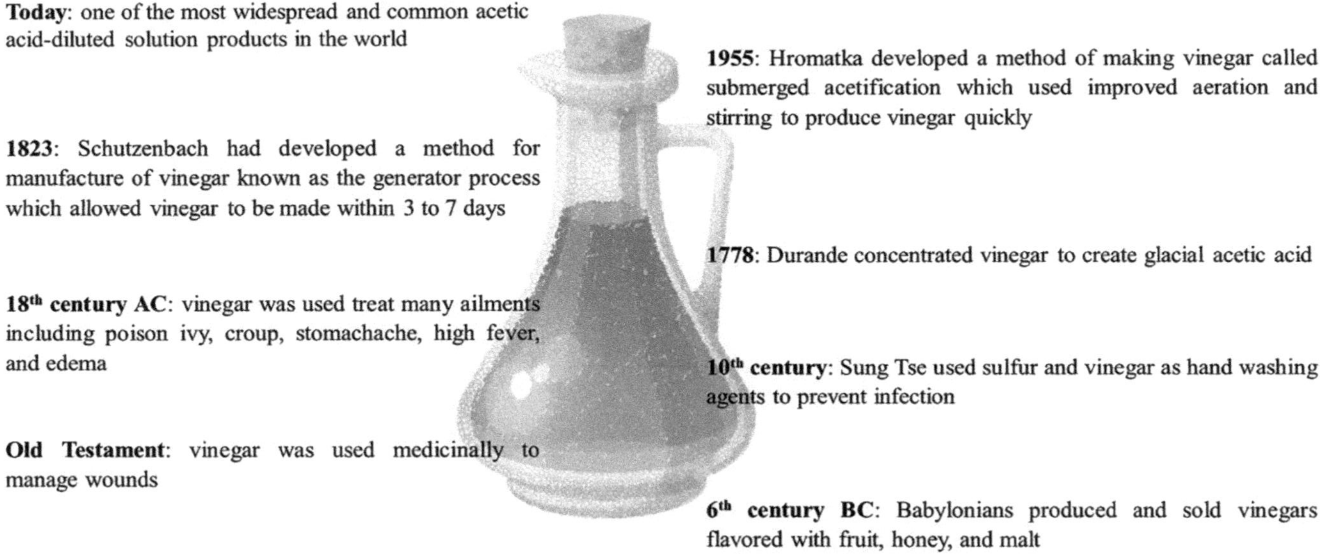

Figure 8.3 Timeline of significant events of vinegar production.

Vinegars can be produced from different raw materials including hydroalcoholic solutions, various fruits, cereals, or amidaceous agricultural products. However, in Europe, the most highly prized quality vinegars derive from grapes and the wines obtained therefrom.

Some vinegars have received protected designation of origin (PDO) or protected geographical indication (PGI).

In particular, the Italian 'Aceto Balsamico Tradizionale di Modena' and 'Aceto Balsamico Tradizionale di Reggio Emilia' and the Spanish 'Vinagre de Jerez' and 'Vinagre de Condado de Huelva' are recognized as PDOs by the European Union, while a PGI has been granted for another Italian vinegar called 'Aceto Balsamico di Modena'. In addition, very recently, a new PDO for 'Vinagre de Montilla-Moriles' has been authorized.[51]

In general, vinegars production requires three steps. The first is starch saccharification, followed by alcoholic fermentation driven by yeasts and finally acetic acid fermentation (AAF), which is the oxidation of ethanol to acetic acid by acetic acid bacteria (AAB).[53] However, sometimes, starch saccharification and alcoholic fermentation can coincide.[54,55]

Very few studies are available about BAs in vinegars. The main study concerning AAB ability to produce BAs has been conducted by Landete *et al.*,[16] who demonstrated that none of the 40 AAB strains studied were able to produce BAs. More interesting are the results achieved by Song *et al.*,[56] who isolated a strain of *Acetobacter aceti* MBA-77 from Bokbunja seed vinegar, which was able to decrease histamine concentration in vinegars.

Ordóñez *et al.*[57] investigated the occurrence of BAs in 26 vinegars including balsamic, apple, red, white, and sherry wine vinegars. The total content of BAs in vinegars ranged from 23.35 to 1445.2 μg L^{-1}, being lower than those reported in wines.

In general, red wine and balsamic vinegars showed the highest concentrations of BAs, ranging from 79 μg L^{-1} to 760 μg L^{-1} and from 491 μg L^{-1} and 1445 μg L^{-1}. This huge variability could be related to different factors such as the variety of grape, the method of winemaking employed, bacterial strains present, and the availability of free amino acids.[25,58] White wine, sherry, and Pedro Ximénez vinegars showed lower levels of BAs ranging from 194 μg L^{-1} to 223 μg L^{-1}, from 23.43 μg L^{-1} to 337.6 μg L^{-1}, and 693 μg L^{-1}, respectively.

Apple vinegars showed the lowest amounts of total BAs (about 24 μg L^{-1}) probably because cider contains a lower concentration of BAs than wine. The main BAs detected were putrescine and histamine, reaching values up to 525 and 309 μg L^{-1}, respectively. These results were expected since these amines are the most abundant in wines.[59]

In general, methylamine and phenylethylamine were not determined in any vinegar.

Chinnici *et al.*[60] analysed 22 vinegars of Jerez, six balsamic vinegars of Modena, and ten traditional balsamic vinegars of Modena at different ageing times and supplied from local producers or purchased in the market.

Spermidine, phenylethylamine, cadaverine, isoamylamine, and tryptamine were not found in any vinegar, while tyramine and putrescine were detected in all the vinegars except traditional balsamic vinegar of Modena, while histamine was recorded in all samples.

These data highlight that the concentration of BAs in vinegars is lower than that found in wines, suggesting that BAs are not produced and even degraded during acetic fermentation. However, further studies are necessary to confirm this hypothesis.

8.5 Liqueur

Liqueurs are highly flavoured sweetened spirit beverages (15–30% alcohol; >10% sugar), based on ethyl alcohol (*e.g.* cognac, rum, or whiskey) or a neutral distillate of agricultural origin flavoured with fruit, cream, herbs, spices, flowers, or nuts and bottled with added sugar, honey, or high-fructose corn syrup.[61,62]

Liqueurs are mainly produced in three basic ways:

- Heat or infusion method, which can be done in the presence of the alcohol base or merely in hot water, with the distilled spirit subsequently added, often preferred when herbs, peels, or roots are used, as heat can add to extract flavouring substances.
- Cold or maceration method, usually done in large oak casks in the presence of distilled alcohol, best when fruits and flowers are used, because heat can destroy some of the most favourable aromatic substances.
- Distillation, in which alcohol and flavouring agents are blended before being distilled.

Very little is known for what concern BAs in beverage spirits, such as liqueurs, whiskey, vodka and brandy. Cunha *et al.*[63] analysed a total of 27 liqueur samples commercialized in Portugal, including 12 from industrial production and 15 homemade liqueurs, from five different types (fruits, herbs, coffee, honey, and milk). The total content of amines in the studied liqueurs was very low, from 0.168 mg L^{-1} to 4.665 mg L^{-1} with an average value of 1.353 mg L^{-1}. Coffee liqueurs have the highest average content (2.117 mg L^{-1}) followed by honey liqueurs (1.932 mg L^{-1}), fruit liqueurs (1.254 mg L^{-1}), and herb liqueurs (1.015 mg L^{-1}). The only sample of milk liqueur studied showed the lowest BAs total content, 0.168 mg L^{-1}. Homemade liqueurs have a significant higher content compared with industrial liqueurs for fruit (1.824 mg L^{-1} *vs.* 0.780 mg L^{-1}) and herb liqueurs (1.225 mg L^{-1} *vs.* 0.663 mg L^{-1}), while in coffee liqueurs, the average content of the two industrial samples analysed (2.378 mg L^{-1}) was slightly higher than the average content of the three homemade samples (1.944 mg L^{-1}). Ethylamine, morpholine, and cadaverine were found in all

Table 8.11 Biogenic amines content (mg kg^{-1}) of boza samples.[a] Data from ref. 64.

Boza samples	MMA	ETH	IPA	PUT	CAD	TRY	SPD	SPM	HIS	TYR
1	nd	nd	nd	3.4 ± 0.2	nd	nd	1.4 ± 0.2	nd	nd	45 ± 26
2	0.3 ± 0.1	0.8 ± 0.3	5.8 ± 1.6	1.2 ± 0.2	1.1 ± 0.8	nd	2.7 ± 0.1	3.0 ± 0.5	2.2 ± 0.1	18 ± 8
3	1.7 ± 0.1	5.0 ± 2.0	6.2 ± 4.8	5.6 ± 0.6	1.5 ± 0.8	3.3 ± 1.1	3.5 ± 0.3	4.0 ± 0.1	6.5 ± 0.4	23 ± 2
4	nd	nd	1.4 ± 1.0	0.6 ± 0.1	nd	nd	1.8 ± 0.7	nd	nd	65 ± 6
5	nd	nd	2.1 ±0.1	0.8 ± 0.1	nd	nd	1.2 ± 0.6	nd	1.8 ± 0.8	19 ± 7
6	0.2 ± 0.1	nd	0.9 ± 0.4	0.4 ± 0.1	nd	nd	0.8 ± 0.2	1.0 ± 0.4	1.6 ± 0.8	55 ± 10
7	0.4 ± 0.1	nd	1.8 ± 1.2	2.8 ± 0.4	nd	nd	1.5± 0.4	nd	7.9 ± 0.3	13 ± 4
8	nd	nd	nd	0.7 ± 0.2	nd	nd	0.6 ± 0.1	nd	nd	52 ± 1
9	0.8 ± 0.1	1.7 ± 1.5	nd	1.0 ± 0.7	nd	nd	2.0 ± 1.0	1.4 ± 0.9	2.1 ± 1.8	33 ± 3
10	0.2 ± 0.1	0.6 ±0.2	1.9 ± 0.5	0.7 ± 0.2	nd	nd	1.2 ± 0.2	0.5 ± 0.3	1.5 ± 0.2	41 ± 17

[a]Abbreviations: MMA, methylamine; HIS, histamine; MET, methionine; ETH, ethylamine; TYR, tyramine; PHE, phenylamine; PUT, putrescine; CAD, cadaverine; AGM, agmatine; ETA, ethanolamine; TRY, tryptamine; SPD, spermidine; SPM, spermine; IPA, isopentylamine; nd, not detected.

27 samples, while methylamine, dimethylamine, 1,3-diaminopropane, putrescine, histamine, and tyramine were found in half of the samples. Diethylamine was not found in any sample. These values agree with the few results for fermented and spirits beverages such as boza and port wine.[64] Boza is a fermented beverage made by yeast and lactic acid fermentation of millet, maize, wheat, or rice. Putrescine, spermidine, and tyramine were detected in boza samples. Tyramine was the prevailing biogenic amine, with values ranging from 13 to 65 mg kg^{-1} (Table 8.11). Differences were observed among boza samples from different manufacturers, but in all cases, the concentrations detected were below the legal limits of 100–800 mg kg^{-1} for tyramine,[65,66] and of 100–200 mg kg^{-1} for the total biogenic amine content.[66]

8.6 Tea

Tea is the second most consumed beverage in the world after water. The tea plant *Camellia sinensis* originates in South Asia and is cultivated in more than 30 countries around the world.[67]

Different types of tea can be produced, and the best-known types of tea are black, green, oolong, dark (pu-erh), white, and yellow. The types of tea essentially differ with regard to the extent of fermentation: green tea (non-fermented), yellow and white teas (lightly fermented, 10–20%), Oolong tea (semi-fermented, 30–60%), black tea (fully fermented, 80–100%), and dark tea (post fermented, $\leq$100%).[68] For green tea production, leaves usually undergo heating or steam treatment and fast drying. On the contrary, during black tea production, the leaves are subjected to a sequence of procedures such as weathering, destruction of plant tissues by various rolling, crushing and/or tearing processes followed by enzymatic maturation, and final drying.

The majority of studies concerning this beverage deals with the positive effects of tea consumption on human health. Bioactive compounds of this beverage, in particular polyphenolic constituents, deeply influence its antioxidative, anti-inflammatory, antimicrobial, anticarcinogenic, antihypertensive, and neuroprotective properties.[69,70] On the other hand, very few studies have investigated the content of BAs in tea.

Palavan-Unsal *et al.*[71] investigated the distribution of putrescine, spermidine, and spermine during processing of *Camellia sinensis*. Before processing, the spermine was detected in higher concentrations than putrescine and spermidine in green tea leaves. However, spermine decreased during processing while putrescine and spermine increased, suggesting that leaf processing strongly influenced BA levels and distribution.

Yang *et al.*[72] reported the presence of four spermidine derivatives on tea flowers that mostly accumulate in anthers.

Brückner *et al.*[73] analysed 14 black teas, five green teas, one oolong tea, and one instant tea. Their results revealed that the aqueous infusions generally contained very low amounts of BAs. In most extracts, only putrescine, spermidine, spermine, and tyramine were detected. In dried green tea leaves, the same BAs occurred in significant amounts. Tyramine concentration in black teas (if present) was larger in comparison with green teas, probably because they are richer in tyrosine.[74]

Spizzirri *et al.*[75] showed that in tea leaves, the main BAs detected were putrescine (1.05–2.25 $\mu g\ g^{-1}$) and spermidine (1.01–1.95 $\mu g\ g^{-1}$), while serotonin, histamine, cadaverine, and spermine were present more rarely, tyramine was not detected. Tea infusions showed the same trend with total BA concentrations never exceeding 80.7 $\mu g\ L^{-1}$. Black teas showed higher amounts of BAs than green teas, probably on the basis of the production process. Finally, decaffeinated black teas and instant green teas generally showed lower amounts of BAs than regular black and green teas, probably in relation to the industrial processes involved in decaffeination and the soluble tea technology.

All these studies indicate that the low amounts of free BAs in tea infusions show no health risks are to be expected, even on consumption of large quantities of tea.

8.7 Coffee

The history of coffee is long. It was introduced during the 15th century and nowadays it is one of the most popular and widely consumed beverages worldwide, producing a great economic impact. The most important commercially types/species of coffee are *Coffea arabica* (Arabicas) and *Coffea canephora* (Robustas).

Coffee brew contains many components such as polysaccharides, melanoidins, volatile aroma compounds, and lipid-like compounds with a range of positive, negative, and neutral health benefits.[76] Biogenic amines can be present in coffee representing an undesirable coffee component.[76] Generally, putrescine is the most abundant BA in both Robusta and Arabica coffee, followed by spermidine, spermine, and serotonin, while cadaverine and tyramine are generally present in smaller amounts.[77–79] Robusta origins are associated with levels of tyramine, and significant differences have been recorded in coffees from Angola and other countries, although Robustas from countries other than Angola are still statistically different from the Arabicas.[78] Restuccia *et al.*[76] studied BAs content in grounded coffees and coffee brews. They found that total BAs content in ground coffee ranged from 25.80 $\mu g\ g^{-1}$, to 88.85 $\mu g\ g^{-1}$. Moreover, they found that with a higher roasting temperature, total BAs were lower. Özdestan,[80] on the contrary, found in Turkish ground coffee, a much higher total BAs level, ranging from 126.0 $mg\ kg^{-1}$ to 352.2 $mg\ kg^{-1}$. This difference should be related to the different agronomic and technological parameters in the coffee production that strongly influence BAs concentration. Moreover, industrial processes have a great impact on BAs formation, facilitating their formation. Also, Casal *et al.*[78] evaluated BAs content in green

Robusta and Arabica coffee grounds. They found that putrescine was the most abundant amine, while cadaverine and tyramine were present in small quantities.

In coffee brews, all BAs are present, even if at very low concentrations (putrescine 2.27–0.43 μg mL^{-1}, spermine, 1.95 μg mL^{-1}, tyramine, 1.89–0.25 μg mL^{-1}, cadaverine, 1.84–0.19 μg mL^{-1}, spermidine, 1.81–0.45 μg mL^{-1}, histamine, 1.62–0.22 μg mL^{-1}).[76] Özdestan[80] found in Turkish brewed coffee higher total amine levels, ranging from 5.67 mg L^{-1} to 48.88 mg L^{-1}. Putrescine, cadaverine, tyramine, and serotonin were detected in all coffee samples. Serotonin was found to be the prevailing amine. Moreover, decaffeinated and instant coffees contain lower amount of BAs, perhaps related to the strictness of the production processes.[76,81]

8.8 Conclusion

Biogenic amines can accumulate in a wide variety of alcoholic beverages that have allowed microbial growth and activity during their manufacture, handling, and/or storage. The production of BAs is a strain-dependent property and can take origin from raw materials and processing practices. In addition, the content and types of BAs seem to have a regional variability and are related to several agricultural and processing practices. Biogenic amine monitoring during fermentation with reliable and sensitive analytical techniques is a worthwhile goal in order to identify problems related to fermentation or raw materials contamination and thus providing solutions to avoid BAs accumulation. However, the most promising approach to reduce BAs occurrence in fermented beverages is the selection of starter cultures unable to form BAs that at the same time are able to conduct the fermentation process, reducing the development of contaminant/spoilage microorganisms.

References

1. V. Ladero, *Curr. Res. Nutr. Food Sci.*, 2010, **6**, 145.
2. M. V. Moreno-Arribas and M. C. Polo, Wine Chemistry and Biochemistry, in *Amino Acids and Biogenic Amines*, ed. M. V. Moreno-Arribas and M. C. Polo, Springer, New York, 2009, pp. 163–189.
3. EFSA Panel on Biological Hazards (BIOHAZ), *EFSA J.*, 2011, **9**, 2393.
4. J. M. Landete, S. Ferrer, L. Polo and I. Pardo, *J. Agric. Food Chem.*, 2005, **53**, 1119.
5. N. P. Jolly, C. Varela and I. S. Pretorius, *FEMS Yeast Res.*, 2014, **14**, 215.
6. A. Lonvaud-Funel, Lactic acid bacteria in the quality improvement and depreciation of wine, in *Lactic Acid Bacteria: Genetics, Metabolism and Applications*, ed. W. Konings, O. P. Kuipers and J. H. J. Huis In't Veld, Springer, Dordrecht, 1999, pp. 317–331.
7. E. J. Bartowsky, *Microbiol. Aust.*, 2017, **38**, 76.

8. M. B. A. Glória and S. M. Vieira, *Food*, 2007, **1**, 258.
9. M. C. Leitao, A. P. Marques and M. V. San Romao, *Food Control*, 2005, **16**, 199.
10. M. Martuscelli, G. Arfelli, A. C. Manetta and G. Suzzi, *Food Chem.*, 2013, **140**, 590.
11. M. García-Marino, Á Trigueros and T. Escribano-Bailón, *J. Food Compos. Anal.*, 2010, **23**, 455.
12. O. Martìnez-Pinilla, Z. Guadalupe, Z. Hernàndez and B. Ayestaràn, *Eur. Food Res. Technol.*, 2013, **237**, 887.
13. H. Wang, Y. Luo, H. Huang and Q. Xu, *Int. J. Food Microbiol.*, 2014, **190**, 66.
14. S. Bover-Cid, M. Iquierdo-Pulido, A. Mariné-Font and M. C. Vidal-Carou, *Food Chem.*, 2006, **96**, 43.
15. M. Caruso, C. Fiorel, M. Contursi, G. Salzano, A. Paparella and P. Romano, *World J. Microbiol. Biotechnol.*, 2002, **18**, 159.
16. J. M. Landete, S. Ferrer and I. Pardo, *Food Control*, 2007, **18**, 1569.
17. P. J. Martín-Álvarez, Á Marcobal, C. Polo and M. V. Moreno-Arribas, *Eur. Food Res. Technol.*, 2006, **222**, 420.
18. D. Restuccia, V. Sicari, T. M. Pellicanò, U. G. Spizzirri and M. R. Loizzo, *Food Res. Int.*, 2017, **102**, 303.
19. V. Del Prete, A. Costantini, F. Cecchini, M. Morassut and E. Garcia-Moruno, *Food Chem.*, 2009, **112**, 474.
20. R. Tofalo, F. Patrignani, R. Lanciotti, G. Perpetuini, M. Schirone, P. Di Gianvito, D. Pizzoni, G. Arfelli and G. Suzzi, *Front. Microbiol.*, 2016, **610**, 1.
21. I. Vigentini, A. Romano, C. Compagno, A. Merico, F. Molinari, A. Tirelli, R. Foschino and G. Volonterio, *FEMS Yeast Res.*, 2008, **8**, 1087.
22. Á. Benito, F. Calderón, F. Palomero and S. Benito, *Molecules*, 2015, **20**, 9510.
23. Á. Benito, D. Jeffares, F. Palomero, F. Calderón, F. Y. Bai, J. Bähler and S. Benito, *PLoS One*, 2016, **11**, e0151102.
24. V. Capozzi, P. Russo, L. Beneduce, S. Weidmann, F. Grieco, J. Guzzo and G. Spano, *Lett. Appl. Microbiol.*, 2010, **50**, 327.
25. C. Ancín-Azpilicueta, A. González-Marco and N. Jiménez-Moreno, *Crit. Rev. Food Sci. Nutr.*, 2008, **48**, 257.
26. A. P. Marques, M. C. Leitao and M. C. San Romao, *Food Chem.*, 2008, **107**, 853.
27. A. Lonvaud-Funel, *FEMS Microbiol. Lett.*, 2001, **199**, 9.
28. Á. Marcobal, M. C. Polo, P. J. Martín-Álvarez and M. V. Moreno-Arribas, *Food Res. Int.*, 2005, **38**, 387.
29. L. Manfroi, P. H. A. Silva, L. A. Rizzon, P. S. Sabaini and M. B. A. Gloria, *Food Chem.*, 2009, **116**, 208.
30. V. Konakovsky, M. Focke, K. Hoffmann-Sommergruber, R. Schmid, O. Scheiner, P. Moser, R. Jarisch and W. Hemmer, *Food Addit. Contam.*, 2011, **28**, 408.
31. A. Lonvaud-Funel and A. Joyeux, *J. Appl. Bacteriol.*, 1994, 77, 401.

32. P. M. Lucas, W. A. Wolken, O. Claisse, J. S. Lolkema and A. Lonvaud-Funel, *Appl. Environ. Microbiol.*, 2005, **71**, 1417.
33. M. C. Vidal-Carou, M. J. Isla-Gavin, A. Mariné-Font and R. Codony-Salcedo, *J. Food Compos. Anal.*, 1989, **2**, 210.
34. C. W. Bamforth, *Annu. Rev. Chem. Biomol.*, 2017, **8**, 1.
35. G. Gasarasi, M. Kelgtermans, K. J. Verstrepen, J. Van Roy, F. Delvaux and G. Derdelinckx, *Monatsschr. Brauwiss.*, 2003, **56**, 58.
36. A. Halasz, Á. Barath and W. H. Holzapfel, *Z. Lebensm.-Unters. -Forsch. A*, 1999, **208**, 418.
37. R. Romero, M. G. Bagur, M. Sanchez-Vinas and D. Gazquez, *Anal. Bioanal. Chem.*, 2003, **376**, 162.
38. A. Slomkowska and W. Ambroziak, *Eur. Food Res. Technol.*, 2002, **215**, 380.
39. M. A. M. Gloria and M. Izquierdo-Pulido, *J. Food Compos. Anal.*, 1999, **12**, 129.
40. M. Izquierdo-Pulido, T. Hernández-Jover, A. Mariné-Font and M. C. Vidal-Carou, *J. Agric. Food Chem.*, 1996, **44**, 3159.
41. P. Kalač and M. Křížek, *J. Inst. Brew.*, 2003, **109**, 123.
42. P. Kalač, J. Šavel, M. Křížek, T. Pelikánová and M. Prokopová, *Food Chem.*, 2002, **79**, 431.
43. S. Buiatti, O. Boschelle, M. Mozou and F. Battistutta, *Food Chem.*, 1995, **52**, 199.
44. O. O. Lasekan and W. O. Lasekan, *Food Chem.*, 2000, **69**, 267.
45. F. Cousin, R. Le Guellec, M. Schlusselhuber, M. Dalmasso, J. M. Laplace and M. Cretenet, *Microorganisms*, 2017, **5**, 39.
46. G. Garai, M. T. Dueñas, A. Irastorza and M. V. Moreno-Arribas, *Lett. Appl. Microbiol.*, 2007, **45**, 473.
47. M. Coton, A. Romano, G. Spano, K. Ziegler, C. Vetrana, C. Desmarais, A. Lonvaud-Funel, P. Lucas and E. Coton, *Food Microbiol.*, 2010, **27**, 1078.
48. G. Garai-Ibabe, A. Irastorza, M. T. Dueñas, P. J. Martín-Álvarez and V. M. Moreno-Arribas, *Int. J. Food Sci. Technol.*, 2013, **48**, 375.
49. J. A. Zee, R. E. Simard, L. L'Heureux and J. Tremblay, *Am. J. Enol. Vitic.*, 1983, **34**, 6.
50. V. Ladero, M. Coton, M. Fernández, N. Buron, M. C. Martín, H. Guichard, E. Coton and M. A. Alvarez, *Food Microbiol.*, 2011, **28**, 554.
51. L. Solieri and P. Giudici, in *Vinegars of the World*, ed. L. Solieri and P. Giudici, Springer, Berlin, 2009, pp. 1–14.
52. N. Budak, E. Aykin, A. C. Seydim, A. K. Greene and Z. B. Guzel-Seydim, *J. Food Sci.*, 2014, **79**, 757.
53. M. Adams, Vinegar, in *Microbiology of Fermented Foods*, ed. B. J. B. Wood, Springer, Berlin, 1997, pp. 1–44.
54. S. Haruta, S. Ueno, I. Egawa, K. Hashiguchi, A. Fujii, M. Nagano, M. Ishii and Y. Igarashi, *Int. J. Food Microbiol.*, 2006, **109**, 79.
55. S. W. Lee, S. R. Yoon, G. R. Kim, S. M. Woo, Y. J. Jeong, S. H. Yeo, K. S. Kim and J. H. Kwon, *Food Sci. Biotechnol.*, 2012, **21**, 453.

56. N. E. Song, H. S. Cho and S. H. Baik, *Braz. J. Microbiol.*, 2016, **47**, 452.
57. J. L. Ordóñez, R. M. Callejón, M. L. Morales and M. C. García-Parrilla, *Food Chem.*, 2013, **141**, 2713.
58. P. Herbert, M. J. Cabrita, N. Ratola, O. Laureano and A. Alves, *J. Food Eng.*, 2005, **66**, 315.
59. A. Peña-Gallego, P. Hernández-Orte, J. Cacho and V. Ferreira, *J. Chromatogr. A*, 2009, **1216**, 3398.
60. F. Chinnici, E. Durán-Guerrero and C. Riponi, *J. Sci. Food Agric.*, 2016, **96**, 3762.
61. Regulation (EC) No 110/2008 of the European Parliament and the Council of 15 January 2008 on the definition, description, presentation, labelling and protection of geographical indications of spirit drinks and repealing Council Regulation (EEC) No 1576/89.
62. J. Murphy, The Principles and Practices of Bar and Beverage Management, in *The Drinks Handbook*, ed. J. Murphy, Goodfellows Publishing Ltd, Oxford, England, 2013, pp. 243–247.
63. S. C. Cunha, R. Lopes and J. O. Fernandes, *J. Food Compos. Anal.*, 2017, **56**, 147.
64. S. Yeğin and A. Üren, *Food Chem.*, 2008, **111**, 983.
65. B. Ten Brink, C. Damink, H. M. L. J. Joosten and J. H. J. Huis In't Veld, *Int. J. Food Microbiol.*, 1990, **11**, 73.
66. M. J. R. Nout, *Food Res. Int.*, 1994, **27**, 291.
67. H. Mukhtar and N. Ahmad, *Am. J. Clin. Nutr.*, 2000, **71**, 1698.
68. W. J. Zheng, X. C. Wan and G. H. Bao, *Phytochem. Rev.*, 2015, **14**, 499.
69. K. Hayat, H. Iqbal, U. Malik, U. Bilal and S. Mushtaq, *Crit. Rev. Food Sci. Nutr.*, 2015, **55**, 939.
70. Q. V. Vuong, *Crit. Rev. Food Sci. Nutr.*, 2014, **54**, 523.
71. N. Palavan-Unsal, E. D. Arisan and S. Terzioglu, *Int. J. Food Sci. Nutr.*, 2007, **58**, 304.
72. Z. Yang, F. Dong, S. Baldermann, A. Murata, Y. Tu, T. Asai and N. Watanabe, *J. Sci. Food Agric.*, 2012, **92**, 2128.
73. H. Brückner, S. Flassig and J. Kirschbaum, *Amino Acids*, 2012, **42**, 877.
74. K. Neumann and A. Montag, *Dtsch. Lebensm.-Rundsch.*, 1983, **79**, 160.
75. G. U. Spizzirri, N. Picci and D. Restuccia, *J. Anal. Methods Chem.*, 2016, **2016**, 8715287.
76. D. Restuccia, U. G. Spizzirri, O. I. Parisi, G. Cirillo and N. Picci, *Food Chem.*, 2015, **175**, 143.
77. S. Casal, M. B. P. P. Oliveira and M. A. Ferreira, *J. Liq. Chromatogr. Relat. Technol.*, 2002, **25**, 2535.
78. S. Casal, E. Mendes, M. R. Alves, R. C. Alves, M. Beatriz, P. P. Oliveira and M. A. Ferreira, *J. Agric. Food Chem.*, 2004, **52**, 6188.
79. M. P. Cirilo, A. F. S. Coelho, C. M. Araújo, F. R. Gonçalves, F. D. Nogueira and M. B. A. Glória, *Food Chem.*, 2003, **82**, 397.
80. Ö. Özdestan, *Food Res. Int.*, 2014, **61**, 167.
81. T. M. L. da Silveira, É. Tavares and M. B. A. Glória, *J. Food Compos. Anal.*, 2007, **20**, 451.

CHAPTER 9

Biogenic Amines in Baby Foods

ANNA CZAJKOWSKA*[a] AND JOANNA LESZCZYŃSKA[b]

[a] Mass Spectrometry Laboratory, PORT Polish Centre for Technology Development, 147 Stablowicka Str., 54-066 Wroclaw, Poland; [b] Faculty of Biotechnology and Food Sciences, Lodz University of Technology, 4/10 Stefanowskiego, 90-924 Lodz, Poland
*Email: anna.czajkowska@port.org.pl

9.1 Introduction – Biogenic Amine Toxicity and Occurrence in Food

It has been acknowledged for a long time that the formation of biogenic amines (BAs) may pose a risk to human health. Since they are formed mainly by decarboxylation of the corresponding amino acids by spoilage or other microorganisms (beside polyamines), BAs can be found in all food products. But only amines present in large amounts in food may be dangerous to human health.

The adverse reactions in food caused by BAs are mainly classified as intolerance reactions, known as pseudoallergies. The most toxic amine is histamine (HIS), as a most common course of food poisonings related to a wide variety of physiological syndromes. These occur in healthy adults when a dose of 50 mg HIS/meal/person NOAEL (no-observed-adverse-effect-level) is consumed. In light of the amine-related risk, a HIS acceptable level was established (200 mg kg^{-1}) in fish products rich in histidine.[1] It is also worthy mentioned that the toxic effect of HIS is potentiated by other amines (and possibly by other substances). Tyramine (TYR) is the second

Food Chemistry, Function and Analysis No. 20
Biogenic Amines in Food: Analysis, Occurrence and Toxicity
Edited by Bahruddin Saad and Rosanna Tofalo

Published by the Royal Society of Chemistry, www.rsc.org

most toxic BA, which when present in food over a critical threshold causes dietary migraines in sensitive individuals. Of other amines, diamines: putrescine (PUT) and cadaverine (CAD), and polyamines: spermine (SPM) and spermidine (SPD), beside the cell regulatory function, are also identified as potential precursors for the formation of cancerogenic *N*-nitroso compounds.[2]

9.1.1 Potentially Aminogenic Foods – Complementary Foods

Appropriate nutrition in early childhood is a cornerstone in the development of life-long health. Nutrition plays a major role in child growth, especially the condition and physical and psychical development of young children during weaning.[3] Proper nutrition may affect the risk of some diseases such as obesity, type-2 diabetes, osteoporosis, atherosclerosis, allergy, cancer or at least condition a milder course in adulthood.[4] Quoting a Professor Lawrence T. Weaver: "Pediatricians should be more interested in adult disease because the child is the father of the man".[5]

Complementary foods are products intended to complete the particular requirements of healthy infants while they are being weaned and of healthy young children as a supplement to their diet and/or for their progressive adaptation to regular food.[6] In the case of infants feeding, the WHO recommends exclusive breastfeeding for the first 6 months of life.[7] However, the introduction of solid foods during infancy (from 4 months of age) is essential for the proper development and growth of a child. But studies on the composition and consequences of introducing solid foods are very scarce in comparison with breastfeeding.[8] Complementary foods such as ready-to-eat food products targeted for infants (from 4 months of age) and young children (1–3 years old) are widely used due to its commercial availability and diversity. In Poland, more than 60% of parents declare using commercially prepared foods. In the UK, two-thirds of mothers provide ready-to-eat baby foods as the first solid products in their baby's diet, and 45% of mothers are giving 8–10 months infants this kind of food at least once a day.[9] Ready-to-eat baby foods are available in many different variants of constituents' origin. Baby foods are mainly accessible as jars, pouches or boxes containing single-ingredient food, which is often puree, mixed dinners, soups or desserts. They are labelled according to a baby's stage development. Commercial baby foods are preserves ready-to-eat after or without heating. In general, baby food products stay fresh in the refrigerator after being opened, and should be used within 48 h.

In the opinion of the Committee of Nutrition (CoN) European Society for Pediatric Gastroenterology Hepatology and Nutrition (ESPGHAN), there is special attention needed to review and evaluate baby foods, also focusing on longer-term safety outcomes. The safety concern is predominantly acute for baby foods because of the infant susceptibility to health effects from the consumption of contaminated food. There is no doubt that the potential adverse reactions among infants and young children could appear after the consumption of food containing small amounts of toxic amines which can also enhance histamine release from basophils and provoke allergy-like responses.[10] In the case of baby food allergies or pseudoallergies, the symptoms most often have a very severe course and are dangerous to a baby's health or even life. This kind of strong mediator of allergic reactions in baby food could be HIS. Since BAs are thermostable compounds, they will be present in baby food even after heat-treating (besides polyamines). Nevertheless, on the basis of the present literature, it is not known whether delaying the introduction of potentially sensitizing foods or causing food intolerance containing high amounts of BAs, *e.g.* fish products, in any way permanently prevents adverse reactions.

In light of that, how could we explain the fact that foods for infants and young children have never been the object of biogenic amine profiling? The primary cause of this is due to the lack of proper analytical methods with appropriate sensitivity, and secondary, that the expected low level of these compounds in food is not hazardous for healthy adults.[11]

9.2 Analytical Methodology of BA Analysis in Baby Foods

For the analysis of BAs in food, several techniques have been developed, mainly chromatographic: high-performance liquid chromatography (HPLC), thin-layer chromatography (TLC), gas chromatography (GC), capillary electrophoresis (CE) and capillary electrochromatography (CEC) or even nuclear magnetic resonance (NMR) spectroscopy.[12–15] The most common technique used for analysis of BAs, due to its sensitivity and selectivity, is HPLC. HPLC methods are based on derivatization in pre- or post-column mode and ultraviolet (UV) or fluorescence (FL) detection,[16] using an evaporative light-scattering detector (ELSD),[17] or more often are performed with mass spectrometric detectors in mass spectrometry (MS), tandem mass spectrometry (MS/MS). Separation of BAs is generally performed on columns with alkyl chain C8 or C18, and may also be conducted in ultra-performance liquid chromatography (UPLC) systems,[18] or on hydrophilic interaction liquid chromatography (HILIC) columns.[19]

Most methodologies consists of two basic steps: extraction from food matrices, with optional clean-up with solid-phase extraction SPE or μSPE,[20] matrix solid-phase dispersion (MSPD),[11] dispersive liquid–liquid microextraction (DLLME)[21] and derivatization to appropriate compounds for the detection technique and type of food to be analysed. The derivatization step can be performed using *e.g.* *o*-phthaldialdehyde (OPA), dansyl chloride (Dns-Cl), benzoyl chloride, dabsyl chloride (Dbs-Cl), 4-chloro-3,5-dinitrobenzotrifluoride (CNBF), 1,2-naphthoquinone-4sulfonate (NQS), 6-aminoquinolyl-*N*-hydroxysuccinimidyl carbamate (AQC) or *N*-hydroxy-succinimide ester (DMQC-Osu).[22] Completing these steps results in low recoveries from food matrices. Therefore, it is indispensable to use different extraction solvents such as perchloric acid, trichloroacetic acid, hydrochloric acid[23] and perform liquid–liquid extraction (LLE) to remove the amino acids from the sample.

Acquired limits of detection (LODs) for the majority of the available analytical methods are generally in the order of μg g^{-1} (μg mL^{-1}) with a few exceptions.[24] In one study, synthesized hydrazine-based ligands were trapped in a silica sol-gel matrix and used for μSPE before HPLC-UV analysis of orange juice (6 BAs, LOD 7–31 ng mL^{-1}, linearity 1–50 ng mL^{-1}).[25] In other study using newly synthesized difluoroboradiaza-*s*-indacene (TMBB-Su), BODIPY-based fluorophore was used as a derivatization agent followed by HPLC-FLD in yoghurt, milk and lake water matrices (8 BAs, LOD 0.1–0.2 nM, linearity 0.001–0.2 μM).[26] LC-MS is one of the most sensitive and selective analytical techniques, notably for routine analysis of food, but it is still not extensively employed. Using tandem MS/MS detectors, the derivatization step could be eliminated, but analysis needs to be performed in HILIC mode, because reverse phase (RP) mode columns promote separation of polar amines and give much better resolution as derivatives. Estimated LODs for analysis of eight underivatized BAs in fish using SPE and LC-MC/MS range from 0.02 to 0.25 μg g^{-1}.[27]

For the determination of histamine content in food, enzyme-linked immunosorbent assays (ELISA) are also used for qualitative analyses or as well as semi-quantitative/quantitative tools.[28] Immunoassays for histamine are commercially available. ELISA tests are validated for application to specific food matrices (fish, wine, cheese) at an appropriate sensitivity level, allowing a rapid quantitative analysis of histamine in samples of up to 0.5 mg kg^{-1}. The most important gain of ELISA kits is a high throughput (from 20 minutes to 2 hours of analysis per many samples), without the need of using an expensive analytical equipment. However, positive results of ELISA tests need to be confirmed by the reference method (HPLC). In the EU, European Commission Regulation (EC) sets the analytical method for the determination of histamine in fishery products (HPLC with precolumn derivatization by dansyl chloride and UV detection at 254 nm).[29] This method can also be used for determination of other BAs. An alternative method of histamine analysis in fish food is also FL method with OPA derivatization (AOAC 977.13) and remains a reference method for the Food and Drug Administration (FDA).

Food products intended for infants and young children are not currently screened for BAs. Biogenic amine profiling in baby foods is undoubtedly a real challenge due to the complexity of food matrices, the presence of free amino acids and compounds that could interfere with the analytes or the low concentration of BAs. The bottlenecks of baby food analysis are the significant differences in the concentrations of individual amines in the amine profile. A methodological requirement is to achieve much lower (1–2 order of magnitude) limit of quantitation (LOQ) in comparison with current methods for quantification of the most frequently occurring BAs in foods (vegetables, fish, meat or fruit), without additional sample preconcentration. Determination of BAs at a low concentration level (ng g^{-1}) requires the use of the most sensitive and selective chromatographic detectors such as the fluorescence detector and the mass spectrometer. The most responsive possibility arises from the connection of separation technique – HPLC with MS detection to overcome all these issues. By using HPLC-APCI-MS with the precolumn derivatization with dansyl chloride, we identified the BA profile in baby food and quantified the putrescine, cadaverine, histamine, tyramine, spermidine and spermine content with an extremely low limit of detection.[30] There is also a possibility of using OPA/2-ME derivatization of the primary amines (putrescine, cadaverine, histamine, tyramine) with HPLC-FLD analysis, but with much higher LOQ in comparison with the HPLC-APCI-MS method or by commercial immunoassays for histamine with the modification of sample preparation (methods in publication) (Table 9.1).

Table 9.1 Comparison of analytical methods applied to BA analysis in baby foods.

	ELISA kits (Histamine ELISA kit, Histamine Food EIA[b], Ridascreen)	RP-HPLC-FLD with OPA/2-ME	RP-HPLC-APCI-MS with Dns-Cl
Using conditions	Needs confirmation by HPLC	—	Reference method
Analysed BAs	HIS	HIS, TYR, CAD, PUT	HIS, TYR, CAD, PUT, SPD, SPM
Analysis time (per 1 sample)	90 min	50 min	30 min
Sample preparation	Water extraction derivatization	Acid extraction derivatization (1 min) LLE extraction	Acid extraction derivatization (60 min) LLE extraction
Equipment	ELISA Microplate reader	High-performance liquid chromatograph with fluorometric detector	High-performance liquid chromatograph with single quadrupole mass detector
LOQ [ng g^{-1}]	30	26, 27, 26, 36	2, 2, 2, 25, 50, 50
The percent of <LOQ results	46% (1 BA)[a]	29% (4 BAs)[a]	2% (6 BAs)[a]
Method advantages	Speed Easy to use Low equipment costs High throughput	Medium sensitivity Good precision High accuracy Medium equipment costs	High sensitivity High precision High accuracy Possibility of BA profiling
Method disadvantages	Only for HIS Results need confirmation	Only for primary amines	High equipment costs

[a]Analysis was performed in 68 ready-to-eat baby foods (vegetable products with fish, vegetable products with meat, vegetable products and fruit products).
[b]EIA = enzyme immunoassay.

The presented methodologies can be successfully applied to determine the content of selected biogenic amines depending on the available equipment. A step forward in baby food analysis would be applying of a tandem MS/MS detector as a confirmatory tool for providing full information/confirmation of BA structures.

9.3 Biogenic Amine Occurrence in Ready-to-eat Baby Foods

BAs presence in baby food is not the consequence of food-associated spoiled microorganisms' activity, but the activity of endogenous decarboxylases and the presence of appropriate amino acids as the substrates for amines production. This indicates that the biogenic amine level in baby foods is an equilibrium between the levels of amines present in the individual product components. We have evaluated the BA profile of commercially available ready-to-eat baby foods in Poland.[31] Analyses were performed in 68 ready-to-eat products *e.g.* soups, dinners and fruit preserves intended for infants and young children produced by ten available on the market manufacturers, divided into fish-based, meat-based, vegetable-based and fruit-based preserves.

Ready-to-eat baby food amine profile consists of BAs commonly found in food: methylamine, ethylamine, putrescine, cadaverine, histamine, agmatine, tyramine, serotonin, spermidine and spermine. Apart from the abovementioned BAs, other psychoactive amines commonly found in food including dopamine and norepinephrine have been detected, but only in fruit-based baby foods. The presence of psychoactive amines such as serotonin, dopamine or norepinephrine in vegetable and/or fruit samples was not surprising. Generally, the amine levels in baby food products vary a lot, from 2 ng g^{-1} (CAD, HIS or TYR) to 53 416 ng g^{-1} (PUT) (Table 9.2).

The obtained amine quantities as mean concentrations in decreasing order were as follows (in ng g^{-1}): putrescine $>$ spermidine $>$ spermine $\gg$ tyramine $>$ histamine $>$ cadaverine. The amine profile of

Table 9.2 BA content in 68 samples of ready-to-eat baby food commercially available in Poland.[31]

Biogenic amine	No. of results $<$LOQ (HPLC-APCI-MS with Dns-Cl method)	Minimum and maximum content [ng g^{-1}]	Average content [ng g^{-1}]	SD[a]
Putrescine	—	704–53 416	5654	8690
Cadaverine	2	2–1263	90	209
Histamine	5	2–2375	186	345
Tyramine	3	2–1668	300	322
Spermidine	—	408–46 680	4907	7689
Spermine	—	34–5619	1278	1111

[a]SD = standard deviation.

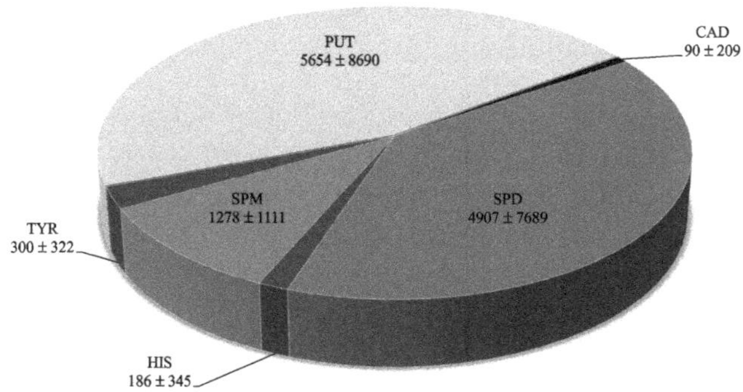

Figure 9.1 Average content of BAs in 68 ready-to-eat baby foods available in Poland [ng g^{-1}].[31]

ready-to-eat baby foods consists mainly of putrescine and spermidine. These amines were the most plentiful ones, representing 86% of the profile (Figure 9.1). Therefore, they could be found in all analysed samples.

Other amines, beginning from spermine, through to tyramine, histamine to cadaverine covered only 14% of the amine profile.

The maximum level of individual amines was obtained for histamine in a vegetable sample with spinach, for tyramine in a fruit product containing banana, for putrescine, cadaverine and spermine in a vegetable product with fish and green peas, and for spermidine in a vegetable sample with green peas.

The most aminogenic baby foods are those containing 43–70% of green peas and 48–50% of banana. The summary amine levels in analysed baby food products were found in the range of 1283–101 421 ng g^{-1}. In general, amine contents were independent of the product manufacturers, which may *e.g.* use lower quality raw materials or apply different production processes, rather by the use of the specific product ingredients.

Among all analysed products, only 21% of the samples contained less than 5000 ng g^{-1} of six BAs. Nearly half of the 68 products contained amines at the concentration range of 5000–10 000 ng g^{-1}. Next, 21% of the samples ranged from 10 000 to 20 000 ng g^{-1} of amines, but in some particular samples, the amine level exceeded over 80 000 ng g^{-1} up to 110 000 ng g^{-1}. This indicates the need of unification of amine level designated for baby foods.

The statistical analysis results indicated a number of important, significant correlations between biogenic amine contents *vs.* sample components and between the individual amine concentrations found in the samples. Regarding individual amines of all baby foods, positive correlation was found between putrescine and spermidine, spermidine *vs.* spermine, putrescine *vs.* cadaverine, cadaverine *vs.* spermine, putrescine *vs.* spermine, cadaverine *vs.* spermidine and tyramine *vs.* putrescine levels. Concerning the diamine (CAD + PUT) and polyamine (SPD + SPM) levels in all analysed samples, a strong positive correlation was found. Consequently, only four out of six BAs have been taken under consideration during further amine-related risk assessment.

Principal component analysis (PCA) was used to assess the similarities and differences between obtained BAs profiles. The PCA analysis exposed the negative correlation between histamine and polyamine contents (SPD, SPM). The most variable level of amine concentrations was assumed for tyramine. The most correlated results were estimated for di- and polyamines, PUT and CAD *vs.* SPD and SPM, respectively. The significantly different amine profiles were found for samples with green peas, with spinach, with rice and with banana. Whereas most similar amine profiles were observed for samples with tuna, containing banana and for vegetable samples.

Based on a one-way ANOVA and Tukey's *post hoc* tests of four food types organized by main ingredients, we indicated the presence of individual components responsible for the variability of BA levels within the food type. In general, the quantity and quality composition of particular food components in the samples was so different, even among selected food types, that it makes their direct comparison with the data in the literature almost impossible.

The 23 vegetable samples with fish contained 8–12% of fish species such as salmon, cod, pangasius, tuna and other sea fishes. The results suggest that the BA profiles were affected by the fish species present in the samples. Fish species such as tuna contain high levels of free histidine (histamine amino acid precursor). Therefore, higher contents of histamine are to be expected. In the case of tyramine, fresh fish contains little or no tyramine, but a large amount can be found in spoiled or fermented fish. The endogenous histamine and tyramine levels in tuna meals could potentially be toxic to sensitive babies. Indeed, it may be necessary to remove tuna as an ingredient of ready-to-eat baby foods, due to elevated level of the most toxic amines found in these meals. The higher mean levels of other amines including cadaverine, putrescine and spermidine were observed in samples containing cod, and spermine in samples with tuna (Figure 9.2).

Among 15 vegetable samples with meat classified by seven kinds of meat (beef, chicken, lamb, veal, rabbit, turkey and pork) results analysis revealed no significant difference between the type of meat and obtained BA profile (Figure 9.3).

However, elevated levels of histamine were obtained for five samples containing beef, and for two out of four samples with chicken. The highest tyramine content was estimated for the sample with lamb, putrescine and spermidine in two products with veal and spermine in meals based on chicken. In relation to the occurrence of histamine in baby food products containing beef on a concentration level of approximately 400 ng g^{-1}, we recommend delaying the introduction of beef beyond 12 months in babies.

Vegetable products (15 samples) were divided into four groups of meals according to the predominant ingredients including spinach, carrot, pumpkin and other vegetables.

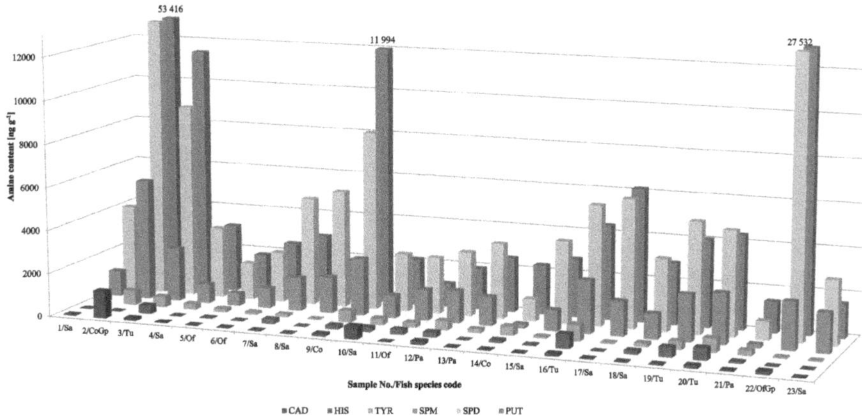

Figure 9.2 Biogenic amine profile estimated for fish-based baby foods ($n = 23$). Samples (No. 1–23) are coded by fish species: salmon (Sa), cod (Co), pangasius (Pa), tuna (Tu) and other sea fishes (Of).[31]

The obtained mean levels of histamine and tyramine in three samples containing over 25.5% of spinach were significantly different in comparison with the results found among other vegetable ingredients (Figure 9.4).

The analysed spinach baby foods have been produced by three different manufacturers. But elevated levels of tyramine (4–32 mg kg^{-1} in frozen spinach puree) and histamine (2–20 mg^{-1} kg in fresh spinach) were also obtained in commercial foods for adults. Certainly, the high level of histamine reaching 2375 ng g^{-1} could not be allowed in baby foods, especially among hyper-sensitive infants from 4 months of age. Therefore, we

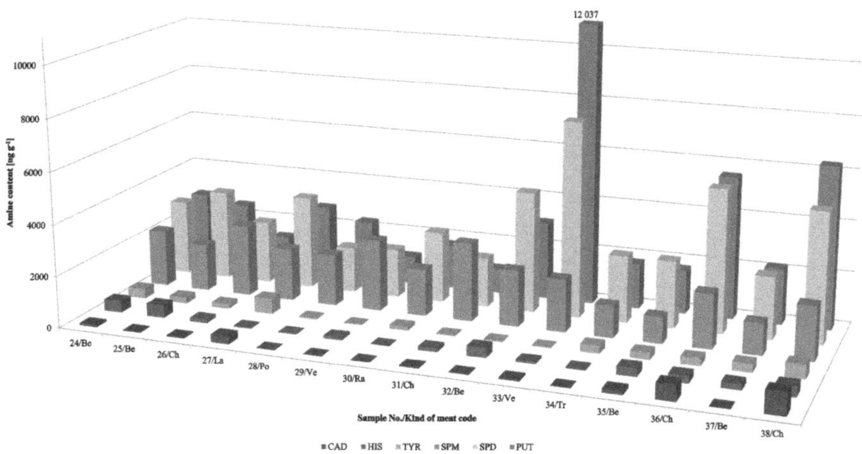

Figure 9.3 Biogenic amine profile estimated for meat-based baby foods ($n = 15$). Samples (No. 24–38) are coded by kind of meat: beef (Be), chicken (Ch), lamb (La), veal (Ve), rabbit (Ra), turkey (Tr), pork (Po).[31]

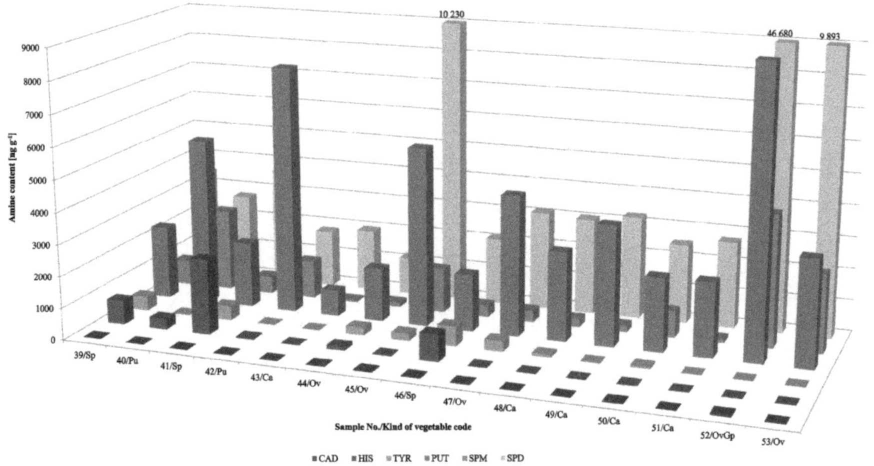

Figure 9.4 Biogenic amine profile estimated for vegetable-based baby foods ($n = 15$). Samples (No. 39–53) are coded by kind of vegetable: spinach (Sp), carrot (Ca), pumpkin (Pu), other vegetables (Ov), green peas (Gp).[31]

recommend removing or reducing (considering high nutritional value) the amount of spinach added to baby foods to *e.g.* 10% of the product. That amount of spinach has not resulted in increasing histamine and tyramine contents and could be found in one of the samples. The BA profile of vegetable baby foods consists mainly of di- and polyamines. The highest level of putrescine with spermidine was found in a sample containing 70% of green peas (intended for 4-month infants). According to the literature, higher concentrations of spermidine in green vegetables are to be expected. Considering the higher concentration levels of di- and polyamines obtained in products with green peas as the main ingredient, we recommend reducing the amount of peas to approximately 10–20% of the product, particularly in foods targeted for infants.

In 15 samples of fruit products intended for babies, only two main ingredients were specified: an apple and a banana. Among baby desserts, the statistically significant differences were found for histamine, tyramine, putrescine, spermidine and spermine between six banana-based samples and the rest of the desserts based on an apple (Figure 9.5).

The assayed amine profile of fruit products targeted for the youngest consumers consists mainly of putrescine, the amine found in two samples at the highest concentration exceeding 20 000 ng g^{-1}. The amine profile obtained for fruit-based baby products differs significantly among other baby food profiles, mainly according to the lowest polyamine levels. Fruit are particularly rich in putrescine. Essentially, among other amines found in commercial banana fruits, putrescine is the predominant one. A higher putrescine level was found in baby preserves with banana, as regards the aforementioned green peas, this level of putrescine is not directly toxic, but may enhance histamine toxicity, strengthened by a higher dose of toxic

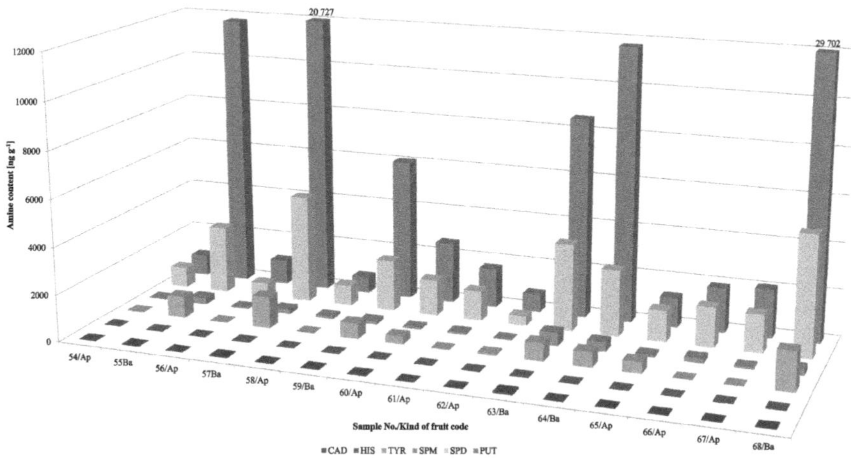

Figure 9.5 Biogenic amine profile estimated for fruit-based baby foods ($n = 15$). Samples (No. 54–68) are coded by kind of fruit: apple (Ap), banana (Ba).[31]

tyramine. Furthermore, analysed baby desserts are often eaten just after the main meal *e.g.* baby food containing meat or fish with an elevated level of dietary histamine. Taking into consideration the obtained BA profiles of fruit-based baby foods, we recommend the consumption of apple-based baby foods with a progressive introduction of products containing banana, especially among infants.

9.4 Toxicity and Risk Assessment of BA Occurrence in Ready-to-eat Baby Foods

On the basis of a limited number of studies regarding a potential no-observed-adverse-effect-level (NOAEL) of histamine for adults, a potential acute reference dose (ARfD = 50 mg per meal) for healthy adults was established. However, it does not include high-risk groups such as infants, young children, pregnant women or the elderly. Especially in the case of healthy young children, where there is an almost total lack of knowledge concerning a potential acute dose of histamine. Taking into account a potential acute reference dose (ARfD) of 5 mg per meal per 70 kg person histamine (10-fold lower than the ARfD suggested by EFSA) for sensitive adult individuals and making calculation of results according to the WHO child weight-for-age standards for boys and girls (WHO, 2006), we assessed the percentage of ARfD for histamine ingested through consumption of one baby meal per day. The highest %ARfD of histamine was estimated for baby food products with tuna (from 11% to 26%ARfD) and spinach (from 21% to 65% ARfD) (Table 9.3).

In case of sensitive young individuals, it can result in so-called pseudoallergies appearing after consumption of products containing tuna or

Table 9.3 Calculated potential doses of HIS according to the ARfD = 5 mg per meal per 70 kg person for sensitive adult individuals in baby foods. Samples (No. 1–23) are coded by fish species: salmon (Sa), cod (Co), pangasius (Pa), tuna (Tu) and other sea fishes (Of); by kind of meat (No. 24–38): beef (Be), chicken (Ch), lamb (La), veal (Ve), rabbit (Ra), turkey (Tr), pork (Po); by vegetable ingredients (No. 39–53): spinach (Sp), carrot (Ca), pumpkin (Pu), other vegetables (Ov), green peas (Gp); and by kind of fruit (No. 54–68): apple (Ap), banana (Ba).

Sample No/Code/Age group	%ARfD estimated for		Sample No/Code/Age group	%ARfD estimated for	
	Girls	Boys		Girls	Boys
1/Sa/1–3 years	4.2–2.7	3.9	35/Be/6 month	8.2	7.6
2/CoGp/9 month	0.1	0.1	36/Ch/6 month	10.5	9.7
3/Tu/6 month	14.5	13.4	37/Be/4 month	9.8	8.9
4/Sa/1 year	0.2	0.2	38/Ch/5 month	15.8	14.5
5/Of/8 month	1.7	1.6	39/Sp/4 month	31.5	28.8
6/Of/11 month	1.0	0.9	40/Pu/4 month	9.3	8.5
7/Sa/5 month	5.4	5.0	41/Sp/4 month	64.9	59.4
8/Sa/15 month	0.2	0.1	42/Pu/4 month	1.8	1.7
9/Co/8 month	8.0	7.3	43/Ca/4 month	0.3	0.2
10/Sa/5 month	8.2	7.6	44/Ov/5 month	4.5	4.1
11/Of/7 month	9.8	9.0	45/Ov/4 month	0.7	0.6
12/Pa/11 month	9.4	8.7	46/Sp/4 month	23.2	21.2
13/Pa/4 month	0.2	0.2	47/Ov/4 month	0.4	0.3
14/Co/8 month	4.4	4.1	48/Ca/4 month	0.3	0.2
15/Sa/5 month	1.7	1.5	49/Ca/4 month	0.2	0.2
16/Tu/1–3 years	26.0–16.6	24.1–16.2	50/Ca/4 month	0.4	0.4
17/Sa/1–3 years	0.6	0.6	51/Ca/4 month	0.1	0.1
18/Sa/6 month	6.3	5.9	52/OvGp/4 month	0.1	0.1
19/Tu/9 month	11.6	10.7	53/Ov/4 month	0.0	0.0
20/Tu/1–3 years	18.1	16.7	54/Ap/8 month	0.1	0.1
21/Pa/11 month	2.9	2.6	55/Ba/1 year	0.2	0.2
22/OfGp/12 month	0.6	0.5	56/Ap/4 month	0.1	0.1
23/Sa/5 month	0.3	0.3	57/Ba/4 month	0.5	0.5
24/Be/1–3 years	16.5–11.4	15.4–11.1	58/Ap/5 month	<0.1	<0.1
25/Be/1–3 years	19.3–12.4	17.9–12.0	59/Ba/4 month	0.5	0.4
26/Ch/1–3 years	5.6–3.6	5.2–3.5	60/Ap/5 month	0.1	0.1
27/La/5 month	1.6	1.5	61/Ap/4 month	<0.1	<0.1
28/Po/6 month	0.8	0.8	62/Ap/6 month	0.1	0.1
29/Ve/5 month	4.2	3.9	63/Ba/8 month	0.2	0.2
30/Ra/5 month	0.3	0.3	64/Ba/5 month	0.3	0.3
31/Ch/7 month	5.7	5.2	65/Ap/6 month	<0.1	<0.1
32/Be/4 month	16.3	14.9	66/Ap/4 month	<0.1	<0.1
33/Ve/5 month	3.3	3.1	67/Ap/4 month	<0.1	<0.1
34/Tr/6 month	1.6	1.5	68/Ba/4 month	0.5	0.5

spinach, especially among infants. The estimated potential doses are related with the consumption of one meal only. The WHO recommendation for infants between 6 and 8 months of age is to introduce complementary foods 2–3 times a day, increasing to 3–4 times daily between 9 and 24 months of age (WHO, 2016). Consequently, the estimated %ARfD of histamine will certainly increase by consumption of the subsequent portion of food. However, even a significant dose of BAs, including toxic histamine, does not necessarily result in adverse reactions. This depends on individual characteristics, health status and many other factors, making it almost impossible to determine the tolerable threshold of histamine concentrations acceptable in children's products. Nonetheless, the obtained % ARfD values for histamine in products with spinach and tuna indicate the need to pay special attention to the proper selection of food ingredients for the youngest consumers in order to protect them from the adverse effects of toxic amine consumption. An additional concern is associated with people suffering from histamine intolerance by diamine oxidase (DAO) deficiency with health problems or gastrointestinal disorders resulting in an escalation of adverse effects.[32] In the case of children, particularly infants at the beginning of the development of an efficient digestion, this concern should not be ignored.

Although there is presently insufficient data related to establishing the NOAEL for tyramine in adults, this amine, considered less toxic than histamine, may be a causative agent of certain food-induced migraines, especially among individuals with lower genetic or physiological activity of monoamine oxidase (MAO) enzymes, and it also acts as a histamine potentiator. Therefore, the presence of tyramine should also be considered as an indicator of the potential toxicity of consumable baby food products, especially desserts with banana. Particularly taking into account results of the *in vitro* analysis, which suggests that tyramine has a stronger and more rapid cytotoxic effect than histamine.[33]

The biogenic amine index (BAI) is often determined for products in which the effects of spoilage (fish, meat) are noticeable as a result of the increasing microbial activity over time, which ultimately results in an increase in the content of BAs. During the aging of the product, we observed the increase of histamine, tyramine, cadaverine and putrescine levels, and slightly decrease polyamines content (six amines in BAI calculations). Because the content of polyamines remains constant, usually only four amines or two (PUT, CAD) are taken into BAI assessment.

The assessment of potential toxicity of ready-to-eat baby foods based on the two most toxic amines (HIS, TYR) is not sufficient. The presence of diamines in baby products may increase the release of histamine into the bloodstream. The obtained maximum total concentration of 55 351 ng g^{-1} for four amines (PUT, CAD, HIS, TYR) would have exposed a 9-month old infant to 7.2 mg of BAs in 130 g of consumable portion, or 157% of 40 mg per meal per person,[34] the value of BAs considered potentially toxic for healthy adults. This may support the application of BAI for amine-related risk assessment of baby foods, according to the equation (eqn (9.1)).

$$\mathrm{BAI}_{\text{baby food}} = \frac{\mathrm{ng\,g^{-1}\,HIS} + \mathrm{ng\,g^{-1}\,PUT} + \mathrm{ng\,g^{-1}\,CAD} + \mathrm{ng\,g^{-1}\,TYR}}{1000\ \mathrm{ng\,g^{-1}}} \quad (9.1)$$

Of all 68 analysed samples in five samples (containing green peas or banana), the BAI reach the range of 20–55. The composition of these products should be immediately re-evaluated to reduce the dietary amine intake. The most interesting is that among all analysed samples, the banana-based baby foods seem to be the most potentially toxic ones (elevated TYR content accompanied by the high amount of PUT). In general, healthy adult tolerance levels are irrelevant to babies. The maximum level of 4 in the most toxic BAs in baby foods should be established at 20 mg kg^{-1} (BAI = 20) in view of the higher vulnerability of babies. Our recommendations of BAI indexes are as follows: BAI < 10 for baby food products safe for consumption by the youngest consumers; 10 < BAI < 20 for acceptable level of BAs in baby foods; and BAI > 20 for products being potentially harmful for infants and young children's health due to adverse effects of exposure to BAs.

The proposed method of BA indexing by the four amines is better adapted to the prediction of the potential toxicity of baby foods than two in the BA index and quickly allows the indication of products with an increased content of BAs, resulting from the use of aminogenic ingredients. However, its use does not exclude the need to eliminate products directly threatening the health of infants and young children due to the high content of toxic histamine. Nevertheless, estimation of the potential toxicity of baby products is extremely multifaceted. This is mainly due to the fact that the content of amines naturally occurring in these products is characterized by a huge variation in both qualitative and quantitative components included in individual products. The obtained BAI results certainly indicate the need to pay special attention to proper components selection in foodstuffs intended for the youngest consumers in order to protect them from the adverse effects of toxic amines intake.

9.5 Future Research Directions in BA Analysis in Baby Foods

The exposure to BAs through the consumption of complementary foods appears to be higher than anticipated. Indeed, further research is required on amine-related risk assessment of newly tested baby food, in particular, histamine in complementary foods including dairy products, especially yoghurt products and also the preparations for the further feeding of infants (initial and follow-up formula milk). A dietary exposure assessment is also needed to investigate the tolerable threshold of BAs in correlation with consumption data of food products intended for infants and young children. In light of estimated results, the re-evaluation of baby food composition is needed, taking into consideration the low amine level to be obtained. It is also necessary to conduct further research on analytical methodologies development for BA analysis in baby foods.

In the light of presented research, our amine-related nutritional recommendations regarding ready-to-eat baby foods are as follows:

1. Tuna should be removed from the list of ingredients allowed in baby foods with regard to the elevated content of histamine and tyramine.
2. Spinach should be removed from baby foods or reduced to the amount of 10% with regard to the very high level of toxic histamine.
3. Green peas content should be reduced to 10–20% of the product with regard to the high concentration of di- and polyamines.
4. Beef usage in baby food should be reduced to babies beyond 12 months with regard to the elevated level of histamine.
5. Banana introduction to a baby's diet needs special attention, especially among infants with regard to the high concentration of toxic tyramine and putrescine.
6. The safest from the point of view of the number of amines provided in baby fruit meals is the consumption of desserts composed of apples and carrots, with the addition of *e.g.* peach or strawberry with a gradual introduction of banana-containing products into the diet.

References

1. European Commission, Commission regulation 1441/2007, 2007.
2. F. Wei, *Meat Sci.*, 2009, **81**, 451.
3. T.-C. Wu and P.-H. Chen, *Pediatr. Neonatol.*, 2009, **50**, 135.
4. M. Domellöf, *J. Pediatr. Gastroenterol. Nutr.*, 2014, **58**, 119.
5. L. T. Weaver, *Clin. Med.*, 2001, **1**, 38.
6. European Commission, European regulation (EU) 609/2013, 2013.
7. N. F. Butte, M. G. Lopez-Alarcon and C. Garza, WHO, 2002. Available at http://www.who.int/child_adolescent_health/documents/9241562110/en/
8. T. C. Wu and P. H. Chen, *Pediatr. Neonatal.*, 2009, **50**, 135.
9. A. L. Garcia, S. Raza, A. Parrett and C. M. Wright, *Arch. Dis. Child.*, 2013, **98**, 793.
10. J. Karovičová and Z. Kohajdová, *Chem. Pap.*, 2005, **59**, 70.
11. A. Czajkowska-Mysłek, PhD Thesis, Lodz University of Technology, 2016.
12. S. Oguri, Y. Maeda and A. Mizusawa, *J. Chromatogr. A*, 2004, **1044**, 271.
13. S. Oguri and Y. Yoneya, *J. Chromatogr. B: Anal. Technol. Biomed. Life Sci.*, 2002, **781**, 165.
14. A. Önal, S. E. K. Tekkeli and C. Önal, *Food Chem.*, 2013, **138**, 509.
15. T. Chatzimitakos, V. Exarchou, S. A. Ordoudi, Y. Fiamegos and C. Stalikas, *Food Chem.*, 2016, **202**, 445.

16. S. Tahmouzi, R. Khaksar and M. Ghasemlou, *Food Chem.*, 2011, **126**, 756.
17. D. Restuccia, U. G. Spizzirri, F. Puoci, G. Cirillo, M. Curcio, O. I. Parisi, F. Iemma and N. Picci, *Talanta*, 2011, **85**, 363.
18. E. Dadáková, M. Křížek and T. Pelikánová, *Food Chem.*, 2009, **116**, 365.
19. R. L. Self, W. H. Wu and H. S. Marks, *J. Agric. Food Chem.*, 2011, **59**, 5906.
20. G. Sagratini, M. Fernández-Franzón, F. De Berardinis, G. Font, S. Vittori and J. Mañes, *Food Chem.*, 2012, **132**, 537.
21. K.-J. Huang, C.-X. Jin, S.-L. Song, C.-Y. Wie, J.-F. Zhang, Y.-M. Liu and J. Li, *J. Chromatogr. B: Anal. Technol. Biomed. Life Sci.*, 2011, **879**, 579.
22. A. Önal, S. E. K. Tekkeli and C. Önal, *Food Chem.*, 2013, **138**, 509.
23. A. Bartócz, *Trends Food Sci. Technol.*, 1995, **6**, 341.
24. F. B. Erim, *Trends Anal. Chem.*, 2013, **52**, 239.
25. C. Basheer, W. Wong, A. Makahleh, A. Tameem, A. Salhin, B. Saad and H. K. Lee, *J. Chromatogr. A*, 2011, **1218**, 4332.
26. P. F. Gao, X. F. Guo, H. Wang and H. S. Zhang, *J. Sep. Sci.*, 2011, **34**, 1393.
27. G. Sagratini, M. Fernández-Franzón, F. De Berardinis, G. Font, S. Vittori and J. Mañes, *Food Chem.*, 2012, **132**, 537.
28. M. Muscarella, L. S. Magro, M. Campaniello, A. Armentano and P. Stacchini, *Food Control*, 2013, **31**, 211.
29. European Commission, Commission Regulation 1441/2007, 2007.
30. A. Czajkowska-Mysłek and J. Leszczyńska, *Chromatographia*, 2018, **81**, 901.
31. A. Czajkowska-Mysłek and J. Leszczyńska, *Food Chem. Toxicol.*, 2017, **105**, 82.
32. M. L. Latorre-Moratalla, O. Comas-Basté, S. Bover-Cid and M. C. Vidal-Carou, *Food Chem. Toxicol.*, 2017, **99**, 78.
33. D. M. Linares, B. del Rio, B. Redruello, V. Ladero, M. C. Martin, M. Fernandez, P. Ruas-Madiedo and M. A. Alvarez, *Food Chem.*, 2016, **197**, 658.
34. A. R. Shalaby, *Food Res. Int.*, 1996, **29**, 675.

CHAPTER 10

Occurrence of Biogenic Amines in Fermented and Non-fermented Soybean Products

N. BATTISTELLI, F. TITTARELLI, G. PERPETUINI, G. SUZZI* AND R. TOFALO

Faculty of Bioscience and Technology for Food, Agriculture and Environment, University of Teramo, via Balzarini 1, 64100 Teramo, TE, Italy
*Email: gsuzzi@unite.it

Soybean belongs to the family of Fabaceae (Leguminosae), in the genus *Glycine*; its scientific name is *Glycine max* (L.) *Merr*. It is an ancient legume and is one of the main protein sources in Asian countries, because it contains approximately 40% protein and 20% oil on an average dry matter base. Production methods of these products have been recorded since ancient times in traditional Chinese and Korean manuscripts, indicating that soy has been consumed for millennia.[1] Since ancient times, in Eastern countries, it has been used to produce a variety of traditional non-fermented and fermented products (Table 10.1).

Most of these products can still be produced *via* traditional methods. Cheonggukjang, for example, traditionally requires 2 or 3 days of fermentation with rice straw added to steamed soybeans. Microbial activity during fermentation gives this product its peculiar flavour and aroma. Doenjang, a typical Korean soybean paste, is prepared by fermentation with *Bacillus subtilis* naturally present in the ingredients. Meju is the main ingredient of

Food Chemistry, Function and Analysis No. 20
Biogenic Amines in Food: Analysis, Occurrence and Toxicity
Edited by Bahruddin Saad and Rosanna Tofalo

Published by the Royal Society of Chemistry, www.rsc.org

Table 10.1 Principal non-fermented and fermented soybean products.

	Product	Description
Non-fermented	Tofu	Non-fermented soybean curd, soft cheese-like consistency made from soybean milk.
	Soy milk	High protein, milky liquid produced from pressing cooked soybeans.
	Soy protein isolates	Highly refined or purified form of soy protein.
	Edamame	Immature soybeans consumed whole.
	Soy flour	Flour derived from ground soybeans, source of high-quality protein.
	Soy chips	Similar in appearance and texture to potato chips but made with soy flour. They are consumed as a high protein snack.
Fermented	Cheonggukjang (Korea)	Traditional soybean paste obtained after natural fermentation *via* rice straws rich in *Bacillus* spp. Modern Cheonggukjang products are manufactured by inoculating soybeans with specific microorganisms such as *B. subtilis*, *B. amyloliquefaciens*, and *B. licheniformis*.
	Doenjang (Korea)	Soybean paste made entirely of fermented soybean and brine. It is rich in flavonoids, vitamins, and minerals. It is traditionally prepared by mixing and fermenting mouldy cooked soybeans with brine in a porcelain pot container.
	Doubanjiang (China)	A spicy, salty paste made from fermented broad beans, soybeans, salt, rice, and various spices.
	Miso (Japan)	Thick paste produced by fermenting soybeans with salt and *A. oryzae*. It is also used to prepare miso soup or as a seasoning for meats and vegetables.
	Natto (Japan)	It is made by soaking whole soybeans, then steaming or boiling them. It is then allowed to ferment over time with *B. subtilis*.
	Soy sauce (China)	Condiment produced from a fermented paste of boiled soybeans, roasted grain, brine, and *A. oryzae* or *A. sojae* moulds. Soy sauce is obtained from pressed fermented paste. The remaining solid product is often used as animal feed.
	Tamari (Japan)	It is the "original" Japanese soy sauce, similar to Chinese soy sauce. The difference is that tamari is made with no, or very little, wheat. It has a darker colour and richer flavour than the common Chinese soy sauce
	Tauchu (China)	Paste made from fermented yellow soybeans and aged.
	Tempeh (Indonesia)	It is made by fermentation with inoculation of *Rhizopus* spp. Traditional tempeh is often produced in Indonesia using *Hibiscus tiliaceus* leaves in which soybeans are pressed and stored. At the end of fermentation its shape is similar to a burger.
	Sufu (China)	Fermented tofu with a strong odor. It is similar to a cheese. Microorganisms used for fermentation belong to *Bacillus* spp. or *Micrococcus* spp.

Doenjang and it is prepared by cooking and processing soybeans into a brick-like shape.[2] Meju is then fermented for 2–3 months and aged. Modern production of Doenjang involves the use of starter cultures, such as *Aspergillus oryzae*. In China, sufu is well known, and it is produced starting from tofu. Moulds and bacteria are inoculated onto tofu which is then salted and ripened. There are different kind of sufu, divided on the basis of the dressing used for ripening: red sufu (red rice koji), grey sufu (without red rice koji) and alcoholic sufu.[3]

Soy is accepted as a healthy ingredient in the United States and Europe. Production of soybeans increased greatly from the end of the 20th century. Among the countries involved in soybean production, the United States are the number one producer with 108 million tonnes in 2014, followed by Brazil (87), Argentina (53), China (12), and India (10).[4] In China, the increase in meat consumption led to the increase of the import of soybeans as feed, and thus the production of soybeans in Brazil and Argentina increased.[5]

Fermented soybean pastes are listed in the Codex Alimentarius Commission (CODEX), a collection of standards and guidelines that ensures food safety for trading; these products are globally recognized for their functionality and distinctive flavours. Nowadays, the consumption of soy-based foods is not only an Asian habit because of its potential health benefits, particularly in preventing different types of cancers, osteoporosis, and cardiovascular diseases.[6] They promote probiotic properties and mediate glucose intake.[7] Consumption of fermented soybean products can also help intestinal regulation and digestion.[8] In the last 20 years, soybean products have been progressively introduced in the European market: at first, they were only present in specific dietetic shops, but now they are available in most food stores and supermarkets.

10.1 Nutritional Value of Soybeans

Soybeans are sometimes referred to as "the meat of the field" because of their high content of excellent quality proteins (65% of recommended daily values) (Table 10.2). They contain ample amounts of dietary fibre, vitamins, minerals, and plant sterols. Moreover, 78% of total fats content is composed of unsaturated fats, with the absence of cholesterol and the presence of lecithin. Soybeans also contains dietary fibres, which help to protect colon mucosa.[9] Moreover, soybeans are among the highest legume sources of folates and, together with vitamin B12, they are one of the essential components for DNA synthesis and cell division.[10] These legumes contain a high amount of antioxidant in the form of isoflavones which have been found to reduce post-menopausal cancers and osteoporosis.[11] Soybeans are very good sources of many B-complex vitamins (riboflavin, thiamine, pyridoxine, pantothenic acid, and niacin). Most of these vitamins work as co-enzymes in carbohydrate, protein, and fat metabolism. Furthermore, they are one of the finest sources of minerals like iron, copper, manganese,

Table 10.2 Nutritional value per 100 g of raw, mature soybean seeds (USDA National Nutrient Database).

Principle	Nutrient value	RDA[a] %
Energy	446 kcal	22.3%
Carbohydrates	30.16 g	23%
Protein	36.49 g	65%
Total fat	19.94 g	99%
Cholesterol	0 mg	0%
Dietary fibre	9.3 g	24%
Vitamins		
Folates	375 µg	94%
Niacin	1.623 mg	10%
Pantothenic acid	0.793 mg	16%
Pyridoxine	0.377 mg	29%
Riboflavin	0.870 mg	67%
Thiamine	0.874 mg	73%
Vitamin A	22 IU	<1%
Vitamin C	6 mg	10%
Vitamin E	0.85 mg	6%
Vitamin K	47 µg	39%
Electrolytes		
Sodium	2 mg	<1%
Potassium	1797 mg	38%
Minerals		
Calcium	277 mg	28%
Copper	1.658 µg	184%
Iron	15.7 mg	196%
Magnesium	280 mg	70%
Manganese	2.517 mg	109%
Phosphorus	704 mg	100%
Zinc	4.89 mg	44%

[a]RDA: recommended daily allowance.

phosphorus, zinc, and magnesium. Manganese is used by the body as a co-factor for the potent antioxidant enzyme, superoxide dismutase.[12]

10.2 Biogenic Amines in Soybean Products

Chao *et al.*[13] and Sun *et al.*[14] explored the microbiota of soy and reported the dominance of lactic acid bacteria (LAB) and species belonging to the genus *Bacillus*.

Most soybean products are commercialized only after fermentation, which is one of the oldest technologies in food preservation and processing. Fermentation of protein-rich foods (fish, meat, soybean) could lead to the formation of secondary and undesirable compounds like biogenic amines (BAs) (histamine, tyramine, β-phenylethylamine, tryptamine, putrescine, and cadaverine) that may cause intoxication in humans. These compounds can be produced by bacteria mostly through enzymatic amino acid decarboxylation or reductive amination of aldehydes and ketones.[8] In 2011, the European Food Safety Authority (EFSA) published a scientific opinion on the risk of BAs formation in fermented food, underlying the difficulty of making a quantitative risk assessment because available information is still lacking.[15] For this reason, there are currently no established standards or regulation for tofu or fermented soybean products to limit BA levels. Even though there is an absence of regulations, previous studies have proposed some toxic limits for BAs: 1000 mg kg^{-1} for total BAs content;[16] 30 mg kg^{-1} for β-phenylethylamine; 100–800 mg kg^{-1} for tyramine; and 100 mg kg^{-1} for histamine.[17]

Aspergillus oryzae or *Aspergillus sojae* and *Saccharomyces cerevisiae* are often used to start fermentation of soy products. Other microorganisms like *Pediococcus halophilus*, *Zygosaccharomyces rouxii*, *Candida* spp., and *Enterococcus faecalis* participate to the other steps of fermentation.[18,19] Adding selected starter cultures is one of the main tools to prevent formation of high levels of BAs in fermented foods. Several studies have shown that BAs in fermented soybean products are most likely formed by the lactic microflora that remain active during fermentation.[20,21]

BAs can be found also in non-fermented soybean products, like tofu, soybean sprouts, and soy milk. In these products, the presence of BAs above a certain level is considered as indicative of undesired microbial activity, underlying a potential microbial spoilage (Table 10.3).

BAs formation is influenced by the high amounts of proteins and free amino acids in soybean used as raw material. Moreover, starter cultures used during fermentation, proteolytic activity of bacteria, and the presence of microorganisms with decarboxylase activity can help to explain the origin of BAs in these food products. Other variables such as pH, moisture content, and NaCl can have an important effect on the production of BAs in soybean food and other food products.

BAs in different varieties of soybean foods have been analysed by several authors as summarized in Table 10.3.

Studies regarding the formation of BAs in soybean products have been conducted mostly by Asian researchers, in fact, information about their content in soybean products in European countries is scarce.

Jeon *et al.*[32] analysing Cheonggukjang found high levels of BAs (tyramine, histamine, and β-phenylethylamine) that exceeded recommended toxicity. These data agree with previous analysis in the same food product, which showed high concentrations of histamine and tyramine.[31] In 2015, Toro-Funes *et al.*[38] analysed BAs in fermented (miso, tamari, soy sauce, natto, tempeh, soybean paste, sufu) and non-fermented (fresh tofu, hard tofu, soy milk, soybean sprouts) commercial soybean products from the Spanish market. In non-fermented products, the bacterial-related BAs (tyramine, histamine, tryptamine, and β-phenylethylamine) were not detected in any sample. Putrescine and cadaverine were found in small amounts and this could be related to naturally occurring amines. In fermented soybean

Table 10.3 BAs content (mg kg^{-1}) in non-fermented and fermented soybean products (data from ref. 1 and 38).[a]

		Biogenic amines								
	Product[b]	TY	TR	HI	PU	PHE	CA	SPM	SPD	No. of samples
Non-fermented	Tofu[22]	ND	1.06–9.46	3.52–5.76	ND–4.73	ND–5.36	ND–3.35	4.05–19.15	14.97–35.30	2
	Soybean sprout[23]	NT	NT	NT	3.8	NT	134.6	3.3	18.2	2
	Soy milk[24]	1.7	20.2	17.5	ND	NT	NT	NT	1.3	1
Fermented	Soy sauce[24–28]	ND–673	NT	ND–592	NT	NT	ND–550	ND–109	ND–486	40
		ND–1.0	ND	2.0–9.6	ND–1.0	NT	NT	NT	ND	2
		3.0–659.9	NT	ND–395.0	ND–58.1	ND–121.9	ND–68.6	NT	NT	42
		0.867	0.094	1.282	11.953	0.477	0.082	0.391	ND	1
		0.139–11.112	ND–0.408	0.249–2.283	0.908–17.832	0.195–2.499	ND–0.429	0.132–0.849	NT	8
		NT	NT	NT	29.79–136.37	NT	NT	0.2–3.84	6.25–16.70	6
	Cheonggukjang[29–32]	0.7–483.1	6.7–236.4	1.3–54.3	4.7–121.3	ND–40.8	2.1–20.2	7.1–14.7	39.6–59.2	7
		4.2–483.1	6.7–76.9	0.2–70.3	4.7–148.5	ND–32.6	4.6–70.6	7.2–23.1	1 33.4–67.7	10
		ND–1913.51	NT	ND–755.40	NT	NT	NT	NT	NT	102
		ND–251.66	2.64–231.98	ND–408.74	0.12–476.30	0.12–562.44	ND–268.16	ND–20.89	ND–79.05	60
	Doenjang[29,33–37]	3.4–1430.7	6.1–234.1	1.5–952.0	9.9–1453.7	ND–529.2	0.3–65.4	ND–10.2	4.2–23.4	14
		12.5–967.6	ND–449.8	1.4–329.2	28.8–1076.6	ND–544.0	2.7–144.1	ND–9.8	ND–30.3	10
		ND–6616.1	ND–2808.1	ND–2794.8	ND–4292.3	ND–8704.6	ND–3235.5	ND–9729.5	ND–8804.0	23
		46.4–1190.7	13.5–45.9	71.1–382.4	46.7–168.2	3.3–65.0	ND–12.9	NT	ND–24.7	7
		3.1–126.9	1.0–11.1	3.3–83.7	4.6–55.2	0.1–20.2	0.9–18.4	NT	0.1–2.8	8
		ND–25.11	ND–192.29	ND–12.81	ND–180.74	ND–71.67	ND	ND–7.21	ND–11.30	14

Tempeh[24,38]	4.3	15.6	4.1	116.9	NT	NT	NT	11.6	1
	ND–10.68	ND	ND	17.53–31.06	ND	1.52–7.21	6.06–21.89	97.3–105.5	3
Natto[39,40]	ND–45.0	ND–301.0	ND–457.0	ND–27.0	ND	ND–42.0	ND–71.0	ND–124.0	39
	ND	ND	ND	5.81–9.49	ND	2.79–5.43	9.17–11.15	56.91–75.21	3
	ND–300.2	ND–45.8	ND–34.4	ND–43.1	ND–51.1	ND–36.8	18.8–80.1	246.5–478.1	21
Miso[28,29,39,41,42]	ND–49.0	ND–762.0	ND–221.0	ND–12.0	ND	ND–201.0	ND–216.0	ND–216	40
	2.0–95.3	21.6–23.7	0.8–1.1	16.4–23.2	0.7–8.1	2.8–3.2	1.3–3.1	9.5–21.9	5
	NT	NT	NT	19.75–34.29	NT	NT	ND–6.27	0.44–5.96	6
	ND–41.9	ND–1.3	ND	ND–12.9	ND–42.0	ND–8.7	ND	ND–35.7	8
	ND–66.66	ND–9.71	ND–24.42	2.69–14.09	2.38–11.76	ND–1.31	ND–2.85	ND–28.31	22
	ND	ND	1.49–4.62	2.73–17.81	ND	ND–3.83	2.54–3.54	7.54–9.91	3
Sufu[38,41,43–48]	NT	NT	NT	1.1–1776.3	NT	ND–883.7	ND–17.9	ND–154.8	11
	ND–48.0	26.0–57.0	ND–40.2	21.2–47.3	ND–4.2	NT	ND–22.9	ND–27.1	10
	ND–446.6	ND–104.1	ND–196.9	0.5–316.9	ND–33.6	0.6–85.8	ND–6.9	ND–4.0	38
	ND–65	ND–81	ND–158	ND–45	ND	ND–371	ND–82	ND	22
	ND–549.34	ND–463.45	ND–415.38	11.38–1147.32	ND–341.03	ND–308.44	ND	ND	33
	910.6–1730	ND	127.6–730.0	9.33–18.76	1.69–4.27	33.39–36.96	0.88–2.21	ND–1.36	3
	13.2–197.1	1.7–31.0	51.9–505.8	32.5–256.7	ND	87.5–310.6	8.4–24.9	ND	8
	11.25–1537.38	ND–469.87	ND–1186.93	8.53–1244.86	ND–110.14	6.84–83.42	ND–7.70	ND–32.87	16

[a]ND: not detected; NT: not tested.

[b]TY: tyramine; TR: tryptamine; HI: histamine; PU: putrescine; PHE: β-phenylethylamine; CA: cadaverine; SPM: spermine; SPD: spermidine.

products, BAs were detected in many samples with a high variability, although only high contents (>50 mg kg^{-1}) were found in sufu, tamari and soybean paste.

In 2013, Byun *et al.*[22] analysed BAs content in doubanjiang and tofu. Doubanjiang is one of the most popular condiments used in Chinese cuisine in the United States and Europe, as well as in Asia. Doubanjiang contains relatively larger amounts of BAs; for example, one sample had β-phenylethylamine higher than the toxic dose of this amine (30 mg kg^{-1}). On the other hand, tofu contains a relatively high level of spermidine which is related to the content of spermidine in the soybean used as a raw material. Byun *et al.*[22] also examined the relationship between BAs content and microbiota. Bacterial strains in tofu samples were isolated and identified (71% of the strains were *B. subtilis*), and the abilities of the strains to produce BAs were further determined and compared with the respective amine content in the corresponding sample, but no apparent relationship was found. The reason is that the strains found in tofu were similarly capable of producing BAs, pointing out that most bacterial contaminants have abilities to form BAs in tofu. Moreover, tofu is obtained from a non-fermenting process and has a very short shelf-life (1–2 days), so, BAs content in tofu is primarily attributed to the raw material, soy, and bacterial contaminants rather than unusual microorganisms.

Due to their toxicological implications, β-phenylethylamine in doubanjiang and spermidine in tofu need to be monitored in both raw material and from foodstuff as well. Production and the global market of tofu are summarized in Box 10.1.

BOX 10.1 Production and Global Market of Tofu

Tofu is a highly versatile and nutritious food that is made from soybean curds. It is important to Asian populations, not only because it has a cheese-like texture, but also because it serves as an important source of calcium and other nutrients. Hence, tofu is considered a healthy food and it is largely consumed worldwide.

To produce tofu, few ingredients are needed: soybeans milk, water, and coagulating agents. To obtain the milk it is important to soak the soybeans in water for 12–14 h so that the beans absorb water and double their size. After soaking, beans are mashed and boiled to inactivate enzymes that hinder digestion. Soy milk is extracted and separated from the pulp, which can be recycled to feed live-stock. Whey is extracted, and coagulating agents are added. These agents could be calcium sulphate, magnesium chloride, or nigari. Nigari is a traditional Japanese ingredient originating from sea water and, therefore, is rich in minerals. The coagulants alter the pH and curds the milk. The soy curds are then processed into tofu in the desired form, primarily in custard-like blocks. Tofu can be produced in a variety of textures, from a dense cheese-like texture to a softer or liquid form.

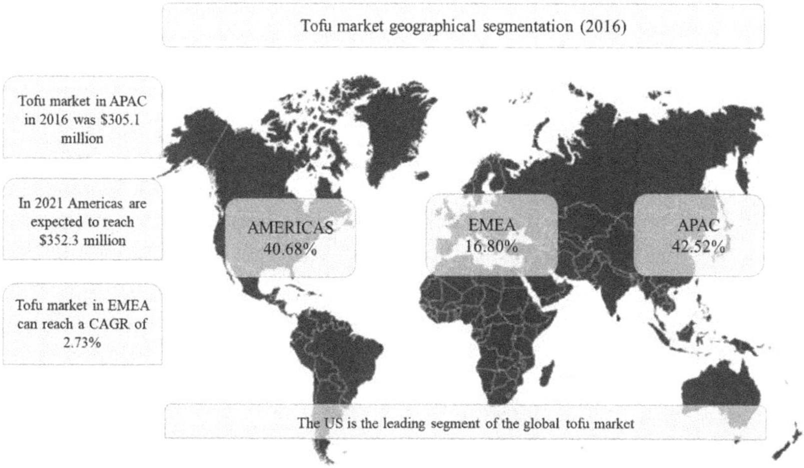

Figure 10.1 Segmentation of tofu global market (data from ref. 49).

Technavio[49] has published a report on the global tofu market from 2017–2021 (Figure 10.1). A growing vegan population base and an increasing number of people with lactose intolerance are driving the growth of the tofu market. The global tofu market size is expected to register a compounded average growth rate (CAGR) of 4.05% during the period 2017–2021. The demand is driven by the United States, which is a leader in the global tofu market.

Among other soy products, natto is a Japanese fermented soybean product made from boiled soybeans fermented for a short period with *B. subtilis* var. *natto*, that gives a unique flavour and viscous texture. In 2012, natto products distributed in Korea were analysed for BAs content.[40] Results showed that 43% of samples had a content of BAs (β-phenylethylamine or tyramine) greater than the toxic dose (30 mg kg^{-1} and 100 mg kg^{-1}, respectively) but no product showed BAs level higher than 1000 mg kg^{-1}, which is the harmful level for human health. High levels of β-phenylethylamine or tyramine were produced by *B. subtilis*, which was the most predominant species in these products.

As regards soy sauce, which is one of the most consumed soybean-based products, the manufacturing process and the BAs content is described in detail in Box 10.2.

Box 10.2 BAs in Soy Sauce

Soy sauce is made from soybean and wheat flour and is a traditional seasoning in China and many other Asian countries. QY Research reported that the soy sauce industry was 23.6 billion USD (United States dollars) in 2017 and is projected to reach 52.8 billion USD by 2023, at a CAGR of 14.32% between 2017 and 2023.[50] Soy sauce is mainly consumed in Asia-Pacific regions, but North American and European markets are growing fast, making soy sauce a common seasoning in those countries. The reason why this market is growing can be found in the growth of consumer's health concerns and interest towards different taste preferences and changing eating lifestyles. Moreover, preparing different multi cuisine foods at home has increased the usage of soy sauces globally. Soy sauce has been used, for more than 2500 years, to improve the flavour and taste of foods, imparting a salty taste and sharp flavour. Recently, soy sauce has also been proved to have health promoting properties.[51–53] Traditionally, soy sauce is prepared by enzymatic brewing of a mixture of soybean and roasted wheat. Soybeans can be steam cooked or defatted and ferment together with roasted wheat flour in the presence of *koji* mould (*A. oryzae* or *A. sojae*). During the fermentation of soy sauce, proteins in the raw materials are hydrolysed into small molecular weight peptides, and amino acids and ammonia by the proteases produced by *A. oryzae* or *A. sojae.* Soy sauce mash obtained by mixing the finished *koji* with salt water is called *moromi* and, at this point, brine fermentation starts and is then subjected to various periods of ageing (Figure 10.2).[54] Microbial communities are responsible of the unique flavour of *moromi.* LAB propagate rapidly in the first stage of *moromi* and produce lactic acid and acetic acid, decreasing pH to <6.0. To improve the quality of soy sauce, *moromi* yeasts like *Z. rouxii*, *Candida versatilis*, and *Candida etchellsii* are necessary to produce characteristics volatile compounds.[55,56] During fermentation and ageing, the flavour develops gradually.

During the fermentation guided by *koji* moulds, soy sauce contains a relatively high amount of free amino acids, which could be a potential source of BA formation.

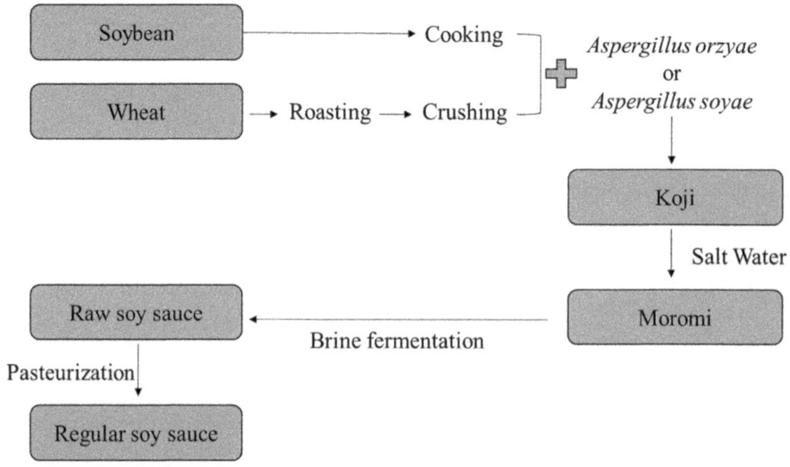

Figure 10.2 Soy sauce manufacturing process.

Literature on the presence and levels of BAs in soy sauce is lacking. High levels of tyramine and histamine in Japanese soy sauces were found by Baek *et al.*[57] Stute *et al.*[58] analysed soy sauce in the German market and found high tyramine levels (up to 5250 mg kg^{-1}) and the presence of histamine, phenylethylamine, putrescine, and cadaverine with lower values. Tyramine and histamine were also above food safety in Chinese soy sauce as observed by Yongmei *et al.*[25]

On the other hand, recently, the BA levels in Thai soy sauce studied by Deetae *et al.*[59] were mostly in the safety zone for consumers. Putrescine and cadaverine were the most common in this product (100% occurrence), followed by tyramine (93.33%) and histamine (86.67%). The knowledge of the levels of BAs in soy sauce is relevant as it can be used as indices of both quality and safety.

The presence of certain amines in soy sauce can indicate poor hygienic-sanitary conditions during processing or the use of low-quality ingredients. Moreover, the presence of high levels of histamine, tyramine, tryptamine, and phenylethylamine in soy sauce can cause adverse effects to human health: histamine can cause histamine poisoning whereas the other amines are implicated in migraines.[60,61]

10.3 Conclusions

It is now clear that soybean foodstuffs produced by fermentation or exposed to microbial contamination during processing or storage may contain BAs. There is some evidence stressing out the high concentrations of BAs in soy products, but these foods are usually consumed in small amounts, so intoxications are rare. For some products, such as miso, natto, sufu, tamari, or soybean paste, BAs are not completely within the safe level for human health. Bacterial contaminants mainly belong to the *Bacillus* group, such as *B. subtilis* and *B. amyloliquefaciens*, and they are responsible for BAs production. Presence of BAs is not only influenced by microbial activity but also by raw materials. The relationship between the accumulation of BAs and the microbial population at different fermentation steps are still not clear. Future investigations are needed to focus on the relationship between BAs production and microbiota at different fermentation steps to allow the selection of less harmful strains with desirable fermentative characters and that are not able to produce BAs. The use of selected starter cultures which do not produce BAs, could be useful to reduce BAs concentration in soybean food products. The control of hygiene and storage conditions is also essential for the reduction of BAs accumulation and should be carefully monitored. Standardization of methods of food production could also be helpful to reduce BAs content in fermented soybean foods because some products are still traditionally manufactured. Improvement of

BA regulations in this field is still needed to allow better monitoring of BAs content and to establish safety limits in soybean products.

References

1. Y. K. Park, J. H. Lee and J. H. Mah, *Food Chem.*, 2018, **278**, 1.
2. K. Y. Park, K. M. Hwang, K. O. Jung and K. B. Lee, *J. Korean Soc. Food Sci. Nutr.*, 2002, **31**, 343.
3. S. Qiu, Y. Wang, Y. Cheng, Y. Liu, M. P. Yadav and L. Yin, *Food Chem.*, 2018, **239**, 1244.
4. FAOSTAT, 2016, Available at http://faostat3.fao.org/
5. K. Kitamura, in *All About Soybeans*, ed. K. Kitamura, Science Forum, Tokyo, 2010, pp. 13–19.
6. A. O. Omoni and R. E. Aluko, *Nutr. Rev.*, 2005, **63**, 272.
7. Y. Youn and Y. S. Kim, *J. Agric. Life Sci.*, 2012, **43**, 20.
8. K. Y. Park, *Food Ind. Nutr.*, 2012, **17**, 1.
9. A. Pituch-Zdanowska, A. Banaszkiewicz and P. Albrecht, *Przegl. Gastroenterol.*, 2015, **10**, 135.
10. N. Delchier, A. L. Herbig, M. Rychlik and C. M. Renard, *Compr. Rev. Food Sci. Food Saf.*, 2016, **15**, 506.
11. Y. Xiao, S. Zhang, H. Tong and S. Shi, *Phytother. Res.*, 2018, **32**, 384.
12. J. Azadmanesh and G. E. Borgstahl, *Antioxidants*, 2018, **7**, 25.
13. S. H. Chao, Y. Tomii, M. Sasamoto, J. Fujimoto, Y. C. Tsai and K. Watanabe, *Int. J. Syst. Evol. Microbiol.*, 2008, **58**, 2555.
14. G. P. Sun, X. J. Zhang, Y. Wang, D. Wang and J. L. Xie, *Mod. Food Sci. Technol.*, 2010, **26**, 1087.
15. EFSA Panel on Biological Hazards, *EFSA J.*, 2011, **9**, 2393.
16. M. H. S. Santos, *Int. J. Food Microbiol.*, 1996, **29**, 213.
17. B. Ten Brink, C. Damink, H. M. L. J. Joosten and J. H. J. Veld Huis in't, *Int. J. Food Microbiol.*, 1990, **11**, 73.
18. W. Shurtleff and A. Aoyagi, *The Book of Tempeh: Professional Edition*, Harper, New York, USA, 1997.
19. E. R. Farnworth, in *Handbook of Fermented Functional Foods*, Functional Foods & Nutraceuticals Series, CRC Press, Boca Raton, FL, 2nd edn, 2008.
20. J. Kirschbaum, K. Rebscher and H. Bruckner, *J. Chromatogr. A*, 2000, **881**, 517.
21. J. E. Stratton, R. W. Hutkins and S. L. Taylor, *J. Food Prot.*, 1991, **54**, 460.
22. B. Y. Byun, X. Bai and J. H. Mah, *Food Sci. Biotechnol.*, 2013, **22**, 55.
23. M. B. A. Gloria, J. Tavares-Neto, R. A. Labanca and M. S. Carvalho, *J. Agric. Food Chem.*, 2005, **53**, 7480.
24. M. Saaid, B. Saad, N. H. Hashim, A. S. Mohamed Ali and M. I. Saleh, *Food Chem.*, 2009, **113**, 1356.
25. L. Yongmei, C. Xiaohong, J. Mei, L. Xin, N. Rahman, D. Mingsheng and Y. Gujun, *Food Control*, 2009, **20**, 593.

26. L. R. Guidi and M. B. A. Gloria, *Food Chem.*, 2012, **133**, 323.
27. J. H. Kim, H. J. Park, M. J. Kim, H. J. Ahn and M. W. Byun, *Korean J. Food Sci. Technol.*, 2003, **35**, 325.
28. N. Nishibori, S. Fujihara and T. Akatuki, *Food Chem.*, 2007, **100**, 491.
29. T. Y. Cho, G. H. Han, K. N. Bahn, Y. W. Son, M. R. Jang, C. H. Lee, S. H. Kim, D. B. Kim and S. B. Kim, *Korean J. Food Sci. Technol.*, 2006, **38**, 730.
30. G. H. Han, T. Y. Cho, M. S. Yoo, C. S. Kim, J. M. Kim, H. A. Kim, M. O. Kim, S. C. Kim, S. A. Lee, Y. S. Ko, S. H. Kim and D. B. Kim, *Korean J. Food Sci. Technol.*, 2007, **39**, 541.
31. Y. J. Ko, Y. H. Son, E. J. Kim, H. G. Seol, G. R. Lee, D. H. Kim and C. H. Ryu, *J. Agric. Life Sci.*, 2012, **46**, 1.
32. R. Jeon, J. H. Lee and J. H. Mah, *LWT–Food Sci. Technol.*, 2018, **92**, 282.
33. H. T. Lee, J. H. Kim and S. S. Lee, *J. Food Hyg. Saf.*, 2009, **24**, 102.
34. S. Shukla, H. K. Park, J. K. Kim and M. Kim, *Food Chem. Toxicol.*, 2010, **48**, 1191.
35. T. K. Kim, J. I. Lee, J. H. Kim, J. H. Mah, H. J. Hwang and Y. W. Kim, *Food Sci. Biotechnol.*, 2011, **20**, 1747.
36. J. I. Lee, Y. K. Oh, J. H. Kim and Y. W. Kim, *Food Sci. Biotechnol.*, 2013, **22**, 1131.
37. S. H. Yoon, M. J. Kim and B. K. Moon, *Appl. Biol. Chem.*, 2017, **60**, 273.
38. N. Toro-Funes, J. Bosch-Fuste, M. L. Latorre-Moratalla, M. T. Veciana-Nogués and M. C. Vidal-Carou, *Food Chem.*, 2015, **173**, 1119.
39. S. Shukla, H. K. Park, J. K. Kim and M. Kim, *Food Sci. Biotechnol.*, 2011, **20**, 851.
40. B. Kim, B. Y. Byun and J. H. Mah, *Food Chem.*, 2012, **135**, 2005.
41. Y. H. Tsai, S. C. Chang and H. F. Kung, *Food Control*, 2007, **18**, 1026.
42. H. F. Kung, Y. H. Tsai and C. I. Wei, *Food Chem.*, 2007, **101**, 351.
43. Y. Byun and J. H. Mah, *J. Food Sci.*, 2012, 77, T216.
44. Z. F. Liu, Y. X. Wei, J. J. Zhang, D. H. Liu, Y. Q. Hu and X. Q. Ye, *Int. J. Food Sci. Technol.*, 2011, **46**, 687.
45. T. Tang, K. Qian, T. Shi, F. Wang, J. Li, Y. Cao and H. Qiongbo, *Food Control*, 2011, **22**, 1203.
46. R. F. Guan, Z. F. Liu, J. J. Zhang, Y. X. Wei, S. Wahab, D. H. Liu and X. Q. Ye, *Food Control*, 2013, **31**, 345.
47. J. Yang, X. Ding, Y. Qin and Y. Zeng, *J. Agric. Food Chem.*, 2014, **62**, 7947.
48. X. Gong, X. Wang, N. Qi, J. Li, L. Lin and Z. Han, *Food Addit. Contam., Part A*, 2014, **31**, 1431.
49. TECHNAVIO, Available at http://www.technavio.com
50. QY research, Available at http://www.qyresearch.com
51. Y. Zhu, Y. Yang, Z. Zhou, G. Li, M. Jiang, C. Zhang and S. Chen, *Food Chem.*, 2010, **118**, 159.
52. B. Yang, H. Yang, J. Li, Z. Li and Y. Jiang, *Food Chem.*, 2011, **124**, 551.
53. M. Jayachandran and B. Xu, *Food Chem.*, 2019, **271**, 362.

54. J. R. Whitaker, *Food Technol.*, 1978, **32**, 175.
55. T. T. Wah, S. Walasiri, A. Assavanig, N. Niamsiri and S. Lertsiri, *Int. J. Food Microbiol.*, 2013, **160**, 282.
56. T. Hamada, M. Sugishita and H. Motai, *Appl. Microbiol. Biotechnol.*, 1990, **33**, 624.
57. I. Baek, C. Weitkamp, T. Erbe, A. Meier, J. Kirschbaum and H. Brückner, 3rd Karlsruhe Nutrition Symposium, European Research Towards Safer and Better Food, Review and Transfer Congress, Congress Centre, Karlsruhe, Germany, 1998, pp. 98–102.
58. R. Stute, K. Petridis, H. Steinhart and G. Biernoth, *Eur. Food Res. Technol.*, 2002, **215**, 101.
59. P. Deetae, P. Jamnong, A. Assavanig and S. Lertsiri, *Int. Food Res. J.*, 2017, **24**, 1575.
60. M. B. A. Glória, Bioactive amines, in *Handbook of Food Science, Technology, and Engineering*, Taylor and Francis Group, Florida, USA, 2005, pp. 13–32.
61. E. Rauscher-Gabernig, R. Grossgut, F. Bauer and P. Paulsen, *Food Control*, 2009, **20**, 423.

CHAPTER 11

Impact of Processing Meat on the Formation of Heterocyclic Amines and Risk of Cancer

AMANAT ALI,*[a] MOSTAFA I. WALY[a] AND SANKAR DEVARAJAN[b]

[a] Department of Food Science and Nutrition, College of Agricultural and Marine Sciences, Sultan Qaboos University, P.O. Box 34, Al-Khoud 123, Muscat, Oman; [b] Nutrition and Food Sciences Program, Department of Human Sciences, University of Arkansas at Pine Bluff, 1200 North University Drive, Pine Bluff, AR 71601, USA
*Email: amanata@gmail.com

11.1 Introduction

Food contains a mixture of hundreds of chemical molecules, which are embedded in a very complex food matrix. Most of the foods are cooked or processed to a varying degree before human consumption. Cooking affects the chemical bonding in the macro- and micro-molecules, their structure, movement of cells, and interactions among various molecules within the complex food matrix. During cooking/processing, these chemical molecules transform their structure and bonding and add taste, flavour, and sensory properties to foods. However, cooking of meat, fish, and poultry at high temperatures, such as pan-frying, grilling, or directly over an open flame or charcoal can result in the formation of certain obnoxious compounds like heterocyclic amines (HCAs), polycyclic aromatic hydrocarbons (PAHs),

Food Chemistry, Function and Analysis No. 20
Biogenic Amines in Food: Analysis, Occurrence and Toxicity
Edited by Bahruddin Saad and Rosanna Tofalo

Published by the Royal Society of Chemistry, www.rsc.org

advanced glycation end products (AGEs), and acrylamides. The formation of HCAs, PAHs, and AGEs can vary depending on the type of meat and method of cooking, as well as the time and temperature *i.e.*, the level of doneness such as rare, medium, or well-done.[1] HCAs belong to a family of mutagenic compounds. The presence of HCAs and PAH in cooked foods, particularly in meat and fish and their indiscriminate consumption may increase the risk of certain types of cancer. In order to increase the quality and safety of foods, it is therefore important to consider reducing the formation of HCAs in foods when optimizing the conditions for various food processes and food processing equipment.

HCAs are formed through the condensation process when proteins, free amino acids (especially tryptophan and glutamic acid), creatine, sugars, and fats present in meat and fish muscles react at high temperatures. All types of muscle meats contain high amounts of creatine, the main substrate for the formation of HCAs. In addition to this, the Strecker's degradation radicals produced during the Maillard reactions can lead to the formation of HCAs.[1,2] As a result of these reactions, some other chemical compounds are also produced, which yield a typical barbecue flavour. The other sources of protein, such as milk, eggs, tofu, and organ meats, contain very little or no HCAs, either in natural or even in cooked form. PAHs are formed when the meat is grilled directly over an open fire and the fat and juices from meat drip on to fire, causing flames.[3] These flames contain the molecules of PAHs, which adhere to the surface of the meat. PAHs can also be formed during other food preparation processes, such as smoking of meats.[2] The food items other than meat and fish, when cooked at high temperatures, do not produce HCAs. On the other hand, some other charred, browned, and blackened food items may contain PAHs. Some other types of carcinogens produced in food during frying and grilling are nitrosamines, aromatic amines, cholesterol oxide products, and so on.[4]

11.2 Factors Affecting the Formation of HCAs and PAHs in Foods

A number of factors affect the formation of HCAs and PAHs in foods. The main factors that influence the formation of HCAs in foods are:

1. Type of food (meat, fish, *etc.*)
2. Fat content of the food
3. Cooking methods
4. Cooking temperature
5. Level and duration of cooking
6. The type of fuel used in cooking/heating
7. Distance from the heat source

The formation of HCAs is dependent on both the physical parameters, such as cooking temperature and time, cooking technique and equipment, heat and mass transport, and on the chemical parameters, especially the presence of substrates or precursors of the HCAs in food.[4] Depending upon the variables, the concentrations of HCAs can vary from 0 to 100-fold in the cooked meat. Cooking for longer time periods as well as the use of high-temperature cooking plates will produce the highest amounts of HCAs.[5] Frying and grilling produce more HCAs as compared with indirect heat methods such as stewing, steaming, or poaching as they are carried out at or below 100 °C. Charcoal grilling produces more HCAs as compared with pan-frying.[6] Cooking temperature is the most important factor in the formation of HCAs. As the temperature increases from 125 °C to 300 °C, there is a proportionate increase in HCA contents. When the meat is cooked at high temperatures (>150 °C) such as during grilling, barbequing, or pan-frying, or for an extended time period, more HCAs are formed. An increase in cooking temperature from 200 °C to 250 °C can lead to a 3-fold increase in the content of HCAs. Tengilimoglu-Metin and colleagues observed that the total number of HCAs increased gradually with an increase in cooking temperatures (150–250 °C).[7] The amounts of HCAs fluctuated between not detectable to 17.60 ng g^{-1} and not detectable to 11.38 ng g^{-1} in chicken breast and beef samples, respectively. As the roasting and baking in an oven is carried out at lower temperatures, there may be lower levels of HCAs formation. During grilling and barbecuing, the meat is exposed to high temperatures, and therefore the highest amounts of HCAs are formed as compared with other cooking methods. Well-done, grilled, or barbecued chicken and steak have been shown to contain high concentrations of HCAs.[4] Common household cooking practices are considered generally safe as they result in lower levels of HCAs. The data from some studies showed that grilled chicken contained higher concentrations of these cancer-causing substances than other types of cooked meat.[8] The browned or burned crusts of meat, which are exposed to high temperatures and are in direct contact with the heat source, as well as the pan drippings that remain after frying, contain much higher concentrations of HCAs.[9] The gravy made from the drippings and/or scrapings of these grilled meats may contain up to 500 ppb of HCAs.[4]

Szterk (2015) demonstrated that free purine and pyrimidine bases as well as their nucleosides are strongly correlated with HCA formation in grilled beef (correlation coefficient 0.78–0.99). It was reported that the most abundant HCA was 9*H*-pyrido-[4,3-*b*]indole (Norharman), which was detected in charcoal grilled chicken breast (32.2 ng g^{-1}), followed by 1-methyl-9*H*-pyrido[4,3-*b*]indole (Harman) and 2-amino-1-methyl-6-phenylimidazo[4,5-*f*]pyridine (PhIP) at 32 and 31.1 ng g^{-1} in charcoal grilled chicken breast, respectively.[10] The pan-fried duck and chicken breasts contained 22 and 18.3 ng g^{-1} of PhIP, respectively. They concluded that processing of meat at high temperatures did not influence the thermal decomposition of amino-imidazo-azaarenes. Methods of cooking that expose meat to smoke or charring, like grilling and broiling over a direct flame, also contribute to the formation of PAHs. PAHs adhere to the surface of food, and the more intense heat is, the higher number of PAHs are formed. Lee and colleagues investigated the effects of grilling procedures on the production level of four PAHs; benzo[*a*]anthracene (B[*a*]A), chrysene (Chr), benzo[*b*]fluoranthene (B[*b*]F), and benzo[*a*]pyrene (B[*a*]P).[11] They concluded that the most important factor contributing to the production of PAHs during grilling was the smoke absorbed by the grilled meat. This smoke is generated as a result of incomplete combustion of fat dripped on to the fire and adhesion and subsequent absorption of the resultant smoke to cooked meat. The other sources of exposures to PAHs and their derivatives in humans are cigarette smoke, emissions from traffic exhausts, and biomass burning.[12]

11.3 Types of HCAs Formed During Cooking

More than 30 different types of these compounds have been identified. The application of new tandem-solid phase extraction techniques in combination with electrospray ionization tandem mass spectrometry (LC/ESI–MS/MS) has led to the identification of many new HCAs in cooked meats at parts-per-trillion range.[13] They are generally classified into two groups. Group 1 includes the compounds that are produced by the pyrolysis of amino acids and proteins through radical reaction and comprise of Trp-P-1, Trp-P-2, Glu-P-1, AαC, and MeAαC. Group 2 contains the compounds (such as IQ, MeIQ, MeIQx, and PhIP) that are generated by heating mixtures of creatine, creatinine, sugars, and amino acids.[14,15] The most commonly studied HCAs are:

1. 2-Amino-3methylimidazo[4,5-*f*]quinolone (IQ)
2. 2-Amino-3,4-dimethylimidazo[4,5-*f*]quinolone (MeIQ)
3. 2-Amino-3,8-dimethylimidazo[4,5-*f*]quinoxaline (MeIQx)
4. 2-Amino-1-methyl-6-phenylimiazo[4,5-*b*]pyridine (PhIP)

These compounds have been listed in the 14th Report on Carcinogens by Department of Health and Human Services, USA, 2016 (https://ntp.niehs.nih.gov/pubhealth/roc/index-1.html, accessed, March 3, 2018). The International Agency for Research in Cancer (IARC) has classified processed meats (hot dogs, sausages, salami, and bacon) as carcinogenic to humans (Group 1), whereas red meat (beef, lamb, and pork) has been classified in Group 2-A (G-2A) as probably carcinogenic to humans.[16,17] The most abundantly formed HCAs in grilled beef, bacon, fish, and poultry are 2-amino-3,8-dimethylimidazo[4,5-*f*]quinoxaline (8-MeIQx) and PhIP.[4,5] The maximum amounts to be present in cooked meat as reported in literature in a 3 ounce serving of ground beef are 136, 145, 1395, and 5783 ng, respectively.[4] The estimated weighted average daily intake throughout the lifetime for the four HCAs ranges between 0.2 and 6.0 ng per kg bodyweight per day.[18] Khan and colleagues analysed the presence of five potential HCAs (IQ, MeIQ, MeIQx, 4,8-DiMeIQx, and PhIP) in home-cooked and fast-food camel meat burgers from Saudi Arabia.[19] They used pan-frying at controlled cooking time and temperature. The level of MeIQx, 4,8-DiMeIQx, and PhIP in the home-cooked and fast-food burgers ranged from 1.52 ng g^{-1} to 3.46 ng g^{-1}, respectively. They however did not detect IQ and MeIQ in home-cooked and fast-food camel meat burgers. They concluded that the variations in HCA contents in the home-cooked and fast-food camel meat burgers might be due to cooking time and temperature. They suggested that the presence of antioxidant food additives in commercial burgers might have played an important role in inhibiting the formation of HCAs.

11.4 Ways to Reduce the Formation of HCAs and PAHs in Cooked Meats and Fish

11.4.1 Simple Modifications in Cooking Methods

There are no specific methods to reduce the amount of HCA and PAH formation in cooked meats and processed foods. However, there is a possibility to inhibit or reduce their production by some simple modifications in cooking methods. The indirect-heat cooking methods can help to reduce the production of HCAs and PAHs in cooked foods. Results from the model systems and cooking experiments suggest many ways to reduce their formation and, thus, reduce human intake, such as simply using lower pan temperatures is a practical way to reduce the heterocyclic amine formation without greatly increasing cooking time.[5] Avoiding direct exposure of meat to an open flame or a hot metal surface, prolonged cooking times, especially at high temperatures over a high heat source, frequent turning and flipping over of meat samples, removing the burned portions of meat as well as abstaining from making the gravy from the meat scarps, can help to reduce HCA and PAH formation.[5,20] Beef patties cooked while frozen did not show any difference in HCAs levels as compared with room-temperature patties.[21] Cooking chicken without the skin can considerably reduce HCA production. Microwaving beef samples prior to exposure to high temperatures such as grilling can help to decrease the formation of HCAs. Microwaving the beef samples before grilling significantly ($P < 0.05$) lowered the HCAs as compared with directly grilling on charcoal. Chicken and beef samples, which were microwaved and deep fried as an alternative method to charcoal grilling, produced significantly fewer HCAs.[22,23] Modifying cooking methods by incorporating these concepts may help to reduce the risk of cancer.

11.4.2 Reducing the Formation of HCAs by Using Herbs and Spices and Marination of Meat

In addition to decreasing the temperature and time of cooking, or changing the cooking method, the use of spices, herbs, and other seasoning agents has also been reported to help in lowering the production of HCAs in cooked foods. Meats and other animal protein foods are generally marinated before frying, barbecuing, and cooking to improve their taste, flavour, and texture. Pretreatments with marinades, together with lower cooking time and temperatures have been shown to be associated with lower HCA production and decreased mutagenic activity, without any compromise on sensory quality of the foods.[24] Mixing, marinating, or rubbing certain herbs, spices, and other food ingredients such as cherries, dried plums, garlic, *etc.* with meat not only lowered the formation of HCAs but also helped to neutralize HCA mutagenicity.[25–27] Shin and colleagues observed that the addition of garlic to hamburger patties reduced the production of HCA > 60%.[26] Plant polyphenols can interfere with the production of HCAs during cooking and can give suitable sensory characteristics to the meat. The marinating products can be applied as direct mix, soaks, or rubs to the foods and can inhibit the formation of HCAs by up to 70%.[20,26,28] Marinating poultry meat before barbecuing with a mixture of brown sugar, olive oil, cider vinegar, garlic, mustard, lemon juice, and salt reduced the total detectable levels of HCAs from 56 to 1.7 ng g^{-1}, from 158 to 10 ng g^{-1}, and from 330 to 44 ng g^{-1} for grilling times of 20, 30, and 40 minutes, respectively.[29] Overnight marination of beef steaks with teriyaki sauce (a local turmeric-garlic sauce), or with commercial honey barbecue sauce resulted in almost 60% less formation of HCAs in the cooked meat as compared with non-marinated. Beef steaks marinated with a standard barbecue sauce containing tomato solids and sugar (containing fructose) however, resulted in a significant increase in the production of HCAs after 15 minutes of cooking.[30] The presence of garlic and onions as well as lemon juice in marinades has also been reported to reduce formation of HCAs in cooked beef patties.[28]

A number of other ingredients have also been reported in the literature that show inhibitory effects on the production of HCAs in cooked meet. The addition of beer and wine in marinades has also been reported to reduce the production of certain types of HCAs in beef steaks.[31] Water-soluble vitamins, because of their antioxidant activities, have been shown to have inhibitory effects against the formation of heterocyclic amine in a chemical model system, with pyridoxamine as the most potent inhibitor. Although heating may destroy some of the antioxidant vitamins like vitamin C, the overall antioxidant capacity of food products is not only maintained but is improved as a result of processing. The addition of vitamin E and rosemary extracts to ground beef patties before frying or grilling inhibited the formation of HCAs by >90%.[32] Antioxidant vitamins may be able to interact with the Maillard reactions and may affect the types and concentrations of reaction products. Such inhibitory activities were found to be in fried beef patties.[33] The cooked meat products may contain some compounds that might impede the carcinogenicity of HCAs. Plant polyphenols and flavonoids have been reported to inhibit the mutagenicity of benzo[*a*]pyrene, which is a polycyclic aromatic hydrocarbon.[34–36] Although in *in vitro* studies, some polyphenolic acids did not inhibit the mutagenicity of HCAs, the flavonoids such as quercetin, morin, and myricetin showed inhibitory effects on IQ, MeIQ, MeIQx, Trp-P-1, and Trp-P-2. However, in *in vivo* trials, the mutagenicity of MeIQ and Trp-P-2 was reported to be higher.[37] Beef burgers when cooked in virgin olive oil contained lower concentrations of HCAs as compared with other oils. They suggested that this effect is probably due to the presence of phenols in virgin olive oil. However, the HCA reducing effect of virgin olive oil decreased during storage, which may be prevented by the addition of rosemary extract in oil.[38]

Spices and herbs, however, differentially inhibit the production of HCAs in cooked meat products. The use of native spices during grilling of meat could minimize the formation of HCAs and may help to reduce their toxic and carcinogenic effects. It was observed that local spices used in Malaysia such as turmeric, lemongrass, torch ginger, and curry leaves were effective in lowering the levels of HCAs in grilled beef samples.[22] Using the best amounts of selected spices, *e.g.* 4 g per 100 g of turmeric and 10 g per 100 g of torch ginger, lemon grass and curry leaves, reduced HCA contents from 40 to 85 ng per 100 g in grilled beef samples. Chili pepper and capsaicin showed the dose-dependent inhibitory effects on the formation of HCAs in roasted beef patties but did not affect ($P > 0.05$) their texture. The highest inhibition (80–98%) was observed at 2 mg of capsaicin per 100 g of beef patties.[22,36] Tengilimoglu-Metin and colleagues showed that hawthorn extract at 0.5% and 1% inhibited the total HCA levels from 12–100% and 19–97% in pan-fried and oven roasted chicken breast and 42–100% and 20–35% in beef, respectively.[7] The addition of microcrystalline cellulose (MCC) or carboxymethyl cellulose (CMC), normally used as fat replacers, has also been shown to be effective in inhibiting the production of HCAs in grilled beef patties.[39] The primitive cooking methodologies are deep-rooted in our cultures and are difficult to change. Introducing simple cooking modifications as mentioned earlier may help to reduce cancer risk. Using modern methods of barbecuing may help to lower the production of HCAs as indirect heat application does not lead to high temperatures. In order to increase the quality and safety of food products, the food manufacturing industry must consider developing effective and optimal processing methods in order to reduce the formation of these carcinogens in final food products.

11.4.3 Mechanisms of Action of HCAs in Cancer Development

HCAs belong to a family of mutagenic compounds, which are believed to play an important role in the aetiology of human cancers. The high fat contents of meat and other animal products increase the hormone production such as oestrogens, which may increase the risk of breast and prostate cancers. Fats require bile salts for their emulsification, digestion, and absorption and nurture the growth of intestinal bacteria, which in turn produce some carcinogenic secondary bile acids to deal with fat digestion. It has been shown that certain HCAs are distributed to mammary glands where they are metabolically activated.[40] In experimental animals and humans, HCAs have been shown to undergo extensive metabolic transformations.

Most of the metabolic transformations of HCAs occur in the liver. The oxidation of HCAs is mainly carried out by CYP1A2 enzyme at the heterocyclic ring and methyl groups to produce the detoxification products. The *in vivo* mutagenic activity of HCAs is activated by some catalases present in the liver.[41,42] A fairly consistent association has been observed between the consumption of grilled, barbecued, broiled, or well-done meats and the risk of various types of cancers such as oesophagus, lung, pancreas, stomach, colorectal, endometrium, and prostate.[43–46]

The data from various *in vitro* and *in vivo* animal model studies shows that these HCAs and PAHs are mutagenic, cause changes in DNA, and increase the risk of cancer.[14] However, the DNA damage only occurs after their bioactivation once metabolized by specific enzymes in the body. The mutagenic potency of HCAs is dependent upon their chemical structure, ability to undergo *N*-oxidation to form the reactive nitrenium ion, and is either markedly boosted or decreased by the simultaneous existence of other dietary components.[14] The metabolic activation of food-borne heterocyclic amines to colon carcinogens in humans is hypothesized to occur through Phase-I and Phase-II reactions in the body. HCAs are oxidized to hydroxyl-amino derivatives by cytochrome P450 enzymes and are converted to ester forms by acetyltransferases and sulfotransferases, producing DNA adducts through the formation of N–C bonds at guanine bases.[14,47] Turteltaub and colleagues reported that the formation of DNA and protein adducts in rodents are dose-dependent and the levels of these adducts in human tissues and blood are generally higher on the basis of equivalent administered doses. They observed that the metabolite profiles differ substantially between humans and rodents for both MeIQx and PhIP, with more *N*-hydroxylation (bioactivation) and there was less ring oxidation (detoxification) in humans.[48] The individual responses towards dietary exposure to HCAs and risk of cancer may vary greatly depending upon their genetics, differences in metabolism, and ability to repair DNA, and may be either 1000 times greater or 1000 times less than the risk for another individual.[4] The individual genetic factors, such as acetylation status, can modify the effects of heterocyclic amines. This may affect the exposure variances, even among people with similar dietary intakes. The individuals with higher genetic susceptibility and heritable background for increased risk of developing cancers could be much more sensitive to HCAs than normal subjects. Age, rapid-rapid phenotype, and consumption of well-done red meat were associated with increased risk of colorectal neoplasia.[49–51]

Depending upon the genetic susceptibility of individuals, the HCAs and PAHs can be bioactivated or inactivated in the body by specific enzymes.[52–54] The compounds in the first group (*e.g.*, Trp-P-1, Trp-P-2, Glu-P-1, AαC, and MeAαC) can be inactivated through the phase-I and Phase-II chemical reactions, whereas the compounds in the second group (*e.g.*, IQ, MeIQ, MeIQx, and PhIP), which are predominantly present in cooked meat, condensate and form adducts with DNA and proteins in the human body in a dose-dependent manner and induce mutations. The genetic mutations induced by MeIQ, IQ, and PhIP have been well characterized in rodents. The chemical structure of HCAs and their ability to undergo *N*-oxidation to form the reactive nitrenium ion (R_2N^+) determines their mutagenic potency (see ref. 55). Exposure to pyrolysis products through consumption of well-done meat increases the risk of CRC, particularly in individuals who smoke and are genetically susceptible as determined by a rapid phenotype for both NAT2 and CYP1A2. A dose-dependent association was observed between the HCA intake estimates and male rectal cancer, with a 2- to 3-fold increase in risk from the low (T1) to high (T3) tertile of intake for each HCA, the strongest association with MeIQx.[56] Cascella and colleagues proposed the underlying mechanism of colorectal cancer (CRC) by red and processed meat intake. They hypothesized that the HCAs and PAHs produced during cooking meat at high temperatures after their metabolic activation can lead to the formation of DNA adducts in the epithelial cells with possible risk of carcinogenesis. The lipid peroxidation as well as endogenous formation of *N*-nitroso compounds (NOCs) may provoke DNA mutations leading to a greater risk of CRC.[17] HCAs and PAHs can however be eliminated from the body through detoxification in the liver. Dietary factors, like antioxidants, bioactive components in food, may induce or inhibit the activities of these enzymes and have been shown to play a significant role in the onset and progression of various types of cancers.[57]

As mentioned earlier, the action of these metabolizing enzymes in the liver can vary among people and therefore may have variable cancer risks related to the dietary intakes of HCAs. Development of tumours in breast, colon, liver, skin, lung, prostate, and other organs including the gastrointestinal tract, have been reported in rodents fed diets supplemented with HCAs and PAHs. The impact of HCAs in mice and rat models has been studied at relatively high doses (10–400 mg per kg of bodyweight), to which the humans are normally not exposed through the regular consumption of cooked and processed foods.[9] The doses of HCAs and PAHs used in these studies were very high and such amounts are generally not consumed by humans from a normal diet.[14,47] PhIP has also been shown to cause cancer of colon and mammary glands in rats in a dose-dependent manner. PhIP can be transferred from mothers to their offspring through the mother's milk. A high intake of PhIP can induce prostate tumours in rodents through the formation of DNA adducts.[58,59] Red and processed meat may be positively associated with the risk of prostate cancer via mechanisms involving heme iron, nitrite, nitrate, and formation of benzo[*a*]pyrene as a result of grilling and barbecuing.[60] It has been demonstrated that both low and high doses of creatine supplementation, given either acutely or chronically, did not stimulate the *de novo* synthesis of carcinogenic HCAs PhIP, 8-MeIQx, IFP, and 4,8-DiMeIQx in healthy subjects.[61] They concluded that it is the diet, rather than creatine supplementation, which is mainly responsible for the formation of HCAs. The intake of these mutagenic compounds such as MeIQx and PhIP from cooked meat may therefore increase the risk of renal cell carcinoma (RCC) in genetically susceptible individuals.[49] Another recent hypothesis is that *N*-glycolylneuraminic acid, a non-human sialic acid sugar present in red meat, becomes incorporated in the cell membrane, triggering the immune response with associated inflammation and reactive oxygen species, which can contribute to DNA damage, tumour promotion, and cancer.[62] The clinical evidence suggests that the type of NOC found after ingestion of red meat in humans consists mainly of nitrosyl iron and nitrosothiols, products that have profoundly different chemistries from certain *N*-nitroso species which have been shown to be tumourigenic through the formation of DNA adducts.[63]

11.4.4 Meat Consumption, Cooking Practices, Dietary Exposure to HCAs, and Risk of Cancer

The amount and type of meat and fish ingested, frequency of consumption, cooking methods, cooking temperature, and duration of cooking all influence the dietary exposure to HCAs. High frying temperature was found to increase colon cancer risk almost 2-fold, and rectal cancer risk by 60%.[43] The people generally prefer to eat well-done and barbecued meats. Some epidemiological studies have tried to correlate the ingestion of HCAs by measuring the consumption of well-done meat and its carcinogenic effects in humans with particular reference to colorectal cancer.[9,51] The consumption of meat and fish was determined based on an extensive food-frequency questionnaires that included questions related to type of meat and fish ingested, frequency of consumption, portion size, cooking methods, and degree of surface browning (with the help of colour photographs at four different temperatures). The number of HCAs ingested was estimated based on the previously carried out laboratory analyses of meat and fish under various cooking conditions. Variable results were observed, which lack in precision and accuracy with respect to intake of HCAs (IQ, MeIQ, MeIQx, DiMeIQx, and PhIP) and risk of cancer. Individuals, who prefer to eat well-done meat had odds ratio of >3 for stomach adenocarcinoma than those who consumed similar quantities of rare or medium rare done.[51,64] Frequent consumption of meat, particularly red meat, increases the risk of colon cancer. In a 6-year prospective study, the relation between diet and incident colon cancer among 32 051 non-Hispanic white cohort members of the Adventist Health Study (California, 1976–1982) was examined, who at baseline, had no documented or reported history of cancer. The overall findings from this cohort identified that both red and white meat intake was an important risk factor for raising the possibility of colon cancer. They however suggested that the risk due to red meat intake reflected a more complex aetiology.[65]

The published data indicates that meat consumption is associated with an increased risk of colorectal cancer and the association was found to be more consistent with intake of red and processed meats. The individuals who ate beef, pork, or lamb daily had approximately three times higher colon cancer risk as compared with those who generally avoided these foods or had lower intakes.[66] High red meat intake was defined as 3 ounces of beef, lamb, or pork for men and 2 ounces for women daily, which is equivalent to a typical hamburger. In a case-control study in the US, Butler and associates examined the intake of meat by cooking method as well as level of doneness and estimated the intake of HCAs in 620 colon cancer patients and 1038 controls. They observed that the red meat intake was positively associated with risk of colon cancer. They observed that the meat prepared through pan-frying was the riskiest as it led to the formation of highest concentration of HCAs in meat.[67] The data from another case-control study conducted at the University of Utah that included 952 subjects with rectal cancer and 1205 controls revealed that men and women with the highest consumption of processed or well-cooked meat had an increased risk of rectal cancer.[68] High consumption of red and processed meat, over longer periods of time, may increase the risk of cancer in the distal portion of the large intestine.[66,69]

High intake of well-done red meat increased the subsequent exposure to HCAs, which was associated with an increased risk of many types of cancers such as colon, rectum, breast, prostate, pancreas, lung, stomach, and esophagus.[56,64,70] Extreme cooking times and temperatures may denote the main risk factors, as the median estimated intake level for HCAs in normal subjects was <100 ng day^{-1} as compared with cancer cases, who had much higher intakes >1900 ng day^{-1}, suggesting that HCAs may induce cancer only at very high intake levels. However, such high amounts are normally not encountered through the usual normal food intake except in those who very frequently consume well-done, barbequed, fried, roasted, and processed meats. The estimated ingested daily amount of heterocyclic amines from cooked/processed foods in humans is, however, relatively too small (~1 ng per kg bodyweight) to make any DNA mutations to exhibit any definite carcinogenic effects in humans. The results of the first European cohort study support data from other case-control studies that there is a positive association between HCA intake and colorectal adenoma risk. The European Prospective Investigation into Cancer and Nutrition (EPIC-Study), Heidelberg estimated the intake of HCAs from meat in 21 462 subjects, examining the associations between meat cooking methods as well as HCA intake from meat and cancer risk in a prospective way.[71] They used a detailed dietary food frequency questionnaire to assess the meat consumption, cooking methods, and degree of browning of respective food items. The median total intake of HCAs from meat was 31 ng day^{-1} (mean 69 ng day^{-1}), which was lower than the results observed in previous studies. The study concluded that 2-amino-1-methyl-6-phenylimidazo[4,5*b*]pyridine was the most common HCA in this cohort (median 17; mean 48 ng day^{-1}). The relative risk for colorectal cancer increased at intakes of >41.4 ng day^{-1}. There was also some indication that a relative risk can occur at intake level (≥ 19.9 ng day^{-1}) of MeIQx but the tendency was not that strong as for PhIP (Rohrmann *et al.*, 2007). An increased risk of colorectal adenoma was associated with PhIP but no statistically significant associations were observed for MeIQx and DiMeIQx intakes.[72]

The data from various other epidemiological studies, which also used comprehensive food frequency questionnaires to estimate the participant's meat consumption as well as the meat cooking methods to assess the HCAs and PAHs intakes, indicated that high consumption of well-done, fried, or barbecued meats was associated with increased risks of various types of cancers.[46,73–75] Joshi and colleagues conducted a comprehensive analysis of CRC risk and red meat and poultry intakes, taking into account cooking methods, level of doneness, estimated intakes of HCAs from cooked meat, tumour location, and tumour mismatch repair proficiency (MMR) status. They observed a significantly positive association between the intake of pan-fried beef steak, oven-broiled short ribs or spare ribs, and risk of CRC that was even stronger among the MMR deficient cases ($P = 0.059$). They, however, observed an inverse association with grilled hamburger intake. They concluded that specific meat types and cooking practices are the possible sources of human carcinogens relevant to CRC risk.[76] Helmus and colleagues observed a statistically significant specific meat type association for red meat-derived MeIQx, DiMeIQx, and mutagenic activity that was not detected for the same mutagens derived from white meat. They reported that the HCAs and PAHs derived from red meat are responsible for colon cancer carcinogenesis and not the white meat derived HCAs and PAHs.[77] The data from two other large cohorts of US health professionals indicated a positive association between meat intake and risk of CRC, particularly for distal cancer. They suggested that future studies with larger cohorts should be conducted to assess the links between the impacts of HCAs and the subsites across the colon to confirm these findings and to explain the underlying mechanisms between the intake of processed and unprocessed red meats and overall risk of CRC.[78]

Le and colleagues prospectively examined the links between estimated intakes of HCAs and meat-derived mutagenicity (MDM) in two cohorts of health professionals. They concluded that the measured intakes of meat mutagens did not show any significant association with CRC risk in the NHS and Health Professionals Follow-up (HPFS) study cohorts during >14 years follow up. They suggested that the results for PhIP from red meat but not from white meat warrant further investigation.[79] The epidemiological studies carried out in Japan and elsewhere suggest that HCAs present in cooked meat and fish are carcinogenic for humans.[80] Chiavarini and colleagues concluded from their meta-analysis that meat-related mutagens (PhIP, MeIQx, 2-amino-3,4,8-trimethylimidazo[4,5-*f*] quinoxaline, DiMeIQx, benzo[*a*]pyrene (B(*a*)P) and MDM) is positively associated with intake of meat and risk of CRA and CRC.[81] Le (2018) raised objections about the results of this meta-analysis and stated that these finding appear confusing and misleading because of the repeated usage of data in calculation of the odds ratio (OR) and relative risk (RR), as well as the missing data from some of the cohort studies, which may have affected the overall wrong outputs of results and thus the skewed findings of the study.[82] Although the data from these epidemiological studies still could not establish a definitive link between the dietary intakes of HCAs and PAHs from cooked meats and the risk of cancer in humans, many of the studies show that a high intake of well-done, fried, or barbecued meats is associated with an increased risk of developing various types of cancers. The major reason for such discrepancies may be because of the difficulties in determining the exact level of intake for HCAs and/or PAHs from cooked meats and other processed foods.[75,81,83] In spite of different confounding factors, the cooked and processed meat intake remains a risk factor for various types of cancers. This may also be associated to the formation of HCAs in meat during cooking and processing at high temperatures.[9,60,81,84]

Jahurul and colleagues assessed the dietary intake of HCAs from various food sources in the Malaysian population through a food frequency questionnaire (FFQ). They appraised that the intake level of total HCAs in the studied samples (meat and fish) ranged from not detected to 38.7 ng g^{-1}. The highest intake was observed for PhIP, followed by MeIQx and MeIQ. They concluded that the major dietary sources of heterocyclic amines in Malaysia are fried and grilled chicken. The estimated average daily intake of HCAs was 553.7 ng per capita per day, which they considered lower as compared with other countries.[85] For the US population, an estimated average intake of 26 ng per kg bodyweight per day has been reported. Dietary fat, saturated fat, dairy products, and meat are regarded as the possible dietary factors contributing to prostate cancer risk. Dietary fats have been hypothesized to increase the production of hormones and promoting the growth of cancer cells in hormone-sensitive organs, particularly in the breast and prostate. Kabat and colleagues assessed the association between meat intake, meat-cooking methods, and meat-mutagen intake and postmenopausal breast cancer risk in the NIH-AARP Diet and Health Study cohort of 120 755 postmenopausal women. The data indicated that the intake of total meat, red meat, meat cooked at high temperatures, and meat mutagens did not show any association with breast cancer risk. The data did not indicate any link between the meat mutagens and the development of postmenopausal breast cancer.[86] High intake of grilled/barbecued and smoked meat may increase mortality after breast cancer.[84] Boldo and colleagues examined the data from a population-based case-control study from Spain (MCC-Spain) to find the relationship between the risk of breast cancer (BC) and meat intake. They concluded that the intake of well-done and stewed red meat was associated with increased risk of breast cancer, particularly in premenopausal women. They suggested that the moderate consumption of well-done, stewed red meat, as well as processed/cured meat, may reduce the risk of developing breast cancer.[45]

Rodriguez and colleagues observed that total red meat intake (such as sausages, bacon, and hot dogs as well as unprocessed red meat) was associated with a higher risk of prostate cancer in the US.[87] The risk of prostate cancer increased gradually with increased consumption of cooked and processed meats and was higher among Black men as compared with White men. The people consuming red meat at least five times per week showed two and half times more relative risk of cancer compared with those, who ate <1 time per week. The results of another Harvard's Health Professionals Follow-Up Study showed that higher red meat intake had a positively significant association with increased risk of prostate cancer.[88] It has been suggested that reducing or eliminating meat from the diet reduces the risk of prostate cancer. Although no association was observed with MeIQx and DiMeIQx, the highest quintile of PhIP was found to be associated with a 1.2-fold increased risk of prostate cancer (95% CI, 1.01–1.48) and a 1.3-fold increased risk (95% CI, 1.01–1.61) of incident disease.[89] This data supports the results of animal studies, which have implicated PhIP as a prostate carcinogen. The consumption of highly processed, charcoaled, very well done, and grilled meats has been positively associated with prostate cancer risk.[90]

Increased meat consumption has also been shown to significantly increase the renal cell carcinoma and pancreatic cancer risk.[46] Grilled red meat intake is not only a risk factor for pancreatic cancer but also for other types of cancer. It has been suggested that the methods of meat preparation as well as the amount of total meat intake are important in assessing the effects of meat consumption in epidemiologic studies. HCAs, particularly benzo[*a*]pyrene (B(*a*)P) from well-done barbecued and pan-fried meats may be associated with increased risk for pancreatic cancer.[73,91] The findings of Taunk and colleagues support the hypothesis that meat intake, particularly the meat cooked at high temperatures may play a role in pancreatic cancer development.[92] Vegetarians have been reported to have 12–40% lower risk of developing cancers as compared with meat eaters, particularly much lower risk (up to 45%) of blood cancers.[93] The people who ate beef $\geq$4 times a week had more than twice the risk of stomach and pancreatic cancer than those consuming beef less frequently.[94,95] The results of a recent meta-analysis indicated that only in case-control studies is there an association between the intake of red and processed meat with increased risk of pancreatic cancer, whereas data from cohort studies showed such association only in men and not in women.[96]

11.4.5 Strategies to Inhibit or Reduce the Dietary Exposure of HCAs and Related Risk of Cancer

Most of the dietary exposure to HCAs and PAHs comes from grilled, fried, barbeque, and charred meats or fish. Fruits and vegetables do not contribute much to HCAs and PAHs unless there is environmental contamination. In addition to dietary sources, the people may also get some exposure to PAHs from other environmental sources, such as pollution, traffic exhausts, and smoking.[12] Data from various scientific studies suggest that meat, dairy products, and energy-dense fatty foods are associated with increased cancer risk, whereas fruits and vegetables may help to reduce cancer risk. The results of *in vitro* studies showed that the carcinogenicity and mutagenicity of HCAs from cooked meats may be inhibited with simultaneous consumption of other foods such as green and black tea, red wine, blue- and blackberries, red grapes, kiwi, watermelon, parsley, and spinach.[97] The cruciferous vegetables contain sulphoraphanes, which have anti-mutagenic effects against the dietary HCAs in varied mixed diet. The results of Canene-Adams and colleagues support the hypothesis that a diet rich in tomato and broccoli can reduce or prevent dietary carcinogen-induced cancers.[98] The probiotic bacteria present in fermented dairy foods may have neutralizing effects on HCAs.[99] Dietary fibre has been shown to decrease the mutagenicity of heterocyclic amines because of their binding effects. Merging such foods, while planning daily meals containing barbecued meat, poultry, and fish, may lower the mutagenic effects of cooked meats. Plant-based diets contain only negligible amounts of HCAs and PAHs and can therefore not only decrease the extent of DNA damage but can also help to fight against the adverse effects of these compounds. In Japan, the breast cancer rates are low because traditional Japanese diets have lower fat content as compared with the typical Western diet. The suitability of urinary HCAs as exposure biomarkers was assessed to measure the dietary intake of HCAs. The results indicated that the concurrent ingestion of raw vegetables may not only inhibit the gastrointestinal absorption of HCAs from well-done cooked beef but also can reduce the formation of DNA adducts by metabolic activation of HCAs.[100] They suggested that concurrent consumption of vegetables with cooked meat may effectively prevent the carcinogenic impact of HCAs. It has been suggested that fresh vegetables and purified phytochemicals can lower the risk of forming the PhIP–DNA adducts and may therefore help in reducing cancer risk.[101]

The total mortality from cancer is steadily on the increase in young Japanese people. The data from epidemiological surveys on the causes of cancer indicated that 50% of cancer cases in Japan can be prevented just simply through controlling infection and abstaining from smoking. This can lead to about 40% reduction in cancer deaths.[80] It is not possible to have a carcinogenic free food environment as the food, which is a complex mixture of many thousands of chemical substances and some of them may contribute to cancer risk. Although the human body has some mechanisms to detoxify heterocyclic amines, one has to be careful in consuming fried, grilled, barbecued meat and fish because of one's genetically predisposed sensitivity to HCA. The body's fight against the risk of cancer can be boosted by consuming a healthy diet that contains $>2:1$ ratio of plant to animal foods. The higher the ratio, the greater will be the protective effect.[102] At present, there are no guidelines concerning the consumption of HCA containing foods and no maximum daily dietary intake levels of HCAs and PAHs. It is generally recommended that people should limit the consumption of foods containing these mutagens. Sander and colleagues reported that their data do not support the hypothesis that HCAs intake as consumed from a regular diet is a risk factor for prostate cancer.[103] IQ, MeIQ, MeIQx, and PhIP are listed in the 'The Report on Carcinogens' by US-DHHS. The World Cancer Research Fund/ American Institute for Cancer Research have issued a report in 2007 with dietary guidelines that recommend limiting the consumption of red and processed meats; however, no recommendations have been provided about the dietary intake of HCAs and PAHs.[46]

Although the consumption of well-done meat can be used as an indirect measure of exposure to HCAs, individual variations in the activity of metabolizing enzymes may affect the exposure differences among people, who ingest the same amount of HCAs and PAHs from dietary sources. The data from various studies have reported that increased consumption of well-done, fried, or barbecued meats is associated with higher risks of colorectal, pancreatic, and prostate cancer. Whereas the other studies did not find any association with risks of colorectal or prostate cancers with intake of meats.[79,104] Bylsma and Alexander, however, observed only a weak positive association for processed meats with prostate cancer.[104] Based on the data from epidemiological as well as from animal model studies, the International Agency for Research on Cancer (IARC) has classified processed meats (hot dogs, cured sausages, salami, and bacon) as carcinogenic to humans (Group 1), whereas red meats (beef, lamb, pork, Group-2A) as probably carcinogenic to humans.[17] However, there is still insufficient evidence to establish a definite conclusion on the role of HCAs in the genesis of human tumours. The results of Carr and colleagues support an association between the consumption of red and processed meat and CRC risk at various sub-sites of CRC.[105] The results from a recent meta-analysis suggest that there is a positive association between the dietary intake of meat mutagens (benzo[*a*]pyrene) and risk of development of CRA and CRC. A significant dose-response effect was observed for PhIP, MeIQx, and mutagenicity index to CRA and CRC.[106] Based on the results of their meta-analysis, Crippa and associates suggested that processed meat may be positively associated with risk of bladder cancer.[107] The data from both case-control and cohort studies have indicated an increased pooled RR for every 50 g increase in intake of processed meat per day. A positive association between red meat and risk of bladder cancer was, however, only observed in case-control studies, while no such association was observed in prospective cohort studies. The results from a large prospective NutriNet-Santé cohort study also suggest that red meat may be involved in the development of cancer risk in many parts of the body.[108]

11.5 Summary and Conclusion

Cooking affects chemical bonding in macro- and micro-molecules, their structure, movement of cells, and interactions among various molecules within the complex food matrix. Cooking of meat, fish, and poultry at high temperatures can result in the formation of HCAs, PAHs, and AGEs. The data from various *in vitro* and *in vivo* animal model studies shows that HCAs and PAHs are mutagenic, which can get bioactivated by specific cytochrome P450 metabolizing enzymes in the body, damaging the DNA and thereby increasing the risk of various types of cancers. The data from various epidemiological studies indicates that high consumption of well-done, fried, or barbecued meats is associated with increased risks of various types of cancers: CRC, pancreatic, prostate, and breast cancer. There is a possibility to reduce the formation of HCAs and PAHs in cooked or barbequed meat through various cooking modifications such as cooking at lower temperatures, avoiding cooking on a direct flame or flipping over meat frequently during preparation, and marinating the meat with herbs and spices before cooking. Although the consumption of red and processed meat is associated with increased risk of various types of cancers, there is still limited scientific evidence based on randomized controlled clinical trials that this risk is related to the dietary intake of mutagenic compounds (HCAs and PAHs) formed during cooking of meat at high temperatures. No dietary guidelines are available about the intake of these mutagenic compounds and no good measures are available to establish the upper intake levels of HCAs and PAHs to reduce the risk of cancer. There is an urgent need to study the associations between meat intake, cooking methods, and dietary human exposure to HCAs and PAHs and risk of cancer.

References

1. G. Z. Liao, G. Y. Wang, X. L. Xu and G. H. Zhou, Effect of cooking methods on the formation of heterocyclic aromatic amines in chicken and duck breast, *Meat Sci.*, 2010, **85**(1), 149–154.
2. A. J. Cross and R. Sinha, Meat-related mutagens/carcinogens in the etiology of colorectal cancer, *Environ. Mol. Mutagen.*, 2004, **44**(1), 44–55.
3. K. Wakabayashi, I. S. Kim, R. Kurosaka, Z. Yamaizumi, H. Ushiyama, M. Takahashi, *et al.*, Identification of new mutagenic heterocyclic amines and quantification of known heterocyclic amines, *Princess Takamatsu Symp.*, 1995, **23**, 39–49.
4. K. I. Skog, M. A. E. Johansson and M. I. Jägerstad, Carcinogenic Heterocyclic Amines in Model Systems and Cooked Foods: A Review on Formation, Occurrence and Intake, *Food Chem. Toxicol.*, 1998, **36**(9), 879–896.
5. M. G. Knize and J. S. Felton, Formation and human risk of carcinogenic heterocyclic amines formed from natural precursors in meat, *Nutr. Rev.*, 2005, **63**(5), 158–165.
6. D. Gu, L. McNaughton, D. Lemaster, B. G. Lake, N. J. Gooderham, F. F. Kadlubar, *et al.*, A comprehensive approach to the profiling of the cooked meat carcinogens 2-amino-3,8-dimethylimidazo[4,5-*f*]quinoxaline, 2-amino-1-methyl-6-phenylimidazo[4,5-*b*]pyridine, and their metabolites in human urine, *Chem. Res. Toxicol.*, 2010, **23**(4), 788–801.
7. M. M. Tengilimoglu-Metin, A. Hamzalioglu, V. Gokmen and M. Kizil, Inhibitory effect of hawthorn extract on heterocyclic aromatic amine formation in beef and chicken breast meat, *Food Res. Int.*, 2017, **99**, 586–595.

8. R. Sinha, N. Rothman, C. P. Salmon, M. G. Knize, E. D. Brown, C. A. Swanson, *et al.*, Heterocyclic amine content in beef cooked by different methods to varying degrees of doneness and gravy made from meat drippings, *Food Chem. Toxicol.*, 1998, **36**(4), 279–287.
9. K. Augustsson, J. Lindblad, E. Overvik and G. Steineck, A population-based dietary inventory of cooked meat and assessment of the daily intake of food mutagens, *Food Addit. Contam.*, 1999, **16**(5), 215–225.
10. A. Szterk, Heterocyclic aromatic amines in grilled beef: The influence of free amino acids, nitrogenous bases, nucleosides, protein and glucose on HAAs content, *J. Food Compos. Anal.*, 2015, **40**, 39–46.
11. J.-G. Lee, S.-Y. Kim, J.-S. Moon, S.-H. Kim, D.-H. Kang and H.-J. Yoon, Effects of grilling procedures on levels of polycyclic aromatic hydrocarbons in grilled meats, *Food Chem.*, 2016, **199**, 632–638.
12. C. A. Alves, A. M. Vicente, D. Custódio, M. Cerqueira, T. Nunes, C. Pio, *et al.*, Polycyclic aromatic hydrocarbons and their derivatives (nitro-PAHs, oxygenated PAHs, and azaarenes) in PM2.5 from Southern European cities, *Sci. Total Environ.*, 2017, **595**, 494–504.
13. R. J. Turesky, Formation and biochemistry of carcinogenic heterocyclic aromatic amines in cooked meats, *Toxicol. Lett.*, 2007, **168**(3), 219–227.
14. T. Sugimura, K. Wakabayashi, H. Nakagama and M. Nagao, Heterocyclic amines: Mutagens/carcinogens produced during cooking of meat and fish, *Cancer Sci.*, 2004, **95**(4), 290–299.
15. M. Jägerstad and K. Skog, Genotoxicity of heat-processed foods, *Mutat. Res., Fundam. Mol. Mech. Mutagen.*, 2005, **574**(1), 156–172.
16. V. Bouvard, D. Loomis, K. Z. Guyton, Y. Grosse, F. El Ghissassi, L. Benbrahim-Tallaa, *et al.*, Carcinogenicity of consumption of red and processed meat, *Lancet Oncol.*, 2015, **16**(16), 1599–1600.
17. M. Cascella, S. Bimonte, A. Barbieri, V. Del Vecchio, D. Caliendo, V. Schiavone, *et al.*, Dissecting the mechanisms and molecules underlying the potential carcinogenicity of red and processed meat in colorectal cancer (CRC): An overview on the current state of knowledge, *Infect. Agents Cancer*, 2018, **13**, 3.
18. K. T. Bogen and G. A. Keating, U.S. dietary exposures to heterocyclic amines, *J. Exposure Anal. Environ. Epidemiol.*, 2001, **11**, 155.
19. M. Rizwan Khan, M. Naushad and Z. Abdullah Alothman, Presence of heterocyclic amine carcinogens in home-cooked and fast-food camel meat burgers commonly consumed in Saudi Arabia, *Sci. Rep.*, 2017, 7(1), 1707.
20. C. P. Salmon, M. G. Knize, F. N. Panteleakos, R. W. Wu, D. O. Nelson and J. S. Felton, Minimization of Heterocyclic Amines and Thermal Inactivation of *Escherichia coli* in Fried Ground Beef, *J. Natl. Cancer Inst.*, 2000, **92**(21), 1773–1778.
21. E. Persson, B. K. Oroszvári, E. Tornberg, I. Sjöholm and K. Skog, Heterocyclic amine formation during frying of frozen beefburgers, *Int. J. Food Sci. Technol.*, 2008, **43**(1), 62–68.

22. S. Jinap, S. Z. Iqbal and R. M. P. Selvam, Effect of selected local spices marinades on the reduction of heterocyclic amines in grilled beef (satay), *LWT–Food Sci. Technol.*, 2015, **63**(2), 919–926.
23. S. Jinap, N. D. S. Hasnol, M. Sanny and M. H. A. Jahurul, Effect of organic acid ingredients in marinades containing different types of sugar on the formation of heterocyclic amines in grilled chicken, *Food Control*, 2018, **84**, 478–484.
24. L. M. Tikkanen, K. J. Latva-Kala and R. L. Heiniö, Effect of commercial marinades on the mutagenic activity, sensory quality and amount of heterocyclic amines in chicken grilled under different conditions, *Food Chem. Toxicol.*, 1996, **34**(8), 725–730.
25. C. Britt, E. A. Gomaa, J. I. Gray and A. M. Booren, Influence of Cherry Tissue on Lipid Oxidation and Heterocyclic Aromatic Amine Formation in Ground Beef Patties, *J. Agric. Food Chem.*, 1998, **46**(12), 4891–4897.
26. I. S. Shin, W. J. Rodgers, E. A. Gomaa, G. M. Strasburg and J. I. Gray, Inhibition of heterocyclic aromatic amine formation in fried ground beef patties by garlic and selected garlic-related sulfur compounds, *J. Food Prot.*, 2002, **65**(11), 1766–1770.
27. M. T. Nunez de Gonzalez, R. M. Boleman, R. K. Miller, J. T. Keeton and K. S. Rhee, Antioxidant properties of dried plum ingredients in raw and precooked pork sausage, *J. Food Sci.*, 2008, **73**(5), H63–H71.
28. M. Gibis, Heterocyclic Aromatic Amines in Cooked Meat Products: Causes, Formation, Occurrence, and Risk Assessment, *Compr. Rev. Food Sci. Food Saf.*, 2016, **15**(2), 269–302.
29. C. P. Salmon, M. G. Knize and J. S. Felton, Effects of marinating on heterocyclic amine carcinogen formation in grilled chicken, *Food Chem. Toxicol.*, 1997, **35**(5), 433–441.
30. P. V. Nerurkar, L. L. Marchand and R. V. Cooney, Effects of marinating with Asian Marinades or Western barbecue sauce on PhIP and MeIQx formation in barbecued beef, *Nutr. Cancer*, 1999, **34**(2), 147–152.
31. A. Melo, O. Viegas, R. Eça, C. Petisca and O. Pinho, Ferreira IMPLVO. extraction, detection, and quantification of heterocyclic aromatic amines in Portuguese meat dishes by HPLC/diode array, *J. Liq. Chromatogr. Relat. Technol.*, 2008, **31**(5), 772–787.
32. K. Puangsombat, P. Gadgil, T. A. Houser, M. C. Hunt and J. S. Smith, Heterocyclic amine content in commercial ready to eat meat products, *Meat Sci.*, 2011, **88**(2), 227–233.
33. D. Wong, K.-W. Cheng and M. Wang, Inhibition of heterocyclic amine formation by water-soluble vitamins in Maillard reaction model systems and beef patties, *Food Chem.*, 2012, **133**(3), 760–766.
34. J. H. Weisburger, M. Nagao, K. Wakabayashi and A. Oguri, Prevention of heterocyclic amine formation by tea and tea polyphenols, *Cancer Lett.*, 1994, **83**(1-2), 143–147.
35. D. X. Lin, P. A. Thompson, C. Teitel, J. S. Chen and F. F. Kadlubar, Direct reduction of *N*-acetoxy-PhIP by tea polyphenols: A possible

mechanism for chemoprevention against PhIP-DNA adduct formation, *Mutat. Res.*, 2003, **523–524**, 193–200.
36. M. Zeng, M. Zhang, Z. He, F. Qin, G. Tao, S. Zhang, *et al.*, Inhibitory profiles of chilli pepper and capsaicin on heterocyclic amine formation in roast beef patties, *Food Chem.*, 2017, **221**, 404–411.
37. A. J. Alldrick, B. G. Lake and I. R. Rowland, Modification of in vivo heterocyclic amine genotoxicity by dietary flavonoids, *Mutagenesis*, 1989, **4**(5), 365–370.
38. E. Persson, G. Graziani, R. Ferracane, V. Fogliano and K. Skog, Influence of antioxidants in virgin olive oil on the formation of heterocyclic amines in fried beefburgers, *Food Chem. Toxicol.*, 2003, **41**(11), 1587–1597.
39. M. Gibis and J. Weiss, Inhibitory effect of cellulose fibers on the formation of heterocyclic aromatic amines in grilled beef patties, *Food Chem.*, 2017, **229**, 828–836.
40. E. D. Stefani, H. Deneo-Pellegrini, A. L. Ronco, P. Boffetta, P. Correa, D. Aune, *et al.*, Meat Consumption, Cooking Methods, Mutagens, and Risk of Squamous Cell Carcinoma of the Esophagus: A Case-Control Study in Uruguay, *Nutr. Cancer*, 2012, **64**(2), 294–299.
41. R. H. Dashwood, Use of transgenic and mutant animal models in the study of heterocyclic amine-induced mutagenesis and carcinogenesis, *J. Biochem. Mol. Biol.*, 2003, **36**(1), 35–42.
42. D. D. Alexander, D. L. Weed, P. E. Miller and M. A. Mohamed, Red meat and colorectal cancer: A quantitative update on the state of the epidemiologic science, *J. Am. Coll. Nutr.*, 2015, **34**(6), 521–543.
43. T. Norat, A. Lukanova, P. Ferrari and E. Riboli, Meat consumption and colorectal cancer risk: Dose-response meta-analysis of epidemiological studies, *Int. J. Cancer*, 2002, **98**(2), 241–256.
44. X. Wang, X. Lin, Y. Y. Ouyang, J. Liu, G. Zhao, A. Pan, *et al.*, Red and processed meat consumption and mortality: dose-response meta-analysis of prospective cohort studies, *Public Health Nutr.*, 2016, **19**(5), 893–905.
45. E. Boldo, A. Castello, N. Aragones, P. Amiano, B. Perez-Gomez, G. Castano-Vinyals, *et al.*, Meat intake, methods and degrees of cooking and breast cancer risk in the MCC-Spain study, *Maturitas*, 2018, **110**, 62–70.
46. M. Wiseman, The Second World Cancer Research Fund/American Institute for Cancer Research Expert Report. Food, Nutrition, Physical Activity, and the Prevention of Cancer: A Global Perspective. Nutrition Society and BAPEN Medical Symposium on 'Nutrition support in cancer therapy', *Proc. Nutr. Soc.*, 2008, **67**(3), 253–256.
47. H. Ohgaki, S. Takayama and T. Sugimura, Carcinogenicities of heterocyclic amines in cooked food, *Mutat. Res., Genet. Toxicol.*, 1991, **259**(3), 399–410.
48. K. W. Turteltaub, K. H. Dingley, K. D. Curtis, M. A. Malfatti, R. J. Turesky, R. C. Garner, *et al.*, Macromolecular adduct formation

and metabolism of heterocyclic amines in humans and rodents at low doses, *Cancer Lett.*, 1999, **143**(2), 149–155.
49. S. C. Melkonian, C. R. Daniel, Y. Ye, N. M. Tannir, J. A. Karam,, S. F. Mti, *et al.*, Gene-environment interaction of genome-wie association study-identified susceptibility loci and meat-cooking mutagens in the etiology of renal cell carcinoma, *Cancer*, 2016, **122**(1), 108–115.
50. J. H. Kim, J. Lee, S. Y. Jung and J. Kim, Dietary factors and female breast cancer risk: A prospective cohort study, *Nutrients*, 2017, **9**(12), 1–20.
51. R. Sinha, U. Peters, A. J. Cross, M. Kulldorff, J. L. Weissfeld, P. F. Pinsky, *et al.*, Meat, meat cooking methods and preservation, and risk for colorectal adenoma, *Cancer Res.*, 2005, **65**(17), 8034–8041.
52. U. Nothlings, J. F. Yamamoto, L. R. Wilkens, S. P. Murphy, S. Y. Park, B. E. Henderson, *et al.*, Meat and heterocyclic amine intake, smoking, NAT1 and NAT2 polymorphisms, and colorectal cancer risk in the multiethnic cohort study, *Cancer Epidemiol., Biomarkers Prev.*, 2009, **18**(7), 2098–2106.
53. T. Cai, L. Yao and R. J. Turesky, Bioactivation of heterocyclic aromatic amines by UDP glucuronosyltransferases, *Chem. Res. Toxicol.*, 2016, **29**(5), 879–891.
54. E. L. Jamin, A. Riu, T. Douki, L. Debrauwer, J.-P. Cravedi, D. Zalko, *et al.*, Combined genotoxic effects of a polycyclic aromatic hydrocarbon (B(*a*)P) and an heterocyclic amine (PhIP) in relation to colorectal carcinogenesis, *PLoS One*, 2013, **8**(3), e58591.
55. F. T. Hatch, M. G. Knize and M. E. Colvin, Extended quantitative structure–activity relationships for 80 aromatic and heterocyclic amines: Structural, electronic, and hydropathic factors affecting mutagenic potency, *Environ. Mol. Mutagen.*, 2001, **38**(4), 268–291.
56. L. Le Marchand, J. H. Hankin, L. M. Pierce, R. Sinha, P. V. Nerurkar, A. A. Franke, *et al.*, Well-done red meat, metabolic phenotypes and colorectal cancer in Hawaii, *Mutat. Res.*, 2002, **506–507**, 205–214.
57. R. F. Tayyem, H. A. Bawadi, I. Shehadah, L. M. Agraib, S. S. AbuMweis, T. Al-Jaberi, *et al.*, Dietary patterns and colorectal cancer, *Clin. Nutr.*, 2017, **36**(3), 848–852.
58. G. Li, H. Wang, A. B. Liu, C. Cheung, K. R. Reuhl, M. C. Bosland, *et al.*, Dietary carcinogen 2-amino-1-methyl-6-phenylimidazo[4,5-*b*]pyridine-induced prostate carcinogenesis in CYP1A-humanized mice, *Cancer Prev. Res.*, 2012, **5**(7), 963–972.
59. M. Bellamri, S. Xiao, P. Murugan, C. J. Weight and R. J. Turesky, Metabolic activation of the cooked meat carcinogen 2-amino-1-methyl-6-phenylimidazo[4,5-*b*]pyridine in human prostate, *Toxicol. Sci.*, 2018, **163**(2), 543–556.
60. R. Sinha, A. J. Cross, B. I. Graubard, M. F. Leitzmann and A. Schatzkin, Meat intake and mortality: A prospective study of over half a million people, *Arch. Intern. Med.*, 2009, **169**(6), 562–571.

61. R. T. Pereira, F. A. Dorr, E. Pinto, M. Y. Solis, G. G. Artioli, A. L. Fernandes, *et al.*, Can creatine supplementation form carcinogenic heterocyclic amines in humans?, *J. Physiol.*, 2015, **593**(17), 3959–3971.
62. R. J. Turesky, Mechanistic evidence for red meat and processed meat intake and cancer risk: A follow-up on the International Agency for Research on Cancer Evaluation of 2015, *Chimia*, 2018, **72**(10), 718–724.
63. C. Kruger and Y. Zhou, Red meat and colon cancer: A review of mechanistic evidence for heme in the context of risk assessment methodology, *Food Chem. Toxicol.*, 2018, **118**, 131–153.
64. W. Zheng and S.-A. Lee, Well-done meat intake, heterocyclic amine exposure, and cancer risk, *Nutr. Cancer*, 2009, **61**(4), 437–446.
65. P. N. Singh and G. E. Fraser, Dietary risk factors for colon cancer in a low-risk population, *Am. J. Epidemiol.*, 1998, **148**(8), 761–774.
66. A. Chao, M. J. Thun, C. J. Connell, M. L. McCullough, E. J. Jacobs, W. D. Flanders, *et al.*, Meat consumption and risk of colorectal cancer, *JAMA, J. Am. Med. Assoc.*, 2005, **293**(2), 172–182.
67. L. M. Butler, R. Sinha, R. C. Millikan, C. F. Martin, B. Newman, M. D. Gammon, *et al.*, Heterocyclic amines, meat intake, and association with colon cancer in a population-based study, *Am. J. Epidemiol.*, 2003, **157**(5), 434–445.
68. M. A. Murtaugh, K. N. Ma, C. Sweeney, B. J. Caan and M. L. Slattery, Meat consumption patterns and preparation, genetic variants of metabolic enzymes, and their association with rectal cancer in men and women, *J. Nutr.*, 2004, **134**(4), 776–784.
69. N. M. Pham, T. Mizoue, K. Tanaka, I. Tsuji, A. Tamakoshi, K. Matsuo, *et al.*, Meat consumption and colorectal cancer risk: An evaluation based on a systematic review of epidemiologic evidence among the Japanese population, *Jpn. J. Clin. Oncol.*, 2014, **44**(7), 641–650.
70. J. H. Weisburger, Comments on the history and importance of aromatic and heterocyclic amines in public health, *Mutat. Res.*, 2002, **506–507**, 9–20.
71. S. Rohrmann, D. Zoller, S. Hermann and J. Linseisen, Intake of heterocyclic aromatic amines from meat in the European Prospective Investigation into Cancer and Nutrition (EPIC)-Heidelberg cohort, *Br. J. Nutr.*, 2007, **98**(6), 1112–1115.
72. S. Rohrmann, S. Hermann and J. Linseisen, Heterocyclic aromatic amine intake increases colorectal adenoma risk: Findings from a prospective European cohort study, *Am. J. Clin. Nutr.*, 2009, **89**(5), 1418–1424.
73. K. E. Anderson, R. Sinha, M. Kulldorff, M. Gross, N. P. Lang, C. Barber, *et al.*, Meat intake and cooking techniques: Associations with pancreatic cancer, *Mutat. Res.*, 2002, **506–507**, 225–231.
74. R. Z. Stolzenberg-Solomon, A. J. Cross, D. T. Silverman, C. Schairer, F. E. Thompson, V. Kipnis, *et al.*, Meat and meat-mutagen intake and pancreatic cancer risk in the NIH-AARP cohort, *Cancer Epidemiol., Biomarkers Prev.*, 2007, **16**(12), 2664–2675.

75. A. J. Cross, L. M. Ferrucci, A. Risch, B. I. Graubard, M. H. Ward, Y. Park, *et al.*, A large prospective study of meat consumption and colorectal cancer Risk: An investigation of potential mechanisms underlying this association, *Cancer Res.*, 2010, **70**(6), 2406–2414.
76. A. D. Joshi, A. Kim, J. P. Lewinger, C. M. Ulrich, J. D. Potter, M. Cotterchio, *et al.*, Meat intake, cooking methods, dietary carcinogens, and colorectal cancer risk: Findings from the Colorectal Cancer Family Registry, *Cancer Med.*, 2015, **4**(6), 936–952.
77. D. S. Helmus, C. L. Thompson, S. Zelenskiy, T. C. Tucker and L. Li, Red meat-derived heterocyclic amines increase risk of colon cancer: A population-based case-control study, *Nutr. Cancer*, 2013, **65**(8), 1141–1150.
78. A. M. Bernstein, M. Song, X. Zhang, A. Pan, M. Wang, C. S. Fuchs, *et al.*, Processed and unprocessed red meat and risk of colorectal cancer: Analysis by tumor location and modification by time, *PLoS One*, 2015, **10**(8), e0135959.
79. N. T. Le, F. A. Michels, M. Song, X. Zhang, A. M. Bernstein, E. L. Giovannucci, *et al.*, A prospective analysis of meat mutagens and colorectal cancer in the Nurses' Health Study and Health Professionals Follow-up Study, *Environ. Health Perspect.*, 2016, **124**(10), 1529–1536.
80. M. Nagao and S. Tsugane, Cancer in Japan: Prevalence, prevention and the role of heterocyclic amines in human carcinogenesis, *Genes Environ.*, 2016, **38**, 16.
81. M. Chiavarini, G. Bertarelli, L. Minelli and R. Fabiani, Dietary intake of meat cooking-related mutagens (HCAs) and risk of colorectal adenoma and cancer: A systematic review and meta-analysis, *Nutrients*, 2017, **9**(5), 514.
82. N. Le, Meat mutagens and colorectal adenoma and cancer: A problem with a recently published systematic review and meta-analysis, *Nutrients*, 2018, **10**(3), 312.
83. Z. Abid, A. J. Cross and R. Sinha, Meat, dairy, and cancer, *Am. J. Clin. Nutr.*, 2014, **100**, 386s–393s.
84. H. Parada Jr., S. E. Steck, P. T. Bradshaw, L. S. Engel, K. Conway, S. L. Teitelbaum, *et al.*, Grilled, barbecued, and smoked meat intake and survival following breast cancer, *J. Natl. Cancer Inst.*, 2017, **109**(6), 1–8.
85. M. H. Jahurul, S. Jinap, S. J. Ang, A. Abdul-Hamid, P. Hajeb, H. N. Lioe, *et al.*, Dietary exposure to heterocyclic amines in high-temperature cooked meat and fish in Malaysia, *Food Addit. Contam., Part A*, 2010, **27**(8), 1060–1071.
86. G. C. Kabat, A. J. Cross, Y. Park, A. Schatzkin, A. R. Hollenbeck, T. E. Rohan, *et al.*, Meat intake and meat preparation in relation to risk of postmenopausal breast cancer in the NIH-AARP diet and health study, *Int. J. Cancer*, 2009, **124**(10), 2430–2435.
87. C. Rodriguez, M. L. McCullough, A. M. Mondul, E. J. Jacobs, A. Chao, A. V. Patel, *et al.*, Meat consumption among Black and White men and

risk of prostate cancer in the Cancer Prevention Study II Nutrition Cohort, *Cancer Epidemiol., Biomarkers Prev.*, 2006, **15**(2), 211–216.

88. E. Giovannucci, E. B. Rimm, M. J. Stampfer, G. A. Colditz, A. Ascherio and W. C. Willett, Intake of fat, meat, and fiber in relation to risk of colon cancer in men, *Cancer Res.*, 1994, **54**(9), 2390–2397.
89. A. J. Cross, U. Peters, V. A. Kirsh, G. L. Andriole, D. Reding, R. B. Hayes, *et al.*, A prospective study of meat and meat mutagens and prostate cancer risk, *Cancer Res.*, 2005, **65**(24), 11779–11784.
90. D. Tang, O. N. Kryvenko, Y. Wang, S. Trudeau, A. Rundle, S. Takahashi, *et al.*, 2-Amino-1-methyl-6-phenylimidazo[4,5-*b*]pyridine (PhIP)-DNA adducts in benign prostate and subsequent risk for prostate cancer, *Int. J. Cancer*, 2013, **133**(4), 961–971.
91. J. J. Anderson, N. D. M. Darwis, D. F. Mackay, C. A. Celis-Morales, D. M. Lyall, N. Sattar, *et al.*, Red and processed meat consumption and breast cancer: UK Biobank cohort study and meta-analysis, *Eur. J. Cancer*, 2018, **90**, 73–82.
92. P. Taunk, E. Hecht and R. Stolzenberg-Solomon, Are meat and heme iron intake associated with pancreatic cancer? Results from the NIH-AARP diet and health cohort, *Int. J. Cancer*, 2016, **138**(9), 2172–2189.
93. T. J. Key, P. N. Appleby, E. A. Spencer, R. C. Travis, N. E. Allen, M. Thorogood, *et al.*, Cancer incidence in British vegetarians, *Br. J. Cancer*, 2009, **101**(1), 192–197.
94. L. D. Boada, L. A. Henríquez-Hernández and O. P. Luzardo, The impact of red and processed meat consumption on cancer and other health outcomes: Epidemiological evidences, *Food Chem. Toxicol.*, 2016, **92**, 236–244.
95. G. Lippi, C. Mattiuzzi and G. Cervellin, Meat consumption and cancer risk: a critical review of published meta-analyses, *Crit. Rev. Oncol. Hematol.*, 2016, **97**, 1–14.
96. Z. Zhao, Z. Yin, Z. Pu and Q. Zhao, Association between consumption of red and processed meat and pancreatic cancer risk: A systematic review and meta-analysis, *Clin. Gastroenterol. Hepatol.*, 2017, **15**(4), 486–493.e10.
97. R. Edenharder, J. W. Sager, H. Glatt, E. Muckel and K. L. Platt, Protection by beverages, fruits, vegetables, herbs, and flavonoids against genotoxicity of 2-acetylaminofluorene and 2-amino-1-methyl-6-phenylimidazo[4,5-*b*]pyridine (PhIP) in metabolically competent V79 cells, *Mutat. Res.*, 2002, **521**(1-2), 57–72.
98. K. Canene-Adams, K. S. Sfanos, C.-T. Liang, S. Yegnasubramanian, W. G. Nelson, C. Brayton, *et al.*, Dietary chemoprevention of PhIP induced carcinogenesis in male Fischer 344 rats with tomato and broccoli, *PLoS One*, 2013, **8**(11), e79842.
99. A. Nowak and Z. Libudzisz, Ability of probiotic Lactobacillus casei DN 114001 to bind or/and metabolise heterocyclic aromatic amines in vitro, *Eur. J. Nutr.*, 2009, **48**(7), 419–427.

100. H. Kataoka, T. Inoue, N. Ikekita and K. Saito, Development of exposure assessment method based on the analysis of urinary heterocyclic amines as biomarkers by on-line in-tube solid-phase microextraction coupled with liquid chromatography–tandem mass spectrometry, *Anal. Bioanal. Chem.*, 2014, **406**(8), 2171–2178.
101. J. K. Kim, M. A. McCormick, C. M. Gallaher, D. D. Gallaher and S. P. Trudo, Apiaceous vegetables and cruciferous phytochemicals reduced PhIP-DNA adducts in prostate but not in pancreas of wistar rats, *J. Med. Food*, 2018, **21**(2), 199–202.
102. M. Kapiszewska, A vegetable to meat consumption ratio as a relevant factor determining cancer preventive diet. The Mediterranean versus other European countries, *Forum Nutr.*, 2006, **59**, 130–153.
103. A. Sander, J. Linseisen and S. Rohrmann, Intake of heterocyclic aromatic amines and the risk of prostate cancer in the EPIC-Heidelberg cohort, *Cancer, Causes Control*, 2011, **22**(1), 109–114.
104. L. C. Bylsma and D. D. Alexander, A review and meta-analysis of prospective studies of red and processed meat, meat cooking methods, heme iron, heterocyclic amines and prostate cancer, *Nutr. J.*, 2015, **14**, 125.
105. P. R. Carr, L. Jansen, S. Bienert, W. Roth, E. Herpel, M. Kloor, *et al.*, Associations of red and processed meat intake with major molecular pathological features of colorectal cancer, *Eur. J. Epidemiol.*, 2017, **32**(5), 409–418.
106. V. Martinez Gongora, K. L. Matthes, P. Rodriguez Castano, J. Linseisen and S. Rohrmann, Dietary heterocyclic amine intake and colorectal adenoma risk: A systematic review and meta-analysis, *Cancer Epidemiol., Biomarkers Prev.*, 2019, **28**, 99–109.
107. A. Crippa, S. C. Larsson, A. Discacciati, A. Wolk and N. Orsini, Red and processed meat consumption and risk of bladder cancer: A dose-response meta-analysis of epidemiological studies, *Eur. J. Nutr.*, 2018, **57**(2), 689–701.
108. A. Diallo, M. Deschasaux, P. Latino-Martel, S. Hercberg, P. Galan, P. Fassier, *et al.*, Red and processed meat intake and cancer risk: Results from the prospective NutriNet-Santé cohort study, *Int. J. Cancer*, 2018, **142**(2), 230–237.

CHAPTER 12

Sample Preparation Methods for the Analysis of Biogenic Amines

BAHRUDDIN SAAD*[a] AND CHANBASHA BASHEER[b]

[a] Fundamental and Applied Sciences Department & Institute of Self-Sustainable Building, Universiti Teknologi PETRONAS, 32610 Seri Iskandar, Perak, Malaysia; [b] Department of Chemistry, King Fahd University of Petroleum and Minerals, 31261 Dhahran, Saudi Arabia
*Email: bahruddin.saad@utp.edu.my

12.1 Introduction to Sample Preparation

Before a sample can be analysed, it must first undergo a procedure that is broadly known as sample preparation. Sample preparation can range from extremely simple operations (*e.g.* filtration to get rid of suspended particulates and undissolved matrix components, grinding and homogenization of solid samples, pH adjustments) to those that are more complicated, requiring hours for the extraction of analytes. Although sample preparation is an important component of the analytical process, it is often the least understood by the majority of practitioners. One of the most common procedure in sample preparation is extraction where the analyte of interest is selectively isolated and removed from the sample, normally using organic extracting solvent(s). Sample preparation is the most time-consuming step in the analytical process, and it is also where most mistakes are made.

Food Chemistry, Function and Analysis No. 20
Biogenic Amines in Food: Analysis, Occurrence and Toxicity
Edited by Bahruddin Saad and Rosanna Tofalo

Published by the Royal Society of Chemistry, www.rsc.org

Thus, it is very important that due attention is given to sample preparation to ensure the quality and credibility of the analytical results.

An ideal sample preparation should fulfil the following criteria:

1. Simple and easy to be implemented, involving minimum steps to minimize loss of analytes.
2. Effective in removing the matrix components to ensure clean extracts.
3. Be able to preconcentrate the analytes.
4. Use a minimum amount of low cost and environmentally friendly solvents.
5. Consumes minimum energy.

12.2 Challenges in Sample Preparation for the Analysis of Biogenic Amines

From the analytical point of view, the determination of biogenic amines (BAs) in food is very challenging.[1–3] This is partly due to the fact that BAs are relatively polar compounds, rendering them readily soluble in aqueous solutions thus making it difficult to be extracted using frequently used organic solvents. The food matrix itself in which the BAs are commonly examined is also often complex. Potentially interfering compounds such as polyphenols, lipids and proteins can be found even in aqueous samples such as beer, wine and dairy products. These compounds must be removed as much as possible prior to the analysis. BAs are also often present in low concentrations (sub-ng mL^{-1} or lower), requiring methods of sufficient sensitivity for the detection. Many BAs lack chromophores in their structures to be detected using common high-performance liquid chromatography (HPLC) detectors such as UV-VIS and spectrofluorimetry. The often occurrence of several BAs together requires an efficient separation method to adequately resolve the components for simultaneous determinations. To achieve the final analytical goals, sample preparation must be carefully chosen to address these aforementioned challenges.

12.3 Classical Sample Preparation Methods for the Determination of BAs

The various sample preparation methods, categorized as either classical or non-classical methods, are shown in Figure 12.1. Classical sample preparation methods for the determination of BAs uses liquid–liquid extraction (LLE) and/or solid phase extraction (SPE).

12.3.1 Liquid–Liquid Extraction

LLE can be accomplished by shaking an aqueous solution (containing the BA of interest) in a separatory funnel with an immiscible organic solvent.

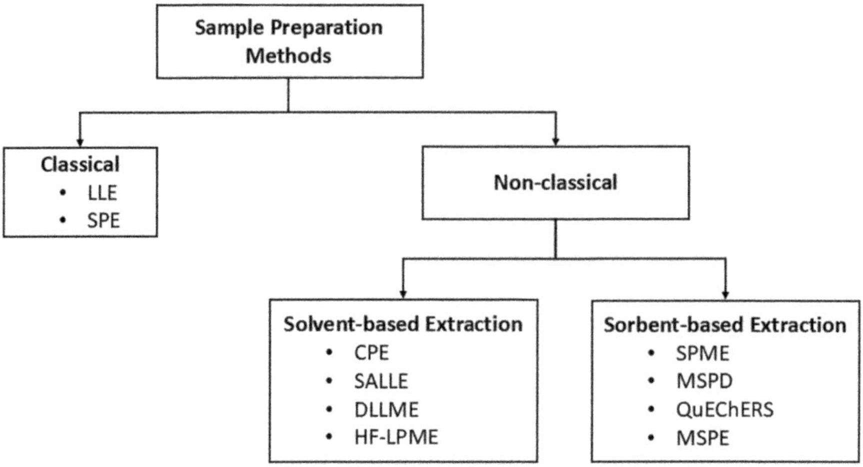

Figure 12.1 Classification of sample preparation methods.

During this process, the analyte distributes itself between the two phases that are governed to a large extent by the distribution ratio, K_d (approximately proportional to its solubility in each phase). After the clear phase separation, the organic extract (containing the BA) is collected. The extraction can be performed several times with fresh solvent in order to extract the analyte quantitatively. To facilitate the extraction, the analytes must be electrically neutral. For basic analytes such as BAs, adjusting the aqueous phase to be sufficiently basic (2 pH units above the pK_a) is necessary to be extracted into the organic solvent.

At the moment, there is no harmonization of the analytical methods among users. A common procedure is shown in Figure 12.2. For the analysis of BAs in solid foods (*e.g.* fish, cheese), the samples are often subjected to an acidic treatment using perchloric acid, trichloroacetic acid, hydrochloric acid *etc.*[4] The charged BAs are next reacted with reagents before the derivatized amines are extracted into suitable organic solvent (*e.g.* butanol, butanol-chloroform) at basic pH. The organic layer is evaporated to dryness, reconstituted with suitable solvent, before finally being injected into the HPLC system.

Sometimes, the LLE technique can be rendered more effective by employing ion-pair techniques to extract the polar ionizable BAs in the non-derivatized form into the organic layer. This is done by adding a suitable ion with opposite charge to form the BA ion-pair complex. Since the complex is electrically neutral, it can be easily extracted into an organic extracting solvent. Taking into account the pK_a values of the BAs, the pH of the aqueous medium needs to be adjusted to achieve high yields of protonated forms of BAs, thereby facilitating the ion-pair extraction of their adducts. This approach was used by Chatzimitakos *et al.*[5] where BAs were extracted into an organic solvent *via* the use of an ion-pairing agent (bis-2-ethylhexylphosphate),

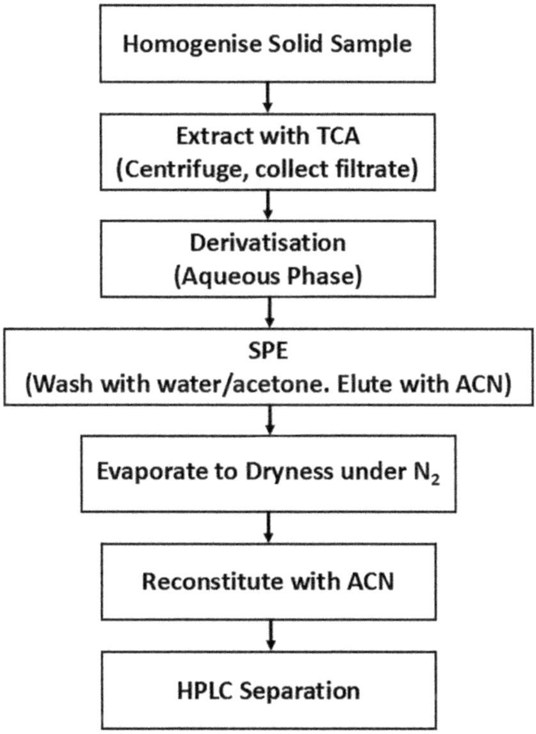

Figure 12.2 Typical procedure for the classical determination of biogenic amines (BAs).[4] Key to abbreviations: TCA, trichloroacetic acid; ACN, acetonitrile; SPE, solid phase extraction.

followed by back extraction in D_2O in order to acquire the proton nuclear magnetic resonance spectra of the underivatized BAs.

From the environmental point of view, LLE is not favoured as the long exposure time towards volatile solvents can pose safety and health problems to the user. The large amount of waste generated adds additional cost for their safe disposal. Furthermore, LLE procedures are labour intensive as automation is not straightforward. Due to these disadvantages, LLE is steadily being replaced with other extraction methods. Common strategies to rectify these problems are:

1. To drastically reduce the amount of extracting solvent used.
2. To substitute for more environmentally friendly solvent.
3. To minimize the number of steps for the extraction procedure (ideally one step).

To assist in the selection of environmentally friendly solvents, a model that uses toxicological and exposure data to obtain total hazard values (tHV) and total analytical hazard values (taHV) was proposed by Tobiszewski and

Namieśnik.[6] The procedure considers toxicity of chemicals to humans and the environment as well as chemical exposure potential. Parameters included in the algorithm are hazards related to oral toxicity, inhalation toxicity, carcinogenicity, other hazardous effects, aquatic acute toxicity, aquatic chronic toxicity and exposure related parameters. Additionally for taHV computation, hazards to inhalation are dually factored as inhalation is the main exposure pathway when solvents are applied in analytical laboratories. According to this model, values of tHV and taHV are the highest for

Table 12.1 Total hazard values (tHV) and total analytical hazard values (taHV) of some solvents. Reproduced from ref. 6 with permission from Elsevier, Copyright 2015.

	Compounds	tHV	taHV
Chlorinated hydrocarbons	Dichloromethane	39.3	59.8
	Chloroform	70.7	103.8
	Carbon tetrachloride	80	109.7
	Trichloroethane	90.9	125.1
	Tetrachloroethane	82.7	107.7
	1,1,1-Trichloroethane	34.7	49
	1,1,2,2-Tetrachloroethane	76.9	97.3
	Chlorobenzene	58	75.9
Hydrocarbons	Pentane	23.4	36.4
	Hexane	53.8	81.4
	Cyclohexane	57.9	79.8
	Heptane	15.2	20.9
	Isooctane	57.1	79.2
	Benzene	84	122
	Toluene	43.9	60.5
	Xylene	51.5	68.2
Aldehydes	Furfural	47.7	69.2
	Benzaldehyde	56.9	77.5
Alcohols	Methanol	8.8	15.7
	Ethanol	4.1	7.2
	Isopropanol	3.9	6.8
	Heptanol	22.9	28.6
	Octanol	38.3	45.7
	Nonanol	39.8	46.7
	Benzyl alcohol	45.2	55.1
	Diethyl ether	8.7	16
	Methyl *tert*-butyl ether	2.7	3.8
	Tetrahydrofuran	9.1	14.9
Organic acids	Formic acid	25.8	43
	Acetic acid	1.3	2.1
	Propionic acid	27	41.2
Ethers	Diethyl ether	8.7	16
	Methyl *tert*-butyl ether	2.7	3.8
	Tetrahydrofuran	9.1	14.9
Ketones	Acetone	1.5	2.6
Esters	Ethyl acetate	5	7.3
Others	Acetonitrile	18.8	26.8
	Carbon disulphide	61.6	89.6

chlorinated hydrocarbons and benzene (Table 12.1) and these solvents are categorized as carcinogens and possess other specific adverse human health effects. They also possess high oral, inhalation and aquatic toxicities. On the other hand, the short chain alcohols, ethers, acetone and ethyl acetate are the least hazardous organic solvents used in analytical laboratories. These polar solvents are of lower toxicities and are biodegradable. The properties of some common solvents according to this model are shown in Table 12.1.[6]

Other non-conventional solvents such as ionic liquids, superheated water, surface active agents, supercritical fluids, deep eutectic solvent and bio-derived solvents have more recently being introduced.[6–8] However, most of these solvents have not been used for the extraction of BAs.

12.3.2 Solid Phase Extraction

The is no clear inventor of the SPE technique but it has been commonly used since the 1980s.[9] SPE is a variation of the liquid chromatographic technique that uses a cartridge that contains a small amount (~50–200 mg) of sorbent material(s) (*e.g.* silica, C18, florosil) to isolate the analyte from the sample solution.

A typical SPE method involves the following steps[9,10] (Figure 12.3): (i) conditioning of the column to activate the active sites; (ii) loading of the liquid sample on the column; (iii) retention of analyte on the column; (iv) washing of column to remove impurities and matrix components; and finally (v)

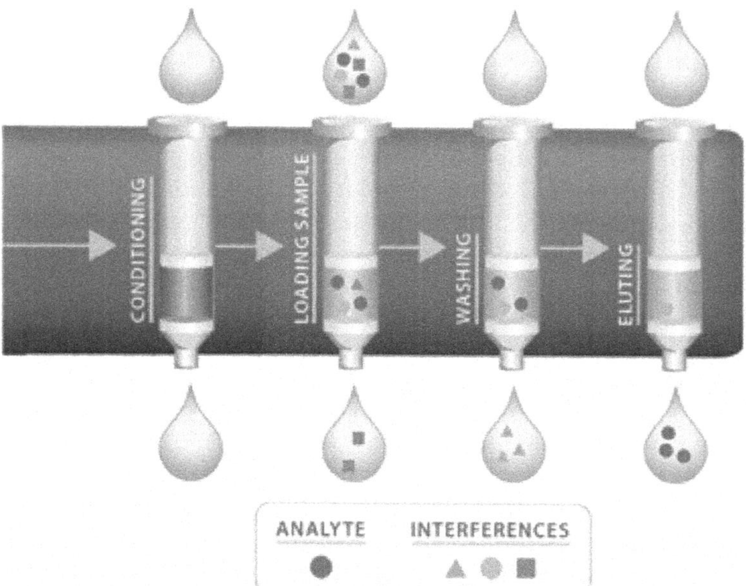

Figure 12.3 Steps in solid phase extraction (SPE). Reproduced from ref. 9 with permission from Elsevier, Copyright 2016.

elution of the analyte from the column using a suitable solvent. Thus, the success of the SPE method does not only depend on the sorbent used but also on the manipulation of the solvents and other variables of these steps. The elution of the more polar amines is normally performed in acidic conditions while less polar amines require organic solvents.[11]

Different commercial sorbents (*e.g.* C18, LiChrolut EN, and ion-exchange resin) are used in SPE work.[12,13] In general, none of these sorbents are able to extract all the BAs effectively. However, the mixed-mode polymeric sorbent (*e.g.* Oasis MCX cartridge), containing reversed phase (C18) and cation exchange (sulfonic) functional groups are superior for the extraction of a wide range of compounds. They are used to adsorb both polar and non-polar, neutral and cationic compounds simultaneously from aqueous media *via* mixed-mode strong cation exchange and reversed-phase mechanisms. This type of sorbent is also used successfully in the determination of BAs.[14,15] In the determination of BAs in wines, Pena-Gallego *et al.*[14] used a mixed-mode sorbent based on Oasis MCX. The pH of the samples is adjusted to be acidic (BAs are converted to cationic species, thus retaining on the cation exchange site of the sorbent) while the neutral compounds are retained on the reversed-phase portion of the sorbent. The authors found that the sequential combination of 2 mL each of three wash solutions (10 mm H_3PO_4 : MeOH (90 : 10, v/v), 10 mM NaOH : MeOH (70 : 30, v/v) and 10 mM $CaCl_2$: MeOH (70 : 30, v/v)) produced the cleanest chromatogram. The second wash was effective to eliminate colouring and phenolic material of the sample. The BAs were finally eluted using 100 mM NaOH : MeOH (65 : 35, v/v). The derivatized BAs when separated on the HPLC column in conjunction with fluorescence (FL) detection with limit of quantitation (LOQ) lower than 0.16 mg L^{-1} for all the BAs (histamine, tryptamine, putrescine, cadaverine) studied by these authors.

A two-step SPE method was used for the clean-up and preconcentration of BAs in red wine.[16] SPE cartridges based on C18, SAX and polyvinylpyrrolidone were first tested mainly to remove polyphenolic compounds followed by a second step where C18 SPE cartridges were used to remove other polar compounds but retain the analyte. Ion-pairing agents such as sodium octanesulfonate, sodium decanesulfonate and sodium dodecanesulfonate were used to enhance the retention of the BAs to the C18 surface of the second step. The optimized procedure enables the derivatized BAs to be detected at μg L^{-1} levels with no interferences when separated on the HPLC-FL unit. In short, the SAX SPE column was found to be effective in removing interferents, especially the polyphenolic compounds (recoveries of 91–111%).

Besides the drive towards miniaturization, another trend in analytical chemistry is towards automation. One of the first attempts to automate the determination of BA was described by Arce *et al.*[17] The method involves online microfilters prior to the introduction to the flow injection manifold. Extraction of amines in the wine was affected by C18 SPE minicolumn in the sample loop. Switching valves were used to direct the flow of solutions for the conditioning, loading and elution of the BAs from the column. After the on-line pretreatments (*i.e.* preconcentration, filtration and

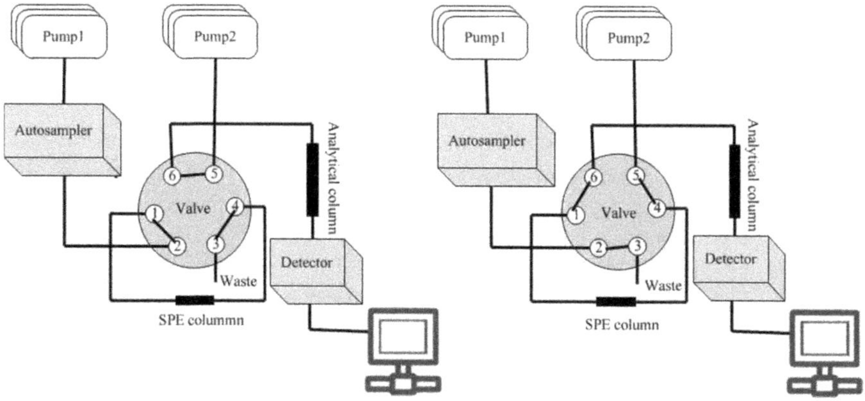

Figure 12.4 On-line solid phase extraction (SPE) coupled with high-performance liquid chromatography (HPLC) for the determination of biogenic amines (BAs) in cheese. (A) On-line preconcentration, (B) elution from the SPE column to the analytical column. Reproduced from ref. 18 with permission from Elsevier, Copyright 2016.

sampling), the sample was introduced into the autosampler of the CE system. Using the appropriate CE background electrolytes, the BAs were separated within 15 min, with LODs in the range 0.05–0.1 μg mL^{-1}. The method was used for the analysis of BAs in synthetic and various types of wine from different regions of Spain. Another on-line SPE pretreatment method that was coupled with capillary HPLC was described by Shan-Shan *et al.*[18] The flow diagram of the set-up is shown in Figure 12.4. The sample was first concentrated on the SPE column using Pump 1 (step A) and transferred to the analytical column for separation using Pump 2 (step B). The direction of movements of the solutions were manipulated by using a six-way switching valve. Limit of detection (LODs) and LOQs for 15 BAs were 0.5–0.25 and 0.15–0.80 mg L^{-1}, respectively. The method was found to be accurate and reliable for the determination of BAs in cheese.

Liquid chromatography-tandem mass spectrometry (LC-MS/MS) has been advocated for its unsurpassed selectivity and sensitivity, making it a wish-list instrument for the modern analytical laboratory. This instrument was used by Sagratini *et al.*[19] for the determination of eight underivatized BAs in fish. After the extraction with trichloroacetic acid (5%), it was subjected to a clean-up and preconcentration using SPE with STRATA X cartridge (conditioned in with methanol and water, adjustment of pH of sample to 11, rinsed with MeOH/water (5 : 95, v/v), eluted with MeOH/acetic acid (99 : 1, v/v). The validated method was compared with a HPLC-FL method based on dansyl derivatives of BAs. As expected, the LC-MS/MS method offered better precision and sensitivity compared with the HPLC-FL. The matrix match calibration (prepared by adding BA standards to fish extract) was found to be effective in overcoming the matrix components that can affect the ionization of the analytes in the MS/MS.

At the moment, the SPE technique has enjoyed wide acceptance as it has been able to reduce the amount of organic solvent used, shorten the analysis time and offer safer work environments for the user. In fact, many standard methods adopted by regulatory agencies are now based on SPE. As an example, many Environment Protection Agency (EPA) USA methods for the analysis of organic compounds in drinking water and wastewater are based on SPE.[9] Other notable advantageous features of the SPE is the convenience of transporting SPE cartridges during transport compared with the bulky glass containers of LLE.[12] Remarkably, analytes sorbed onto the sorbents can be stored for a rather long duration without any change in its concentration. Although not cheap, SPE methods can be automated.

Despite its obvious advantages over the LLE technique, SPE involves multiple steps where loss of analytes is possible at each step. The time required to drive off the extracting solvent after the elution step can take hours. Furthermore, the numerous manipulations of solvents and pH conditions make the method development a daunting task for most users.[12,13] Due to these reasons, there are active efforts to explore alternative methods. Important targets of these methods are the reduction of organic solvent used and the simplicity of procedures.

12.4 Non-classical Sample Preparation Methods

The last two decades witnessed a plethora of studies focussing on alternative sample pretreatments to those of the traditional methods.

12.4.1 Sorbent-based Extractions

The lack of selectivity offered by conventional SPE sorbents such as C18 motivated the search for alternative materials. One strategy is to incorporate molecular recognition compounds onto a suitable support. As an example, Saaid *et al.*[20] chemically immobilized three crown ethers, namely 1-aza-18-crown-6 (A18C6), 1,4,10,13-tetraoxa-7,16-diazacyclo octadecane (DA18C6) and 4-aminobenzo-18-crown-6 (AB18C6) onto the surface of mesoporous silica as support (Figure 12.5). The ability of crown ethers to form selective host–guest complexes with alkali, alkaline-earth, amines and amino acids has been documented. The selectivity of these sorbents is due to the formation of the inclusion complex between crown ethers with the protonated cations such as BAs that was made possible by hydrogen bonding and electrostatic interactions between the cations and the unshared pair of electrons from oxygen, sulphur or nitrogen donor atoms. The authors found that these sorbents exhibit highest selectively towards spermidine in the presence of tryptamine, putrescine, histamine and tyramine.

Other ligands such as those based on benzohydrazide[21] were also tested. With a view for repeated applications, these ligands were physically immobilized in a silica sol–gel matrix. The sorbent could be used for six extraction cycles without any noticeable deterioration in response.

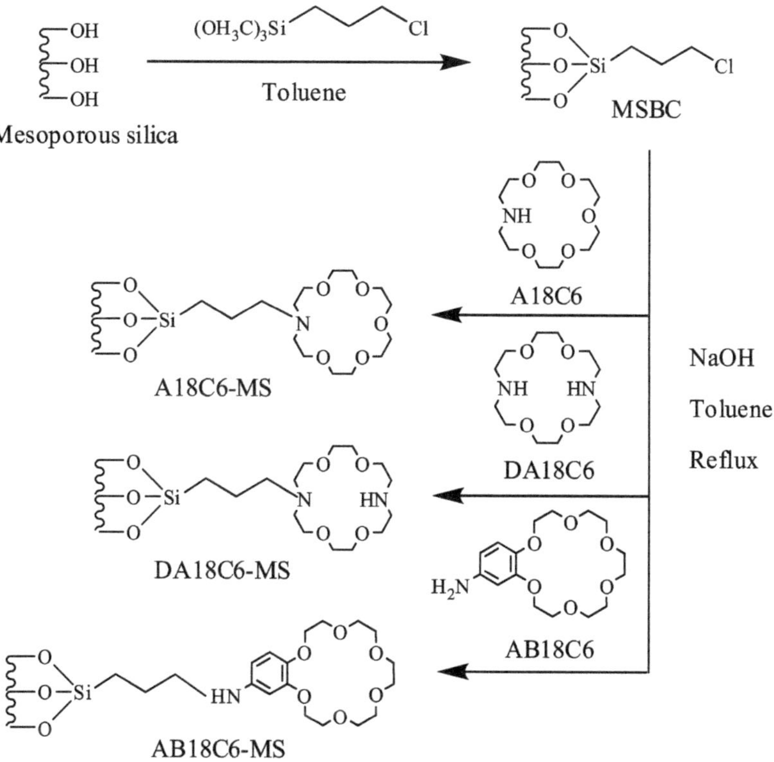

Figure 12.5 Reaction scheme for the chemical immobilization of crown ether ligands onto mesoporous silica support. Reproduced from ref. 20 with permission from Elsevier, Copyright 2010.

Interestingly, these sorbents showed preferential selectivity towards the aliphatic (putrescine, cadaverine, spermidine) and the heterocyclic (histamine) over the aromatic (tyramine, 2-phenylethylamine, tyramine) BAs. When applied for extraction, acceptable recoveries were found on spiked ketchup, soy sauce and juice. However, the recoveries for local shrimp pastes ('budu') were not satisfactory. An important property of sorbents is capacity (defined as mmol of analyte extracted per g of sorbent). The capacity of the benzohydrazide-based ligand is higher than the crown ether-based ligands. Amghouz *et al.*[22] prepared modified zirconium phosphate. After the characterization, the sorbent was tested as sorbent for histamine, putrescine, cadaverine and tyramine in synthetic wine. Histamine and putrescine were the highest adsorbed BAs and the adsorption is postulated to ion-exchange of the cationic BAs with Na^+ cations on the sorbent.

12.4.1.1 Solid Phase Micro Extraction

One of the most significant developments in sample preparation has been, without doubt, solid phase microextraction (SPME), first described by

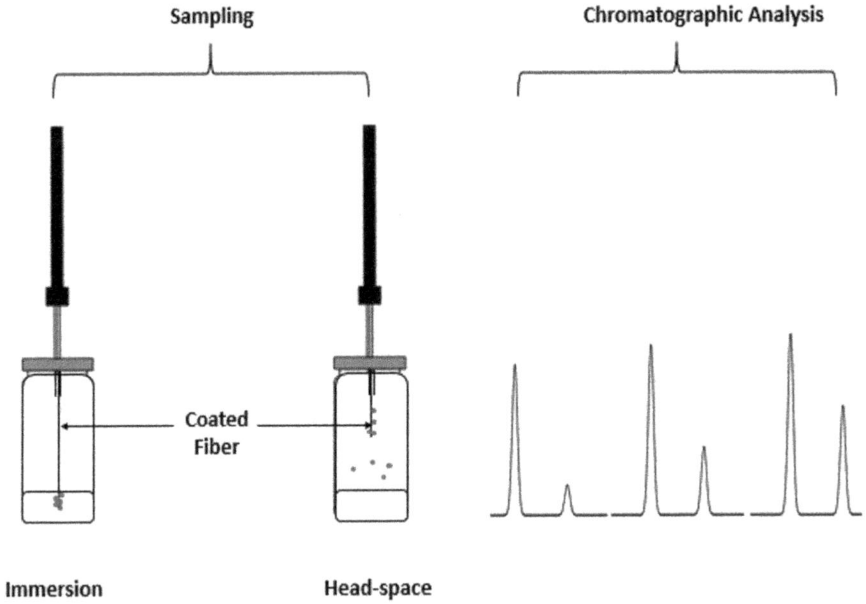

Figure 12.6 Solid phase microextraction (SPME) involving either immersion or head-space sampling techniques, followed by chromatographic analysis.

Arthur and Pawliszyn in 1990.[23] This technique offers a new paradigm shift in sample preparation as for the first time a solvent-free sample preparation technique that combines sampling, analyte isolation and enrichment into one step was demonstrated. An important component of SPME is a fused silica rod that is coated with a suitable stationary phase. The silica rod is either exposed directly to the sample solution or *via* headspace extraction (HS-SPME) (Figure 12.6).

During the extraction, the analytes are adsorbed to the stationary phase sorbent, made possible by interactions such as hydrogen bonding, pi–pi interactions *etc.* After the extraction, the fibre (containing the extracted BAs) is retracted into the SPME housing. The device can then be taken to the instrumentation laboratory for the final analysis. At the lab, the fibre is directly injected into the gas chromatography (GC) or high performance liquid chromatography (HPLC) unit where the BAs are desorbed by thermal or solvent effects. Quantification is based on the amount of analyte extracted under the adopted experimental conditions. Due to its convenience and speed, the SPME technique was quickly adopted by researchers worldwide in food and drugs analysis, clinical chemistry, forensic medicine, environment and other fields.[24,25]

The HS-SPME technique was used for the extraction and determination of histamine, putrescine, cadaverine and tyramine in canned fish samples by ion mobility spectrometry (IMS) without any derivatization process. Histamine and cadaverine have the same mobilities in nitrogen as buffer gas and their corresponding peaks are severely overlapped in ion mobility spectrum. Thus, peak separation was affected by adding 18-crown-6 vapour as complexation reagent into carrier gas and modified ionization region of IMS at optimum flow rate. The interaction between 18-crown-6 and the mentioned amines forms nanocluster product ions with different cross section areas and ion mobilities. Relative standard deviations (RSD%) of the BA determination at 50 g L^{-1} concentration level were in the range 5.7–6.3% while LODs ranges from 0.6 to 1 ng g^{-1}. Nanostructured polypyrrole fibre was used by these authors.[26]

Derivatization can also be conveniently done using the SPME technique. A popular approach is the on-fibre derivatization where the fibre is first dipped into a mixture that contains the organic solvent and the derivatization reagent. The fibre is next taken out of the mixture and air dried. The dipping and air-drying steps can be repeated a few times until the required thickness is achieved. The modified fibre is next immersed into the sample solution to extract the analytes. Finally, the fibre is directly inserted into a GC chromatographic unit for the analysis. This strategy, *i.e.* the simultaneous on-fibre derivatization and extraction of the BAs using HS-SPME has been described by Ali Awan *et al.*[27] The derivatization (using trifluoroacetylacetone) and extraction of putrescine and cadaverine occurs simultaneously at elevated temperature (90–120 °C). The method not only simplifies the procedure, but only requires μg quantities of sample extract and derivatization reagent. Satisfactory recoveries (93.9 and 103.3%) of BAs in the food samples tested were obtained. A PDMS/DVB fibre was used by these authors. Huang *et al.*[28] also used the on-fibre derivatization coupled with GC-MS for the determination of putrescine, cadaverine, histamine and tyramine in fish samples. LODs of the method were 2.98–45.3 μg kg^{-1} with recoveries ranging 78.9–110%.

A key component of the SPME device is the sorbent coating on the fibre substrate. A novel crown ether functionalized ionic liquid-based organic-inorganic hybrid SPME coatings was reported by Zhou *et al.*[29] The unique properties of both ionic liquid (IL) and benzo-15-crown-5 functional group was responsible for the high selectivity and capacity for the extraction of medium polar to polar compounds, including aromatic amines.

The early SPME fused silica fibre is fragile, but this has since been replaced with a metal support to increase its lifetime.[30] The limited range of coating materials commercially available which are more suited for non-polar compounds has prompted individual homemade coatings for task-specific applications.[31] Sometimes, carry-over between runs are also observed.[32]

12.4.1.2 Matrix Solid Phase Dispersion

The matrix solid phase dispersion (MSPD) technique was first introduced by Barker *et al.*[33] in 1989 for the extraction of drug residues from bovine tissues. The technique involves the simultaneous disruption, extraction, preconcentration and clean-up of solid, semi-solid and highly viscous samples. The main steps of the technique are shown in Figure 12.7.[34] The extraction

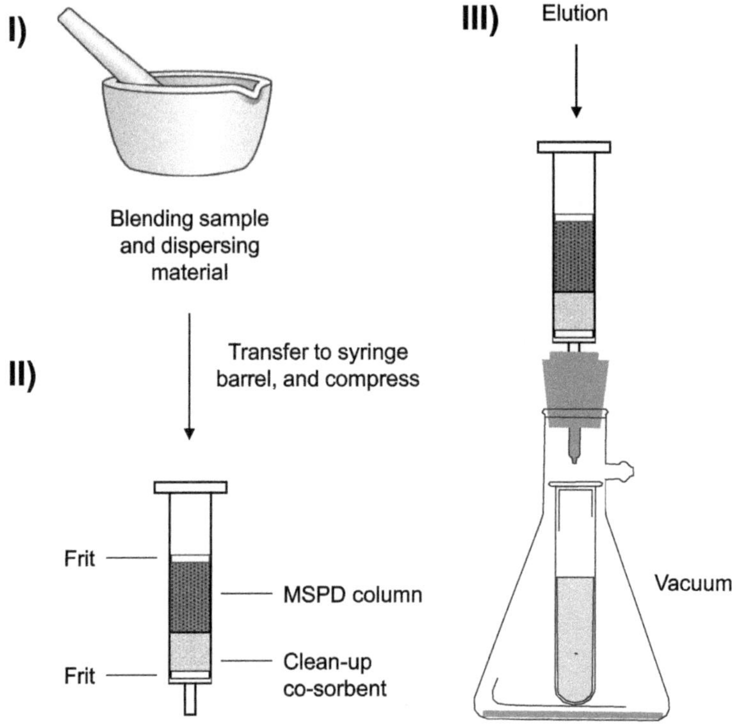

Figure 12.7 Schematic diagram of the matrix solid phase dispersion (MSPD) technique. Reproduced from ref. 34 with permission from Elsevier, Copyright 2015.

can be assisted by vortex and ultrasound. It eliminates most of the complications of the SPE technique and at the same time helps to increase the sensitivity by the direct mechanical blending of sample matrix with an appropriate sorbent. C18- and C8-silica bonded particles are used in most reports, although new sorbents such as carbon-based materials and nanoparticles offer interesting options.[34]

The MSPD technique was explored for the extraction of agmatine, cadaverine, histamine, phenylethylamine, putrescine, tryptamine, tyramine and urocanic acid in canned and frozen tuna. The extracts were analysed by ultrahigh performance hydrophilic interaction chromatography (UHPLC-HILIC) with orbitrap MS detection. The method was validated with respect to accuracy, repeatability, linearity, and LOD and LOQ for all eight BAs. Recoveries were in the 80–109% range, percent relative standard deviation (RSD) ranged from 0.2 to 9.5%, and correlation coefficients for linear regressions were not less than 0.9951 for all samples and analytes.[35]

12.4.1.3 QuEChERS

QuEChERS is an acronym for quick, easy, cheap, effective, rugged and safe method, first reported in 2003 by Anastassiades, Lehotay and co-workers for the extraction of multi-class pesticide residues from fruits and vegetables.[36] After homogenization of the sample, it was subjected to two steps: (i) LLE where the separation of phases is facilitated by salting-out effects (see Section 12.4.2.3), followed by (ii) dispersive solid phase extraction (DSPE) clean-up. The main aim of the second step is to eliminate the possible interfering compounds (*e.g.* organic acids, polar pigments and sugars) from the food extracts. Although the technique was not completely new (*e.g.* solid–liquid extraction/partitioning, including the SALLE were well known. So is the DSPE step, where sorbents were applied for sample clean-up mostly in SPE cartridges is also not new), it is the unique combination of procedures (extraction and clean-up), solvents, salts and sorbents that was found to be very effective, up to the point that it is currently the standard sample-preparation procedure for pesticides analysis in fruits and vegetables[36] with potential for other applications in environmental, agrochemicals and bioanalytical matrices.[37,38] To make it user-friendly, a number of vendors currently sell QuEChERS kits to make the procedure faster, to avoid weighing the salts, and adapted to each application.

A rapid and sensitive UHPLC-MS/MS method after subjecting samples to a QuEChERS method was developed for the simultaneous determination of seven BAs in soy sauce.[39] Samples were firstly diluted by water and pH values were adjusted using ammonia water. The mixture was vortexed for 30 s and acetonitrile (5 mL) and sodium chloride (2 g) were further added. Then the pooled extraction solution was diluted to 10 mL with acetonitrile and mixed evenly. The extraction solution (2.0 mL) was moved to a 2.5 mL high speed plastic centrifuge tube which contains sorbent mixture of C18 (100 mg) and primary-secondary amine (100 mg). Analysis of seven BAs was completed within 4 min. Under the optimized conditions, linear relations were found over the selected concentration ranges of 3.0–600 μg L^{-1} with correlation coefficients greater than 0.999. The average accuracy was between 84% and 115%. Intra- and inter-assay precisions for six replicates ranged from 2.2% to 8.8% or below 12%, respectively. The method LOD and the method LOQ were 4–8 μg kg^{-1} and 15–30 μg kg^{-1}, respectively.

12.4.1.4 Micro-SPE

Porous membrane protected, also known as micro-solid phase extraction (mSPE) was developed in 2006 by Basheer *et al.*[40] mSPE has been widely accepted as powerful sample preparation technique to counter the challenges faced in conventional LLE and SPE.[41,42] This novel sample preparation approach was used to extract and preconcentrate analytes in different matrices. In this technique, sorbent (about 5–10 mg) is placed inside a sheet of porous polymer membrane and its edges are heat sealed to convert it into a small bag. This bag containing the sorbent (known as mSPE device), is then cleaned and conditioned in a suitable solvent and stored in the same solvent until further use. For extraction, the mSPE device is taken out from the conditioning solvent and is air dried. After that, the device is

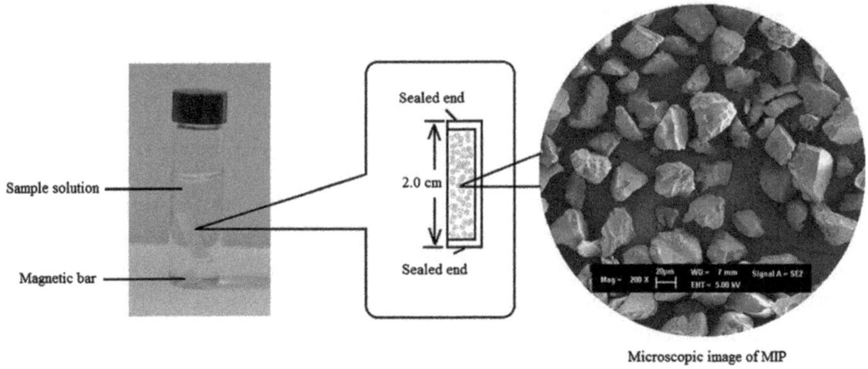

Figure 12.8 Schematic diagram of the micro-solid phase extraction (mSPE) technique. A small amount of sorbent is trapped in an envelope before placing into the sample solution. Reproduced from ref. 41 with permission from Elsevier, Copyright 2012.

added into the sample solution and the device is stirred for an optimum period of time. After the extraction, the device is taken out from the sample solution and is rinsed with ultra-pure water and dried with lint-free tissue. The device is next transferred to small glass or plastic vial and analytes are desorbed by adding a small volume of solvent. The desorption is normally assisted by ultra-sonication. The extract is then analysed using instrumental methods such as CE, GC or HPLC. An mSPE system is shown in Figure 12.8.[41]

Undoubtedly, the success of mSPE lies in the choice of sorbent. With this in mind, Basheer *et al.*[43] has synthesized eight hydrazone ligands. They were next physically trapped in sol–gel matrices before being use as the mSPE device and tested for the dansyl chloride derivatives of histamine, tyramine, tryptamine, putrescine, spermidine and phenylethylamine. Of the ligands that were prepared, benzophenone 2,4-dinitrophenylhydrazone was the best. The recognition was contributed mainly due to the π–π interaction between the BAs and the ligand. Another contributor to the success of the technique is the exclusion of macromolecules by the porous membrane, thus increasing the selectivity of the method. Another major advantage is the high enrichment factors (94–460) that were obtained. The method was applied to the determination of BAs in orange juice samples where high levels of putrescine (550–2210 $\mu g\ L^{-1}$) were found.[43]

12.4.2 Solvent-based Extractions

Solvent-based extractions and microextractions capitalize on the type of extracting solvent and the protocols used to achieve selectivity. A simple strategy adopted to achieve good enrichments is to use a relatively large volume of the sample relative to the volume of the extracting phase.

12.4.2.1 Cloud Point Extraction

Surfactant-based extraction methodologies such as the cloud point technique are becoming widespread for the preconcentration of analytes present at trace levels. This technique capitalizes on the phenomenon that when an aqueous solution containing the analyte and some non-ionic surfactant in amounts above their critical micellar concentration is heated above a critical temperature (the cloud point temperature, CPT), the solution becomes cloudy and the surfactant separates into a second phase. This surfactant phase, constituting a very small portion of the original volume of the mixture, usually contains all the analytes. The micellar phase is next separated from the aqueous phase and diluted with HPLC compatible solvents, such as acetonitrile or methanol, enabling it to be injected into an HPLC system. One obvious drawback of this method is that it is not suitable for the extraction of heat-labile analytes if the surfactant used has a high CPT or if much higher temperatures than the CPT must be maintained for a relatively long period of time in order to allow for maximum extraction.[44,45] Alternatively, another surfactant-based methodology is the phase-transfer method which does not require the use of heat for extraction. After the vigorous shaking of an aqueous salt solution with an added immiscible organic solvent (*e.g.* heptane) and the appropriate time interval for enabling a separation of the resulting two clear phases, reverse micelles are formed in the organic phase. A surfactant such as sodium bis-[2-ethylhexyl]sulphosuccinate (AOT) has been widely employed for micellar extraction of metals, amino acids and proteins from aqueous salt solutions.[44–46]

The cloud point technique was applied for the first time for the extraction of BAs (putrescine, cadaverine, agmatine, tyramine, tryptamine, phenylethylamine, spermine, spermidine and histamine) in aqueous samples after their derivatization with benzoyl chloride.[47] The derivatives were separated with gradient elution micellar liquid chromatography. The combination of these two techniques resulted in a simple, rapid and effective alternative HPLC method for the simultaneous determination of the BAs in trout samples.

Romero *et al.*[48] compared the results from the phase-transfer methodology with AOT in heptane as the extracting phase with those obtained by an optimized cloud point procedure with Triton X-114 for the measurement of tryptamine, cadaverine, tyramine, putrescine, spermine, spermidine, histamine and phenylethylamine. The techniques were then employed for the extraction of BAs from a beer sample. It was concluded that the reverse micelles of AOT in heptane represent a better extraction system than the cloud point extraction with Triton X-114 since lower detection limits and better preconcentration factors are obtained, and the reproducibility is better. The optimized extraction procedure with AOT is faster, more sensitive and more reproducible than extractions with Triton X-114. Additionally, the proposed procedure is extremely simple and requires no heating of the sample.

12.4.2.2 Ionic Liquid LLME

Ionic liquids (ILs) are liquids that are entirely composed of organic cations and inorganic or organic anions at or close to room temperature. Their unique characteristics of high thermal stability, non-flammability and good solubility for inorganic and organic compounds lend themselves as replacements for conventional organic solvents in some extraction processes, such as LLE, liquid phase microextraction, SPE and aqueous two-phase systems extraction.[49,50] Recently, ILs when used in ultrasonic-assisted extraction methods have emerged as an attractive alternative for sample preparation because they showed high recovery and enrichments, simplicity, rapidness and low cost.[49–51] Room temperature ILs have been used as novel media for derivatization, extraction and preconcentration of BAs in wines. Dansyl chloride was selected as the derivatizing reagent. The derivatizations of BAs were conducted in a water-IL two-phase system and the dansylated amines were separated by HPLC with ultraviolet (UV) detection. The use of IL significantly improved the efficiency of both the derivatization reaction and the preconcentration.[52]

A sensitive method for the determination of octopamine, tyramine and phenethylamine in beer samples using the fluorescent reagent 2,6-dimethyl-4-quinolinecarboxylic acid *N*-hydroxysuccinimide ester followed by ionic liquid-based ultrasound-assisted liquid–liquid microextraction was described by Huang *et al.*[53] The room temperature ionic liquid 1-butyl-3-methylimidazolium hexafluorophosphate was used as extractant. The extraction time was very short (1 min) and the HPLC analysis time was 4 min. Reduction of exposure to toxic solvent and the evaporation of the extractant is not required are other significant advantages of the method.

12.4.2.3 Salting-out LLE

The salting-out LLE (SALLE) technique was first proposed by Matkovich and Christian in 1973.[54] A typical SALLE technique uses a binary solvent system consisting of water and another organic solvent (*e.g.* acetonitrile, THF, ethanol, acetone). After the two phases are mixed together and after the addition of salt (*e.g.* sodium chloride, sodium hydroxide, calcium chloride, zinc sulphate, ammonium sulphate, potassium phosphate) phase separation occurred and the analytes are transferred towards a small volume of organic phase which is next collected and injected into HPLC or GC units.

Miscible organic solvent is used to provide a large interface area. The addition of salts decreased the solubility of the solvent and removal of one solvent *via* hydration processes.[54] The phase separation is due to the reduction of mutual miscibility of two liquids by addition of electrolytes, the whole process being a consequence of disruption of weak intermolecular forces between organic solvent and water. At the same time, the salt should have high solubility in water, insoluble in the organic solvent and has the ability to precipitate hydrophilic substances.[54]

A SALLE technique for the determination of BAs in fruit juices and alcoholic beverages involving the reaction of BAs with 1-naphthylisothiocyanate followed by extraction of 1-naphthylthiourea derivatives with water-miscible organic solvent (acetonitrile) was described. Addition of ammonium sulphate was found to be not only effective for the phase separations but it also improves the extraction of molecular species into the organic phase.[55] A major advantage of the method is that the extraction can be simultaneous executed with derivatization, thus reducing the number of stages before the chromatographic analysis.

A simple and fast methodology for the determination of BAs in wines (white, red and rose) based on SALLE was proposed by Ramos *et al.*[56] The proposed methodology uses acetonitrile as organic solvent for the extraction, and typical of SALLE techniques, the procedure is very straightforward. Additionally, the extraction can be accomplished in 25 min. In comparison with the common extraction procedure (SPE), SALLE has a lower solvent consumption and requires fewer steps; therefore, there is the lower probability of analytes loss and is simpler to execute. Other hallmarks of the method are simplicity and that it uses readily available glassware, thus enabling it to be adopted even in laboratories with basic infrastructure.

12.4.2.4 Hollow Fibre Liquid Phase Microextraction

The hollow fibre liquid phase microextraction (HF-LPME) technique was introduced by Pedersen-Bjergaard and Rasmussen in 1999.[57] In this technique, the extracting solvent (a few μL) is securely trapped within a porous polypropylene membrane (hollow fibre, HF) typically of 0.2 μm pore size, 200 μm wall thicknesses and 600–1200 μm internal diameter. The HF is highly porous, allowing the rapid transfer of analytes from the donor to the acceptor phase.

Before the extraction, the HF is washed with volatile solvent to remove any contaminants and it is allowed to evaporate completely. After that, the HF is dipped into the organic solvent to obtain the supported liquid membrane (SLM). The HF is next washed with water to remove the excess organic solvent, before filling the lumen of the HF with the acceptor phase and immersed into the sample solution. After the extraction, the acceptor phase is collected for instrumental analysis (*e.g.* GC, CE and HPLC). The analytes are transferred from the donor phase (aqueous) through the SLM into the acceptor phase. The pores of the HF can reduce interferences from macromolecules by size exclusion effects.

Based on the type of acceptor phase, HF-LPME can be categorized into either the two-phase HF-LPME or three-phase HF-LPME.[58,59] In the two-phase HF-LPME, the acceptor phase in the lumen of the HF is filled with organic solvent (similar solvent used to immobilize within the pores of the HF), while in the three-phase HF-LPME technique, the acceptor phase could either be acidic or alkaline aqueous solution (Figure 12.9).[60]

The organic solvent should be of low volatility, immiscible with water and provide high solubility of analytes.[61] This technique is suitable for the

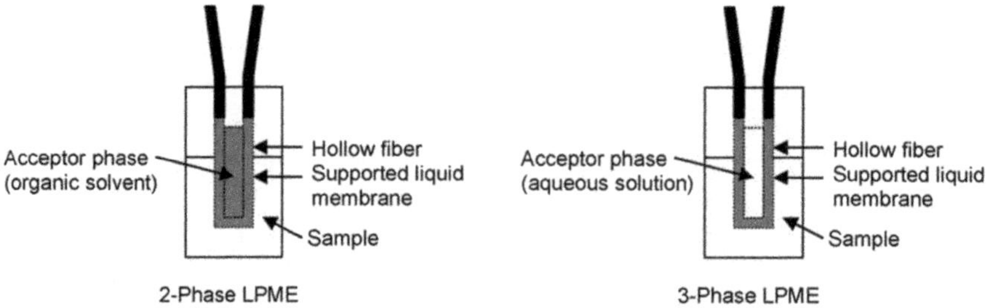

Figure 12.9 Schematic diagram of two-phase and three-phase hollow fibre liquid phase microextraction (HF-LPME) setups. Adapted from ref. 60 with permission from Elsevier, Copyright 2008.

extraction of hydrophobic compounds with high log partition coefficient (P) values. As the acceptor phase is organic, the extract is suitable for direct GC analysis. Further manipulation of the microextracts such as evaporation or reconstitution is required if they are to be analysed using HPLC or CE.

In the three-phase HF-LPME, the analytes are extracted from aqueous phase (donor) through the organic solvent (immobilized in the pores of the HF) into another aqueous phase (acceptor) in the lumen of the HF. The three-phase LPME is suitable for HPLC and CE analysis as the extract is aqueous. This technique is targeted for the analysis of ionizable compounds with high $\log P$ values. pH plays key role in the three-phase HF-LPME technique. As an example, for the extraction of acidic analytes, the pH of the sample should be adjusted to low pH to decrease the analyte solubility in the sample and ease the transfer into the SLM, while at the same time, the pH of the acceptor phase should be alkaline to enhance the analytes solubility.[58] Under these conditions, the analyte will be extracted into the SLM and back-extracted to the acceptor phase. The opposite conditions are used for the extraction of basic compounds.

Dziarkowska *et al.*[61] developed a LPME method for the extraction of the polyamines, putrescine, cadaverine, spermine and spermidine in biological samples. The extracted analytes were subjected to a precolumn derivatization procedure using tosyl chloride followed by the HPLC-UV analysis. These findings triggered Saaid *et al.*[62] to develop the HF-LPME for food applications. For the first time, simultaneous *in situ* derivatization (with dansyl chloride) and HF-LPME extraction was carried out in a single step for the determination of BAs in food samples. The analytes in the sample solution were derivatized and then extracted into the organic phase residing in the pores and the lumen of the HF. The method was applied for the analysis of BAs in shrimp sauce and tomato ketchup samples. A notable advantage of the method is the good enrichment factors (47–456) that were realized. This is due to the transfer of analytes from a relatively large sample volume into small volumes of the lumen. The clean HPLC chromatograms from the complex food samples suggest the HF membranes were effective in excluding endogenous matrix components during the extraction. The HF is cheap and is used once, overcoming carry-over problems in real sample analysis.

A major drawback of the HF-LPME is the long extraction times (about 20–40 min) and the manual preparation of the fibre can lead to poor precision. However, the extraction time has been reduced using the electromembrane approach,[63,64] where a potential is used to accelerate the movement of the analytes. An on-chip electromembrane extraction coupled to HPLC was described by Zarghampour *et al.*[64] for the determination of histamine, tryptamine, putrescine, cadaverine and spermidine in meat product samples including sausage and kielbasa (Iranian food).

12.4.2.5 *Dispersive LLME*

Dispersive LLME (DLLME) was introduced by Assadi and co-workers in 2006 as a new LLE technique for the determination of polyaromatic hydrocarbons.[65] This technique is based on a ternary component solvent system comprising extraction solvent, disperser solvent and an aqueous sample. When an appropriate mixture of extraction and dispersive solvents is injected into an aqueous sample containing the analytes of interest, a cloudy solution is formed. The aim of the cloudy suspension is to markedly increase the contact area for maximum extraction. When the cloudy solution is centrifuged, the extraction solvent is normally sedimented at the bottom of the tube (if the density is above that of water) and removed with a microsyringe for instrumental analysis. A schematic diagram of the technique is shown in Figure 12.10.[66] The hallmarks of the DLLME technique are the simplicity of operation, rapid, low cost, high recovery, high enrichment factor and short extraction time.[50,65–67] Since the initial introduction

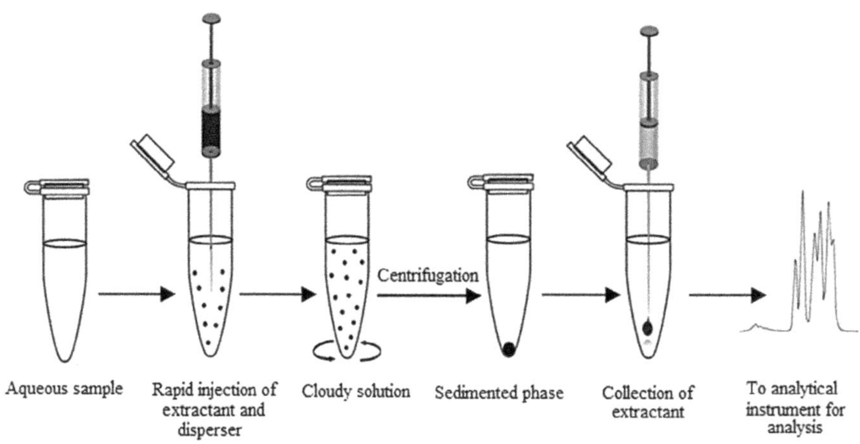

Figure 12.10 Steps in dispersive liquid–liquid microextraction (DLLME) protocols. Reproduced from ref. 66 with permission from Elsevier, Copyright 2015.

of the DLLME technique, several modifications have been made including use of solvents with density lower than that of water; extraction without the need for centrifugation; and use of ionic liquids as extraction solvent. Organic solvents that are denser than water are not only more toxic (*e.g.* halogenated solvents) but they are also more difficult to be collected after the centrifugation. To overcome the difficulty in collecting the extracted solvent, Jia *et al.*[68] described an *in situ* derivatization of nine BAs using benzoyl chloride based on solidification of floating organic droplets. Using solvents of low density, the organic extractant floats on the sample surface. The floating droplet can be solidified using an ice-bath and can be transferred out of the sample matrix. The method was successfully applied in the analysis of white wine, red wine, rice wine and beer.

The combination of microextraction and derivatization has been proven to be a convenient method for the determination of BAs. Obvious benefits by the integration of derivatization and sample preparation steps are the improvements of recovery and sensitivity, as well as the reduction in analysis time. The polarity of BAs is significantly reduced by derivatization and thus extraction efficiencies are dramatically enhanced. As an example, Huang *et al.*[69] uses ultrasound to increase the DLLME for the HPLC-FL determination of octopamine, tyramine and phenethylamine in rice wine samples; 2,6-Dimethyl-4-quinolinecarboxylic acid *N*-hydroxysuccinimide ester was used as the derivatization reagent. Recoveries of 95.42–104.6% for spiked samples were obtained by the authors. The simultaneous DLLME with derivatization of 13 BAs followed by GC-MS analysis was described by Płotka-Wasylka *et al.*[70] Methanol (215 μL) and chloroform (400 μL) were used as disperser and extractive solvent, respectively, while isobutyl choloroformate (90 μL) was the derivatization reagent. The LODs of the BAs ranged from 1.1 to 4.1 $\mu g\ L^{-1}$ while the LOQs ranged from 3.3 to 11.7 $\mu g\ L^{-1}$. Rather high concentrations of putrescine, tyramine, histamine, dimethylamine and diethylamine were found when the method was used for the determination of 13 BAs in homemade wine samples. Other applications of the DLLME technique are shown in Table 12.2.

The extraction efficiency of DLLME is influenced by several factors (*e.g.* types and volumes of extraction and disperser solvents, extraction time, sample amount, pH and salt addition). DLLME is a simple and cheap tool and requires small volumes of organic solvents. In addition, dispersing the organic extractant in the aqueous sample increases the contact surface between both phases, improves the extraction kinetics and enhances the efficiency. However, using manual shaking to induce dispersion is not effective. Thus, vortexing and ultrasonic effects were found to be effective.

12.5 Sample Preparation in the Context of Green Analytical Chemistry

Recently, there have been significant developments in methodology to minimize and reduce the deleterious effects of analytical activities, in line

Table 12.2 Examples of analysis of biogenic amines (BAs) using dispersive liquid–liquid microextraction (DLLME) technique.[a]

Analytical technique	Disperser solvent and volume	Extractive solvent and volume	Derivatization reagent and volume	Type of sample and volume	BAs	LOD	LOQ	Recoveries, %	Assistance	References
GC-MS	ACN, 1000 μL	Toluene, 325 μL	Isobutyl chloroformate, 25 μL	Beer, 5 mL	18	0.3–2.9 μg L^{-1}	1.0–9.5 μg L^{-1}	72–113	—	71
GC-MS	Methanol, 215 μL	Chloroform, 400 μL	Isobutyl chloroformate, 90 μL	Homemade wine, 5 mL	13	1.1–4.1 μg L^{-1}	3.3–12.3 μg L^{-1}	77–105	—	70
HPLC-FL	ACN, 400 μL	Octanol, 50 μL	2,6-Dimethyl-4-quinolinecarboxylic acid *N*-hydroxysuccinimide ester	Rice wine, 100 μL	3	0.02–5 ng mL^{-1}	—	96.7–105	Ultrasound	69
HPLC-UV	Methanol, 450 μL	1-Dodecanol, 50 μL	Benzoyl chloride, 125 μL	Alcoholic beverages, 2 mL	9	0.005–0.01 μg mL^{-1}	0.05 μg mL^{-1}	44–147	Solidification of floating organic droplets	68
HPLC-UV	ACN, 300 μL	1-Octanol, 60 μL	Dansyl chloride, 1 mL	Turkey breast meat, 10 mL	3	0.8–1.4 ng g^{-1}	0.8–1.4 ng g^{-1}	95–105	Microwave	72
HPLC-FL	—	1-Butyl-3-methylimidazolium hexafluorophosphate, 30 μL	2,6-Dimethyl-4-quinolinecarboxylic acid *N*-hydroxysuccinimide ester, 400 μL	Beer, 100 μL	3	0.25–50 ng mL^{-1}	0.8–167 ng mL^{-1}	90–116	Ultrasound, ionic liquid	73

[a]Abbreviations: ACN, acetonitrile; GC-MS, gas chromatography-mass spectrometry; HPLC-FL, high-performance liquid chromatography-fluorescence; HPLC-UV, high-performance liquid chromatography-ultraviolet; LOD, limit of detection; LOQ, limit of quantification.

with the aspirations of green analytical chemistry. The main thrust of green analytical chemistry is the reduction of environmental impacts of the analytical methodology by introducing alternative eco-friendly approaches without sacrificing the analytical performance. In lieu of this, more attention should be paid to sample preparation and separation techniques as they are the biggest consumers of reagents and solvents in analytical chemistry. Focus should be given to: (i) reduction of solvent volume for the extraction; (ii) reduction of reagents used, and derivatization should be avoided where ever possible; (iii) minimizing sample preparation steps; (iv) availability of extract for direct analysis; and (v) enrichment of analyte in the final extract to facilitate the use of low-energy consumption instruments (*e.g.* using HPLC with UV detector instead of MS).

Thus, microextraction techniques being specifically targeted to the reduction of extractant (μL range for liquids and less than 100 mg for solid) complies with green analytical chemistry principles. It must be pointed out

Table 12.3 Analytical Eco-Scale penalty points for selected reagents and instruments.[a] Reproduced from ref. 75 with permission from Elsevier, Copyright 2012.

Reagent		
Type		Hazard PP
Acetonitrile		4
Methanol		6
Tetrahydrofuran		6
1-Octanol		1
DEHPA		4
NaOH		2
$(NH_4)_2SO_4$		0
		Total PP for reagents = hazard PP × amount PP
		Amount PP
Amount	<10 mL (g)	1
	10–100 mL (g)	2
	>100 mL (g)	3
Instruments		
		PP
Technique energy	HPLC	1
	GC	1
Waste amount	None	0
	<1 mL (g)	1
	1–10 mL (g)	3
	>10 mL	5
Occupational hazard	Emission of vapours and gases to the air	3
	No emission	0

[a]Abbreviations: HPLC, high-performance liquid chromatography; PP, penalty point.

that most of these microextraction techniques, unlike LLE and SPE, are not an exhaustive extraction since only a small portion of the analytes are extracted. Several models have been proposed to compare and rank the greenness of an analytical method.[2,74,75] One such approach is the Analytical Eco-Scale as proposed by Galuszka *et al.*[75] According to this approach, the ideal green method must has the following characteristics: (i) reagents are not hazardous; (ii) no waste is produced; and (iii) energy consumption per sample is >0.1 kW h. A value of 100 is assigned for an ideal green method. Penalty points (PP) are deducted for each of the analytical methodology components (*i.e.* reagents, hazards, waste and energy consumption). The Analytical Eco-Scale total is calculated according to the following equation:

$$\text{Analytical Eco-Scale} = 100 - \text{total penalty points}$$

Three categories of the Analytical Eco-Scale are assigned: excellent green analysis (>75); acceptable green analysis (>50); and inadequate green analysis (<50). Details are shown in Table 12.3.[75]

12.6 Conclusion

The importance of sample preparation cannot be over emphasized. Traditional methods of sample preparation based on LLE and SPE are still being extensively used. The demand for green analytical methods has resulted in a thriving area of research based on microextractions. These methods support the green analytical chemistry principles, especially in terms of simplicity and minimization of organic solvents used. Although the applications of microextractions to BA analysis is rather small at the moment, it can be expected that they will become increasingly important in the coming years. Of the many microextraction methods proposed, it is forecasted that DLLME will be predominant due to its versatility and furthermore requires normal laboratory infrastructure and glassware.

References

1. G. I. Mohammed, A. S. Bashammakh, A. A. Alsibaai, H. Alwael and M. S. El-Shahawi, *TrAC, Trends Anal. Chem.*, 2016, **78**, 84.
2. M. Papageorgiou, D. Lambropoulou, C. Morrison, E. Kłodzińska and J. Namieśnik, *TrAC, Trends Anal. Chem.*, 2018, **98**, 128.
3. J. L. Ordóñez, A. M. Troncoso, M. D. C. García-Parrilla and R. M. Callejón, *Anal. Chim. Acta*, 2016, **939**, 10.
4. M. Saaid, B. Saad, N. H. Hashim, A. S. M. Ali and M. I. Saleh, *Food Chem.*, 2009, **113**, 1356.
5. T. Chatzimitakos, V. Exarchou, S. A. Ordoudi, Y. Fiamegos and C. Stalikas, *Food Chem.*, 2016, **202**, 445.
6. M. Tobiszewski and J. Namieśnik, *Ecotoxicol. Environ. Saf.*, 2015, **120**, 169.

7. J. Płotka-Wasylka, M. Rutkowska, K. Owczarek, M. Tobiszewski and J. Namiesnik, *TrAC, Trends Anal. Chem.*, 2017, **91**, 12.
8. M. Tobiszewski and J. Namiesnik, *Curr. Opin. Green Sustain. Chem.*, 2017, **5**, 1.
9. A. Andrade-Eiroa, M. Canle, V. Leroy-Cancellieri and V. Cerdà, *TrAC, Trends Anal. Chem.*, 2016, **80**, 641.
10. P. Lucci, D. Pacetti, O. Núñez and N. G. Frega, in *Chromatography*, IntechOpen, London, 2012, ch. 5, p. 127.
11. C. Molins-Legua and P. Campins-Falco, *Anal. Chim. Acta*, 2005, **546**, 206.
12. A. Andrade-Eiroa, M. Canle, V. Leroy-Cancellieri and V. Cerdà, *TrAC, Trends Anal. Chem.*, 2016, **80**, 641.
13. J. Płotka-Wasylka, N. Szczepańska, M. de la Guardia and J. Namiesnik, *TrAC, Trends Anal. Chem.*, 2015, **73**, 19.
14. A. Pena-Gallego, P. Hernández-Orte and J. Cacho, *J. Chromatogr. A*, 2009, **1216**, 3398.
15. T. Tang, K. Qian, T. Shi, F. Wang, J. Li, Y. Cao and Q. Hu, *Food Control*, 2011, **22**, 1203.
16. O. Busto, J. Guasch and F. Borrull, *J. Chromatogr. A*, 1995, **718**, 309.
17. L. Arce, A. Rios and M. Valcarcel, *J. Chromatogr. A*, 1998, **803**, 249.
18. Y. Shan-Shan, Y. Ya-Nan, L. Xue-Lin and Z. Yang, *Chin. J. Anal. Chem.*, 2016, **44**, 396.
19. G. Sagratini, M. Fernández-Franzón, F. De Berardinis, G. Font, S. Vittori and J. Mañes, *Food Chem.*, 2012, **132**, 537.
20. M. Saaid, B. Saad, I. Ab Rahman, A. S. M. Ali and M. I. Saleh, *Talanta*, 2010, **80**, 1183.
21. A. A. Tameem, B. Saad, A. Makahleh, A. Salhin and M. I. Saleh, *Talanta*, 2010, **82**, 1385.
22. Z. Amghouz, C. Ancin-Azpilicueta, K. K. Burusco, J. R. Garcia, S. A. Khainakov, A. Luquin, R. Nieto and J. J. Garrido, *Microporous Mesoporous Mater.*, 2014, **197**, 130.
23. C. L. Arthur and J. Pawliszyn, *Anal. Chem.*, 1990, **62**, 2145.
24. C. Xu, G. Chen, Z. Xiong, Y. Fan, X. Wang and Y. Liu, *TrAC, Trends Anal. Chem.*, 2016, **80**, 12.
25. Q. Zhang, L. Zhou, H. Chen, C. Wang, Z. Xia and C. Yuan, *TrAC, Trends Anal. Chem.*, 2016, **80**, 57.
26. R. Parchami, M. Kamalabadi and N. Alizadeh, *J. Chromatogr. A*, 2017, **1481**, 37.
27. M. Ali Awan, I. Fleet and C. L. P. Thomas, *Food Chem.*, 2008, **111**, 462.
28. J. Huang, N. Gan, F. Lv, Y. Cao and C. Ou, *Sep. Sci.*, 2016, **39**, 4384.
29. X. Zhou, P. Xie, J. Wang, B. Zhang, M. Liu, H. Liu and X. Feng, *J. Chromatogr. A*, 2011, **1218**, 3571.
30. H. Piri-Moghadam, M. N. Alam and J. Pawliszyn, *Anal. Chim. Acta*, 2017, **984**, 42.
31. M. Lashgari and Y. Yamini, *Talanta*, 2019, **191**, 283.
32. J. Nawala, B. Dawidziuk, D. Dziedzic, D. Gordon and S. Popiel, *TrAC, Trends Anal. Chem.*, 2018, **105**, 18.

33. S. A. Barker, A. R. Long and C. R. Short, *J. Chromatogr.*, 1989, **475**, 353.
34. A. L. Capriotti, C. Cavaliere, P. Foglia, R. Samperi, S. Stampachiacchiere, S. Ventura and A. Laganà, *TrAC, Trends Anal. Chem.*, 2015, **71**, 186.
35. R. L. Self and W. Wu, *J. Food Compos. Anal.*, 2012, **27**, 169.
36. M. Anastassiades, S. J. Lehotay, D. Štajnbaher and F. J. Schenck, *J. AOAC Int.*, 2003, **86**, 412.
37. M. L. Schmidt and N. H. Snow, *TrAC, Trends Anal. Chem.*, 2016, **75**, 49.
38. M. Á. González-Curbelo, B. Socas-Rodríguez, A. V. Herrera-Herrera, J. González-Sálamo, J. Hernández-Borges and M. Á. Rodríguez-Delgado, *TrAC, Trends Anal. Chem.*, 2015, **71**, 169.
39. H. Dong and K. Xiow, *Food Chem.*, 2017, **229**, 502.
40. C. Basheer, A. A. Alnedhary, B. S. M. Rao, S. Valliyaveettil and H. K. Lee, *Anal. Chem.*, 2006, **78**, 2853.
41. T. P. Lee, B. Saad, W. S. Khayoon and B. Salleh, *Talanta*, 2012, **88**, 129.
42. M. Sajid, *Anal. Chim. Acta*, 2017, **965**, 36.
43. C. Basheer, W. Wong, A. Makahleh, A. A. Tameem, A. Salhin, B. Saad and H. K. Lee, *J. Chromatogr. A*, 2011, **1218**, 4332.
44. E. K. Paleologos, D. L. Giokas and M. I. Karayiannis, *TrAC, Trends Anal. Chem.*, 2005, **24**, 426.
45. K. Madej, *TrAC, Trends Anal. Chem.*, 2009, **28**, 436.
46. K. Pytlakowska, V. Kozik and M. Dabioch, *Talanta*, 2013, **110**, 202.
47. E. K. Paleologos, S. D. Chytiri, I. N. Savvaidis and M. G. Kontominas, *J. Chromatogr. A*, 2003, **1010**, 217.
48. L. Romero, J. Osorio Grisales and M. Reta, *Talanta*, 2010, **81**, 1431.
49. J. Nawała, B. Dawidziuk, D. Dziedzic, D. Gordon and S. Popiel, *TrAC, Trends Anal. Chem.*, 2018, **105**, 18.
50. A. Berthod, M. J. Ruiz-Angel and S. Carda-Broch, *J. Chromatogr. A*, 2018, **1559**, 2.
51. I. Rykowska, J. Ziemblińska and I. Nowak, *J. Mol. Liq.*, 2018, **259**, 319.
52. H. L. Jiang, L. Y. Ying, S. C. Zhou, M. Ying, W. Shen and D. H. Qiu, *J. Sep. Sci.*, 2011, **34**, 1055.
53. K. J. Huang, C. X. Jin, S. L. Song, C. Y. Wei, Y. M. Liu and J. Li, *J. Chromatogr. B*, 2011, **879**, 579.
54. C. E. Matkovich and G. D. Christian, *Anal. Chem.*, 1973, **45**, 1915.
55. A. Jain, M. Gupta and K. Verma, *J. Chromatogr. A*, 2015, **1422**, 60.
56. R. M. Ramos, I. M. Valente and J. A. Rodrigues, *Talanta*, 2014, **124**, 146.
57. S. Pedersen-Bjergaard and R. K. Rasmussen, *Anal. Chem.*, 1999, **71**, 2650.
58. V. Sharifi, A. Abbasi and A. Nosrati, *J. Food Drug Anal.*, 2016, **24**, 264.
59. V. Salvatierra-stamp, R. Muñiz-Valencia, J. M. Jurado and S. G. Ceballos-Magaña, *Microchem. J.*, 2018, **140**, 87.
60. S. Pedersen-Bjergaard and K. E. Rasmussen, *J. Chromatogr. A*, 2008, **1184**, 132.
61. K. Dziarkowska, J. Åke Jönsson and P. P. Wieczorek, *Anal. Chim. Acta*, 2008, **606**, 184.
62. M. Saaid, B. Saad, A. Salhin, M. Ali, M. I. Saleh, C. Basheer and H. K. Lee, *J. Chromatogr. A*, 2009, **1216**, 5165.

63. E. Carasek, J. Merib, G. Mafra and D. Spudeit, *TrAC, Trends Anal. Chem.*, 2018, **108**, 203.
64. F. Zarghampour, Y. Yamini, M. Baharfar and M. Faraji, *J. Chromatogr. A*, 2018, **1556**, 21.
65. M. Rezaee, Y. Assadi, M.-R. Milani Hosseini, E. Aghaee, F. Ahmadi and S. Berijani, *J. Chromatogr. A*, 2006, **1116**, 1.
66. W. Ahmad, A. A. Al-Sibaai, A. S. Bashammakh, H. Alwael and M. S. El-Shahawi, *TrAC, Trends Anal. Chem.*, 2015, **72**, 181.
67. M. Rezaee, Y. Yamini and M. Faraji, *J. Chromatogr. A*, 2010, **1217**, 2342.
68. S. Jia, Y. Ryu, S. Won Kwon and J. Lee, *J. Chromatogr. A*, 2013, **1282**, 1.
69. K. J. Huang, C. Y. Wei, W. L. Liu, W. Z. Xie, J. F. Zhang and W. Wang, *J. Chromatogr. A*, 2009, **1216**, 6636.
70. J. Płotka-Wasylka, V. Simeonov and J. Namiesnik, *J. Chromatogr. A*, 2016, **1453**, 10.
71. C. Almeida, J. O. Fernandes and S. C. Chunha, *Food Control*, 2012, **25**, 380.
72. M. Bashiry, A. Mohammadi, H. Hosseini, M. Kamankesh, S. Aeenehvand and Z. Mohammadi, *Food Chem.*, 2016, **190**, 1168.
73. K. J. Huang, C. X. Jin, S. L. Song, C. Y. Wei, Y. M. Liu and J. Li, *J. Chromatogr. B*, 2011, **879**, 579.
74. J. Płotka-Wasylka, *Talanta*, 2018, **181**, 204.
75. A. Galuszka, P. Konieczka, Z. M. Migaszewski and J. Namiesnik, *TrAC, Trends Anal. Chem.*, 2012, **37**, 61.

CHAPTER 13

Derivatization Strategies for the Determination of Biogenic Amines with Chromatographic Techniques

TAUQIR AHMAD,[a] CHANBASHA BASHEER*[a] AND BAHRUDDIN SAAD[b]

[a] Chemistry Department, King Fahad University of Petroleum and Minerals, Dhahran, Saudi Arabia; [b] Fundamental and Applied Sciences Department & Institute of Self-Sustainable Building, Universiti Teknologi PETRONAS, 32610 Seri Iskandar, Perak, Malaysia
*Email: cbasheer@kfupm.edu.sa

13.1 Introduction

Biogenic amines (BAs) are naturally produced in many food products like vegetables, fish, fruits, meat, dairy products, wine and beer.[1,2] They are formed by microorganisms such as bacteria, moulds or yeast from metabolism of amino acids, especially from decarboxylation of amino acids. A typical spoiled food contains high level of BAs.[3] BAs are classified in different ways, *e.g.*, based on the chemical structure such as aliphatic (*e.g.*, cadaverine, agmatine, putrescine, spermidine), aromatic (*e.g.*, tyramine) or heterocyclic (*e.g.*, tryptamine, histamine). They can also be classified based on their functional groups, *e.g.*, morpholine, pyrrolidone, methyl-, ethyl-,

Food Chemistry, Function and Analysis No. 20
Biogenic Amines in Food: Analysis, Occurrence and Toxicity
Edited by Bahruddin Saad and Rosanna Tofalo

Published by the Royal Society of Chemistry, www.rsc.org

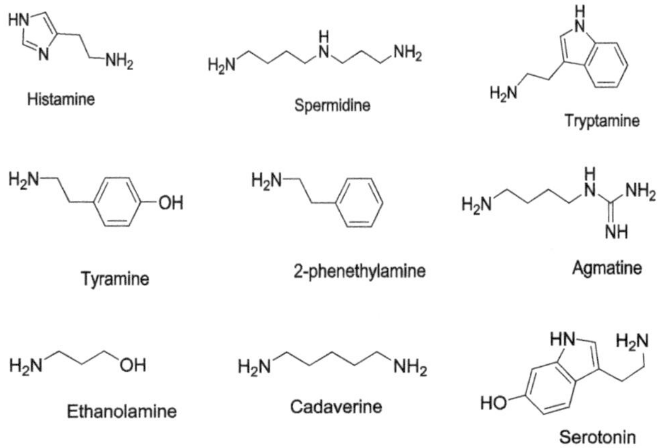

Figure 13.1 Structures of important BAs.

isoamyl-, isobutyl-, isopropyl-, hexyl-, dimethyl-, 3-methyl- and *n*-propylamine are some of the BAs that are volatile. They can also be categorized as monoamines (*e.g.*, tyramine, diamine), two amines (*e.g.*, cadaverine) and polyamines (*e.g.*, spermine and spermidine[4–6]).

The chemical structures of the common BAs are shown in Figure 13.1. BAs can also be classified based on their physiological effect which they act on the human body. On this basis, they are classified into psychoactive amines (*e.g.*, histamine, cadaverine and putrescine), *i.e.*, act upon the nervous system. On the other hand, amines (*e.g.*, tryptamine, tyramine) that act on the vascular system are categorized as vasoactive amines.[7,8]

Microorganisms producing BAs are not always harmful, some beneficial microorganisms, such as lactic acid bacteria, are also involved in the synthesis of BAs. Although a large number of BAs are formed as a result of microbial metabolism in different fermented products like fish, fermented vegetables, beer, wine and cheese, a small quantity is also produced biosynthetically in plant and animal cells. The concentration and contents of BAs are always different in different food samples, which mostly depend upon fermentation conditions, composition and microbial strain. Their effects will also be different depending on their composition and concentrations. They are partially responsible for the bad aroma of food. It has also been reported that ingesting food containing excessive amounts of BAs cause nausea, blood pressure change, headache, respiratory disorders, palpitations and even anaphylactic shock.[9–12] So, in order to evaluate the toxicological consequences and find the level of freshness or spoilage of food, the determination and quantification of BAs are necessary.

13.2 Analysis of BAs and Derivatization

Different analytical techniques such as high-performance liquid chromatography (HPLC), capillary electrophoresis (CE) and gas chromatography-mass spectrometry (GC–MS) have been used for the analysis of BAs.[13–16] HPLC is the most widely used technique and is coupled with different detectors such as UV/FLD and mass spectrometry (MS or MS/MS). Although analysis of BAs is essential for healthy living, it is challenging to analyse them because most of them neither have natural fluorescence nor absorption, which makes their detection very difficult. Therefore, structural modification is necessary before their analysis. Derivatization modifies the structure, converting BAs into their derivatives with improved fluorescent properties, enhancing their analytical sensitivity and selectivity. Different derivatizing agents, *e.g.*, dabsyl chloride (DBS-Cl), dansyl chloride (DNS-Cl), benzyl chloride, 9-fluorenylmethoxycarbonyl chloride (FMOC), *o*-phthaldialdehyde (OPA), fluorescein isothiocyanate (FITC), 6-amino-quinolyl-*N*-hydroxysuccinimidyl carbamate (AQC), 1,2-naphthoquinone-4-sulfonate (NQS) and diethyl ethoxy methylene malonate (DEEMM) are available. However, the choice of derivatizing reagent depends on the analytical method for the analysis. Experimental conditions that are responsible for the stability of the derivatized product must be carefully studied. Furthermore, derivatization also contributes to improvements in separation by chromatographic methods.[17,18] Chemical structures of common derivatizing agents are shown in Figure 13.2. In this chapter, derivatizing reagents will be classified under the heading of HPLC derivatizing reagents and GC derivatizing reagents. Conditions for the optimum formation and stability of derivatizing products of the different derivatizing reagents will also be discussed.

Derivatization may be carried out using offline pre-column, online pre-column or potentially post-column. In offline pre-column derivatization, the sample is mixed in a flask or a vial with a small volume of derivatizing reagent and a buffer solution (most derivatization reactions are pH sensitive) for a specific time. Almost 90% of derivatizations are done this way.

Some offline pre-column derivatizations have some disadvantages, *e.g.*, the derivatized products are not stable at high temperatures and are pH sensitive. Besides, the formation of multiple amines derivatives under non-optimized conditions can occur as BAs may have several amine groups. As a result, amine derivatives with different proportions will be formed.

To overcome the drawback of low stability of derivatized products, the online pre-column derivatization approach was introduced. In this method, derivatization is done in a continuous flow system which is connected to the chromatographic instrument.

Further drawbacks of instability of the derivatized product can also be overcome with the online pre-column derivatization method. One main advantage of this method is that it can easily be automated. Problems related to pre-column derivatization, especially the formation of multiple amine derivatives is circumvented with the post-column derivatization[19,20] approach.

Recently, *in situ* derivatization was developed by several workers which makes the analysis easy and fast. It couples sample preparation and derivatization into one step, *e.g.*, derivatization combined with DLLME or SPME. Such methods were reported by several researchers for the determination of

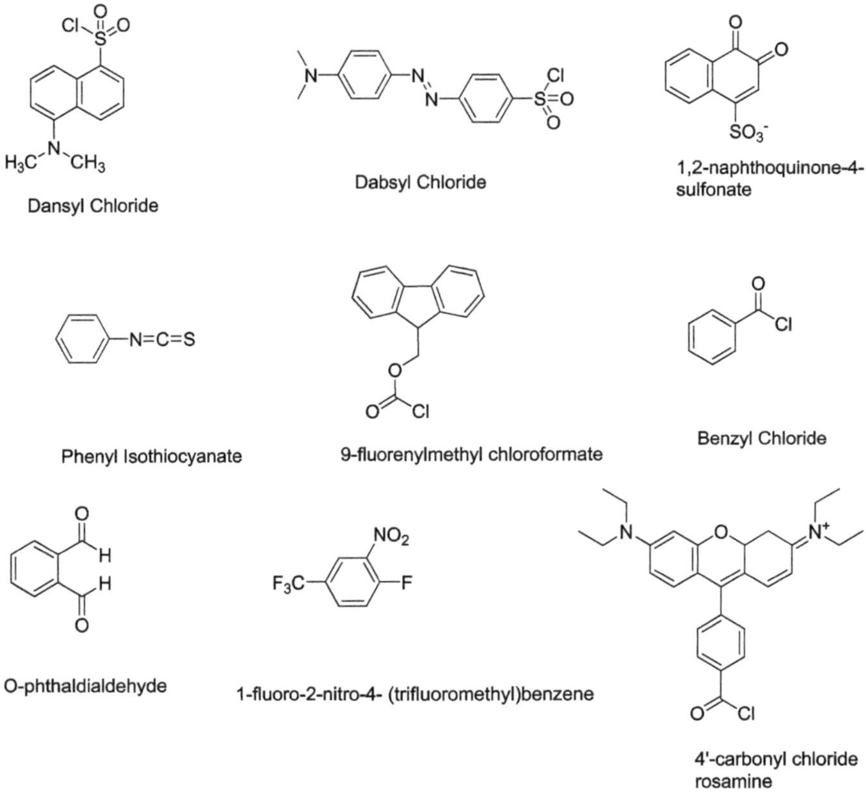

Figure 13.2 Structures of common derivatization reagents for BAs.

BAs in different alcoholic drinks and fish. These methods were very effective in term of efficiency and reproducibility, resulting in accurate determination and quantification of many BAs. Moreover, the *in situ* derivatization has made analysis time faster without sacrificing the method sensitivity.[21,22]

13.3 Derivatization Reagents for HPLC Analysis

As highlighted earlier, HPLC enjoyed widespread acceptance in the analysis of BAs due to its excellent sensitivity and resolution. For HPLC analysis, the BAs are either derivatized before (pre-column) or after (post-column) the separation. Also, pre-column derivatization has been shown to reduce the polarity of BAs, making it more compatible with the HPLC column used.[23]

13.3.1 Derivatization with DNS-Cl

Pre-column derivatization with four derivatizing agents, namely DNS-Cl, 9-fluorenylmethoxycarbonyl chloride, benzyl chloride and dabsyl chloride, for the determination of BAs in cheese and sausages before their analysis with UV and HPLC was reported.[24] The authors found that dabsyl chloride and 9-fluorenylmethoxycarbonyl were less stable when compared with dansyl and benzoyl derivatization reagents. The pre-column dansyl derivatization-HPLC-UV analysis of BAs in sausages and cheese samples showed a wider linear range and higher sensitivity, accuracy, precision and stability of the derivatives than the other derivatization regents[24] studied.

Lee *et al.* analysed BAs including histamine, putrescine, cadaverine, tyramine, tryptamine, β-phenylethylamine, spermine and spermidine in fermented soybean paste by derivatizing with DNS-Cl. The results showed that conditions for optimum derivatization were pH 9, at 40 °C and for 45 min.[25] Ke *et al.*, in 2018, while analysing BAs in wine, showed that the derivatized product of DNS-Cl is stable at 4 °C.[26] Silver carp is an important freshwater fish species in China. Samples stored for different intervals were derivatized with DNS-Cl prior to HPLC analysis. Again, the results showed that the optimum conditions for the derivatized product was when incubated at 40 °C for 45 min.[27] DNS-Cl is also effectively applied as the derivatizing reagent for the HPLC-FL determination of ten kinds of BAs in rat plasma cells.[28,29]

For optimizing the derivatization conditions, 1 mL of extract or standard solution was mixed with 200 μL of 2 N NaOH and 300 μL of saturated $NaHCO_3$ solution and vortexed. Next, 2 mL of dansyl chloride solution was added (2 mg dansyl chloride per mL in acetone) and vortexed again. The mixture was shaken for 5 min and then left in the incubator at 40 °C for 45 min. Extra DNS-Cl was removed by adding 100 μL NH_4OH 25% (v/v) and vortexed for 30 min and left for another 30 min at room temperature. Finally, a mixture of ammonium acetate and acetonitrile (1 : 1; v/v) was added.[30] The reaction involved is shown in Figure 13.3.

13.3.2 Derivatization with Dabsyl Chloride

Dabsyl chloride is another important pre-column derivatization agent which gives a stable derivatized product. Its superiority over other derivatizing reagents is the stable product formed, fast derivatization and good reproducibility of the reaction. The products can be easily separated when analysed with HPLC. Similar to other amine derivatizing agents, it can form mono and diamine derivatives when various amines are present.[31]

The optimum conditions for the dabsylation process are in basic conditions. As the reaction is pH sensitive, sodium bicarbonate buffer under

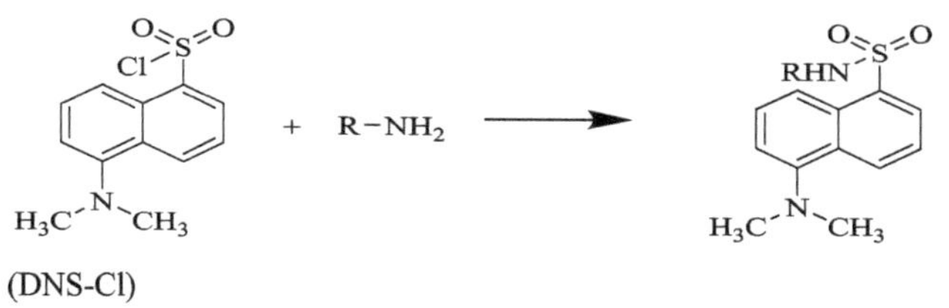

Figure 13.3 Derivatization reaction of BAs with DNS-Cl.

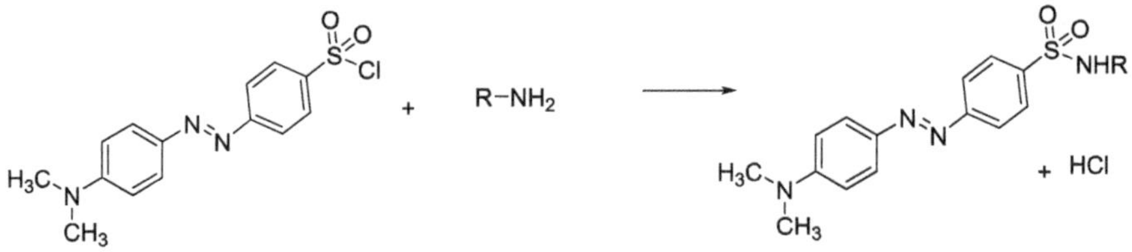

Figure 13.4 Dabsylation reaction between dabsyl chloride and a primary amine.

basic pH is added to the reaction mixture. Typically, 50 μL of sample and 10 μL of 1.5 mol L^{-1} of $NaHCO_3$ buffer were mixed in a vial. The vial was then cooled down and 20 μL was injected into HPLC for analysis.[32] The reaction involved is shown in Figure 13.4.

13.3.3 Derivatization Using 1,2-naphthoquinone-4-sulfonate (NQS)

NQS is used for the derivatization of both primary and secondary BAs. It can be used both for pre-column and post-column derivatizations. BAs derivatized with NQS can be separated by HPLC and CE[33–36] methods. To optimize the derivatization process for the quantification of BAs in wine, 250 μL of each, *i.e.*, wine sample, derivatizing reagent (0.07 M NQS in 0.1 M HCl) and buffer solution (0.125 M $Na_2B_4O_7$ in 0.1 M NaOH) were mixed in a reaction vial. The reaction was allowed to proceed for 5 min at 65 °C at pH 9.2. After completion of the reaction, the derivatized product was extracted with CH_3Cl and solvent was removed under N_2 flow.[37] The general derivatization of BAs with NQS is shown in Figure 13.5.

13.3.4 Derivatization with Phenyl Isothiocyanate (PIT)

Phenyl isothiocyanate (PIT) is a suitable derivatizing reagent for primary and secondary amines, forming their phenylthiourea derivatives which can be conveniently detected using ultraviolet-visible (UV-VIS). Chinese rice wine was analysed for seven BAs and 17 amino acids by derivatizing the sample with PIT before the RP-HPLC separation and UV detection at 254 nm.[38]

Cri-du-chat syndrome (CDCS) is a childhood disease due to depletion of the 5p chromosome. Blood samples from the affected person and healthy

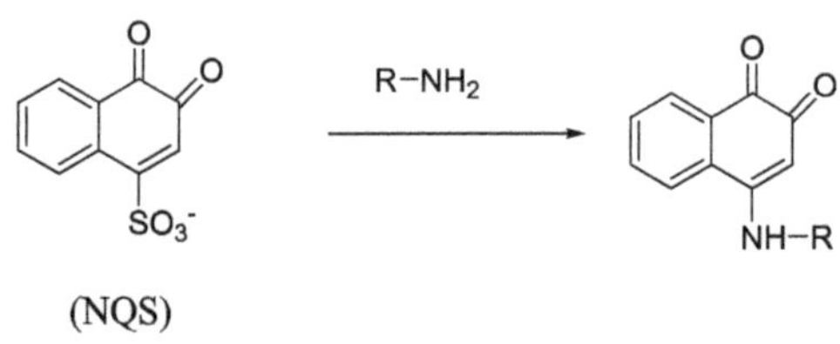

Figure 13.5 The derivatization of BAs with NQS.

Figure 13.6 Derivatization reaction of BAs with PIT.

humans were investigated for amino acid and BA content using UHPLC/MS. The samples were derivatized with PIT[39] and was successfully analysed using LC with amperometric detection.[40]

A typical condition for the derivatization were: 40 µL of sample solution in acetonitrile and 40 µL of PITC were added with 20 µL of 20% v/v solution of pyridine and acetonitrile. The BAs were allowed to derivatize by heating the mixture up to 60 °C for 20 min. After the derivatization reaction, 40 µL of HCl and 60 µL of water were added, and the products were ready for analysis with LC/MS.[41] The general reaction for the derivatization of BAs with PIT is shown in Figure 13.6.

13.3.5 Derivatization with 6-aminoquinolyl-*N*-hydroxysuccinimidyl Carbamate (AQC)

Extraction and pre-treatment of the sample is normally a tedious and time-consuming process. There is always an enormous interest to find methods which can analyse BAs without sample preparation. Pe *et al.* in 2006 reported a direct method for the analysis of target BAs in wine without any sample preparation. The sample was directly derivatized with AQC and then analysed with HPLC.[42]

Another advantage of using AQC as a derivatizing agent is its suitability for derivatization of polyamines. Furthermore, the derivatized product is UV active and can thus be detected by UV-VIS technique.[43] AQC was successfully used as a derivatization agent for HPLC analysis of BAs in different wines.[44]

Changes in the concentration of BAs with time in a few opened wine bottles were studied. Young white wine, red wine and standard were chosen for the study. The effects of temperature and type of stopper were also investigated. The samples were derivatized with AQC and analysed using HPLC. No significant changes in the concentration of BAs due to different temperature conditions[45] were found.

Typical conditions for the optimum derivatization using AQC were: 20 µL of sample was mixed 60 µL of sodium borate buffer (0.2 M, pH 8.8) and 20 µL of AQC was added. The mixture was heated to 50 °C to ensure all the BAs were derivatized.[44]

13.3.6 Derivatization Using Fluorenyl Methyl Chloroformate (FMOC)

FMOC is seldom used in routine work. Nevertheless, it was used for the derivatization of spermine, spermidine and other BAs in wine samples.[46]

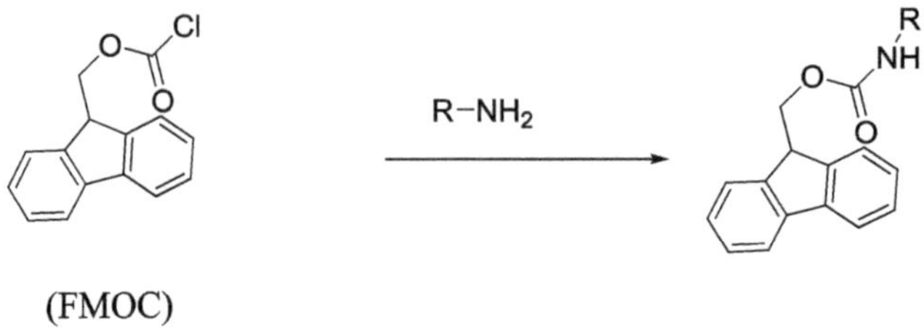

Figure 13.7 The reaction between BAs and FMOC.

A common procedure to derivatize BAs using FMOC involves the addition of 150 μL amino acid solution in water (5.4×10^{-8} M) into 150 μL borate buffer (0.4 M, pH 9) and with 300 μL of FMOC reagent dissolved in CH_3CN (3×10^{-7} M FMOC). The reaction was completed in 20 min and stopped by adding 50 μL ADAM (9-anthrylazomethane. The derivatized samples were ready for chromatographic analysis.[47,48] The reaction between BAs and FMOC is shown in Figure 13.7.

13.3.7 Derivatization with Diethyl Ethoxymethylenemalonate (DEEMM)

DEEMM is a useful derivatizing reagent as, unlike other derivatization reagents, can react with proline and hydroxyproline. The derivatized products can be detected by photodiode array and UV detectors.

Derivatization using DEEMM can be done by taking 1.75 mL of 1 M borate buffer (pH 9), 750 μL methanol, 20 μL internal standard (*e.g.*, L-2-aminoadipic acid, 1 g L^{-1}) and 30 μL of the derivatizing agent in a beaker and reacted in an ultrasonic bath for 30 min. By-products were degraded by heating the derivatized sample for 20 h at 70 °C.[49] The reaction involving this reagent is shown in Figure 13.8.

13.3.8 Derivatization with Benzoyl Chloride

Derivatization with benzoyl chloride (Bn-Cl) is attractive because it needs moderate and short reaction conditions. However, sample preparation is required before the derivatization reaction.[50]

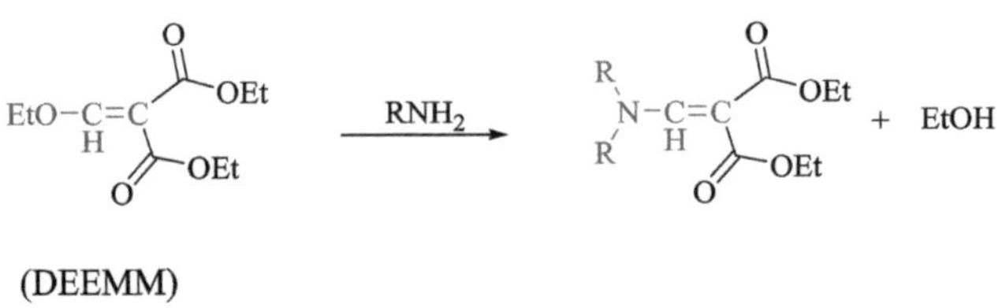

Figure 13.8 Derivatization reaction between DEEMM and BAs.

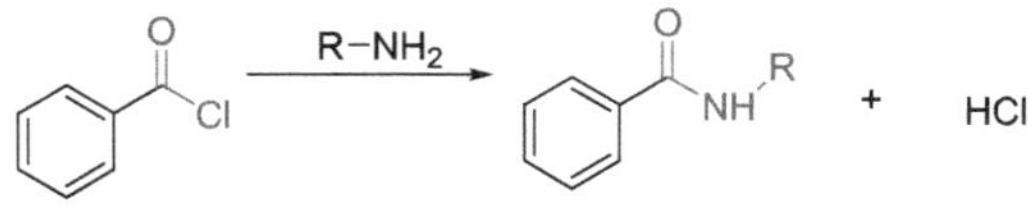

Figure 13.9 Derivatization reaction between benzyl chloride and BAs.

Fu *et al.* derivatized fish samples by reacting 45 µL of Bn-Cl solution in acetonitrile (1% v/v) with 45 µL of buffer solution (sodium tetraborate, pH 8.3) with 10 µL of sample. In order to complete the reaction, the mixture was stirred at 25 °C for 10 min. The derivatized product was removed by centrifugation for 10 min at 4 °C at a speed of 14 000 rpm.[51] The general reaction of derivatization of BAs with Bn-Cl is shown in Figure 13.9.

13.3.9 Derivatization with 1,3,5,7-tetramethyl-8-(*N*-hydroxysuccinimidyl butyric ester)-difluoroboradiaza-*s*-indacene (TMBB-Su)

TMBB-Su is a BODIPY (boron-dipyrromethene) based fluorophore derivatizing agents which can be used for derivatization of BAs. The advantage of using this reagent is the short time and moderate temperature for the reaction. To obtain higher derivatization efficiency, a 75 mL buffer solution of $H_3BO_3/Na_2B_4O_7$ of pH 7.2 was mixed with pretreated sample in a vial. About 30 µL of 1×10^{-3} M derivatizing reagent (10 times to that of sample, w/w) was added to the mixture. Methyl cyanide was added until the overall volume reached 0.5 mL. The mixture was then kept at 20 °C for 20 min to complete the derivatizing reaction.[52] The general reaction of derivatization between TMBB-Su and BAs is shown in Figure 13.10.

13.3.10 Derivatization with 2-chloro-1,3-dinitro-5-(trifluoromethyl)benzene (CNBF)

Piasta *et al.* used CNBF for the derivatization of tryptamine, tyramine, phenethylamine and histamine. The products obtained were pure and was easily analysed with NMR.

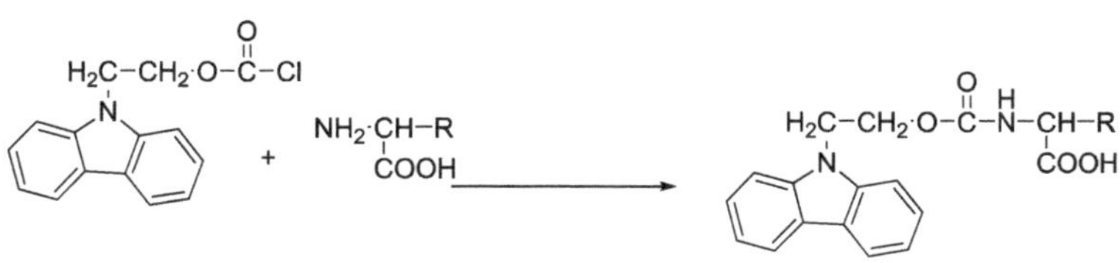

(TMBB-Su)

Figure 13.10 Derivatization reaction between TMBB-Su and BAs.

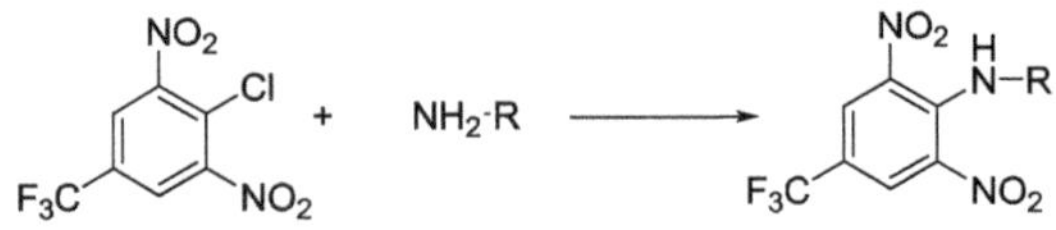

(CNBF)

Figure 13.11 Derivatization reaction of BAs with CNBF.

The authors derivatized the selected amines with CNBF by refluxing the amines together with the derivatizing reagent for 2 h in ethanol. The derivatized products obtained were separated using column chromatography with a solvent system of hexane : water (20/80 v/v).[53] The general derivatization reaction of BAs with CNBF is shown in Figure 13.11.

13.3.11 Derivatization with 1-fluoro-2-nitro-4-(trifluoromethyl)benzene (FNBT)

Jastrzebska *et al.*[54] derivatized wine samples of with 1-fluoro-2-nitro-4-(trifluoromethyl)benzene before the HPLC analysis. The advantage of using this reagent is the fast derivatization and it undergoes fast nucleophilic substitution reaction in basic medium.[54]

The derivatization reaction is similar to that of CNBF, as shown in Figure 13.12.

13.3.12 Derivatization with 4′-carbonyl Chloride Rosamine (CCR)

He *et al.*[55] in 2016 synthesized a new derivatizing reagent called CCR. It is a valuable reagent as it can be used for *in situ* derivatization where derivatization efficiency of 99% can be obtained. Besides, this reagent is suitable for MS analysis. Furthermore, it can be used for nearly all the commonly found food samples.

He *et al.* described a simultaneous extraction and derivatization method of BAs in food by taking 20 µL of the food sample and 20 µL of an IS standard in a test tube; 1 mL of buffer solution (pH 9.2) of sodium carbonate and sodium bicarbonate was also added to the test tube. Then, 20 µL of

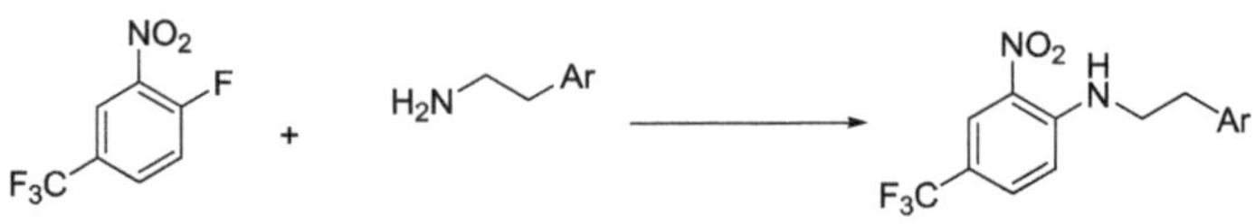

(FNBT)

Figure 13.12 Derivatization of BAs with FNBT.

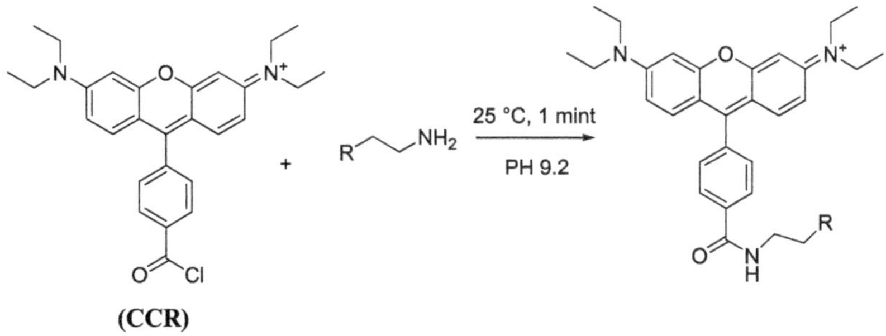

Figure 13.13 Derivatization reaction of CCR with BAs.

derivatizing reagent and 90 μL of extractant (bromobenzene) was injected using a syringe. The tube was first manual shaken for 10 s to mix the phases and then placed in an ultrasound bath for 1 min. The derivatized product was separated from the solution by centrifugation.[55] The general reaction of BAs with CCR is shown in Figure 13.13.

13.3.13 Derivatization of Amino Groups with 2-(9-carbazole) Ethyl Chloroformate (CEOC)

Fluorescence detection is preferred for routine analysis because of its high sensitivity. Thus, the use of fluorescence derivatizing agents such as CEOC is attractive for the analysis of BAs that are not naturally fluorescent.

The following conditions were recommended for the derivatization of BAs using CEOC: 50 μL of the BAs sample that was extracted from shrimp sauce was mixed with 100 μL of CEOC solution in acetonitrile, and 200 μL of borate buffer of pH 10 was added. The mixture was allowed to react for 5 min and the excess of CEOC was removed by extracting twice with pentane, and the aqueous portion was ready for analysis. The product was stable for 24 h and can be increased further if the solution was neutralized at pH 6–7 with quenching reagent. Derivatized tyramine and histidine were stable for less than 24 h.[56] The general reaction of BA derivatization with CEOC is shown in Figure 13.14 and Table 13.1.

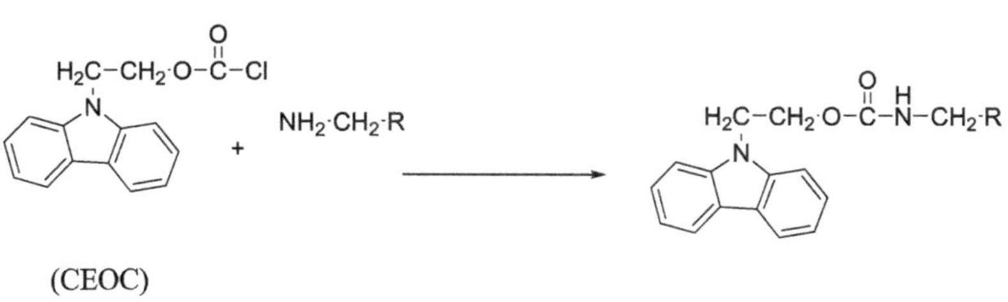

Figure 13.14 Derivatization of amino groups with CEOC.

Table 13.1 Details of the analysis of some BAs using various derivatization reagents.[a]

Target compound	Sample type	Derivatizing reagent	Derivatization time	Sample preparation method	Instrument type	LOD/LOQ	References
His, Tyr, Put, Cad, Tryp and β-Phe	Sausage and cheese	DNS-Cl, FMOC, Bn-Cl and DBS-Cl	60 min, 3 min, 30 min and 20 min, respectively	Vortex-mixing	HPLC	0.03–0.38 mg kg^{-1}/ 0.11–1.26 mg kg^{-1}	24
His, Tyr, Tryp and 2-Phe	Wine	3,5-Bis-(trifluoromethyl) phenyl isothiocyanate	24 h	Stirring on a hot plate	LC-MS/MS	—	57
His, Tryp, Put, Cad and Spe	Sausage and kielbasa	DBS-Cl	10 min	On-chip electromembrane for extraction and vortexed for derivatization	HPLC	3–8 μg L^{-1}/ 10–25 μg L^{-1}	32
Tryp, Phe-Eta, Put, Cad, His, Tyrp, Spe and Sper	Wines	DNS-Cl	60 min	Vortex	HPLC	LOD 0.91–33.6 μg L^{-1}	26
Tryp, 2-phe, β-phe, Phe, Put, Cad, His, Tyr, Spe and Sper	Soybean	DNS-Cl	45 min	Thermo-hygrostat	HPLC	—	58

Put, Cad, Tryp, Phe, Put, His, Tyr and Spe	Sausage	DBS-Cl	45 min	Centrifuged (3000 rpmfor 10 min under 4 °C)	HPLC	—	30
Tryp, Phe Put, β-Phe, His, Tyr, Put, Cad and Spe	Probiotic yogurt	Bn-Cl	40 min	Vertex	HPLC	0.65–4.01 mg L^{-1}/ 3.0–5.0 mg L^{-1}	59
Phe Put, His, Tyr, Cad, Tryp and Spe	Wine	AQC	—	—	HPLC	—	45
Tryp, Put, Cad, Phe Put, His, Tyr, Cad, β-Phe and Spe	Wine	AQC	—	—	HPLC	0.991–0.999 μg L^{-1} /17–51 μg L^{-1}	44
His, Tyr, 2-pe, Ser, Tryp, Oct, Dop, Cad, Put, Agm, Spr, Spre and Put	Fresh fish	OPA		Solvent extraction	UHPLC	0.05–0.2 mg L^{-1} /0.1–0.3 mg L^{-1}	60

[a]Abbreviations: LOD, limit of detection; LOQ, limit of quantification.

13.4 Pros and Cons of Derivatization Reagents

As discussed earlier, the analysis of BAs needs pre- or post-column derivatization before their detection. Different derivatization reagents are available that react with the amine part of BAs, thus exclusively used for the analysis of BAs.[61] Due to the presence of complex mixture of BAs, selection of suitable derivatization reagents becomes necessary. The choice of the derivatization reagent is dictated by sample condition, availability of chromatographic and detection techniques and analytical conditions (temperature, *etc.*).[62] Here, the pros and cons of commonly used derivatization reagents for BAs is presented.

The most commonly used derivatization reagent, DNS-Cl, produces a derivatized product that is not only relatively stable but can be easily detected by DAD, FL and MS. The active group of the DNS-Cl is sulfonyl chloride which can react with primary and secondary BAs.[63] The product is stable for 48 h at 4 °C and even at 25 °C. The product can be detected at 250 nm using UV detection. DNS-Cl had been successfully used for the analysis of many samples such as wine, beer, sausage, *etc.* However, some disadvantages include the need for sample preparation before the derivatization reaction (*e.g.*, SPE and LLE) because it can react with alcoholic and phenolic compounds. Excess reagent must be removed, otherwise, it will also cause absorption in UV-VIS and thus cause interference. Also, longer reaction time, *i.e.*, from 20 to 60 min and higher temperature, *i.e.*, 40–70 °C are the other disadvantages when using this derivatization reagent.[64] However, the problem of high temperature reaction can be solved by using ionic liquids as a media which enables the derivatization reaction to take place at room temperature.[65]

Another sulfonyl chloride reagent is DBS-Cl which can react with both primary and secondary amines. The optimum pH for the derivatization reaction is 8.2 and was successfully applied for the determination of different amines.[66] The derivatives are relatively stable and can be detected at 446 nm.[67]

OPA is a very imported derivatization reagent which eliminates the practice of the tedious sample preparation step. Its analysis time is less compared with DNS-Cl, and the derivatization reaction occurs at room temperature. Another critical aspect of its use is that it can be used for both pre and post-column derivatization. OPA reacts only with primary amines and the product obtained is separated by HPLC and detected by fluorescence detector near λ excitation at 330 nm and λ emission at 445 nm. It has good sensitivity and does not generate interferences with reagents.[68,69] The reaction happens at pH 10, but the products are not stable. To solve this problem, on-column derivatization is introduced. Sample pH is made 10 and reagent is added, shaken for 2 min. Another way to overcome the problem of instability is that it is used with, *e.g.*, 2-marceptophenol, FMOC. This not only solves the problem of instability but also it can react with secondary BAs which makes possible the simultaneous derivatization of primary and secondary amines.[70]

Another derivatization reagent which is widely used is benzoyl chloride. It has many useful properties such as the formation of stable derivatized product and it is less expensive as compared with other derivatizing reagents. The benzoyl derivative of Phe, Tyr, His and Cad from cheese and sausage sample were stable for 48 h even at 25 °C. Further, its analysis is quick and needs moderate reaction conditions, *i.e.*, pH 8.2, 25 °C and 10 min. The main disadvantage of using this reagent is that it is not specific like DAN-Cl, so it needs sample preparation before derivatization reaction. Even sometime after the derivatization reaction, liquid extraction is required. This reagent had been used for different samples of wine, beer and rice.[71]

AQC is another important reagent used for the derivatization of primary and secondary amines. The derivatized BAs are stable for 1 week at room temperature. HPLC is a suitable chromatographic technique for the separation of BAs after derivatization. Reaction time is just 2 min to produce suitable fluorescence product. However excess reagent will cause interference when the UV detector is used for detection. Sample preparation is simple and has a low limit of detection.[61,72] It can also be used with CE.[73]

Diethyl ethoxymethylenemalonate is another important derivatization reagent which reacts with both primary and secondary amines. The product is stable, and the reaction time is almost 30 min but needs a long time (up to 2 h) heating at an elevated temperature, *i.e.*, 70–80 °C for removal of excess reagent. This reagent is also used for the determination of ammonium ions, BAs and amino acids. There is no interference from excess reagents and the products can be detected at 280 nm using UV-VIS spectroscopy. However, it has a problem of low specificity and lowers sensitivity as compared with other derivatizing reagents.[74–76]

Another derivatization reagent is phenyl isothiocyanate (PITC) which reacts favourably in alkaline conditions. The product of PITC which is thiocarbamoyl can be detected at 254 nm. It can also be used for simultaneous extraction and derivatization when used with suitable extraction reagent. Furthermore, it can also be used for the simultaneous determination of amino acids and BAs in wine. The product is stable for 10 days and can be extracted with SPME or ultrafiltration. Low sensitivity was obtained when derivatization is done with DAB-Cl but the results are comparable with NQS. However, the problem of this reagent is that it cannot detect the main BAs.[77–79]

FMCO is another important derivatizing reagent which forms carbamine derivatives from the reaction of BAs with carbonyl fluorides or chlorides. It is used for the introduction of the fluoride group; reaction time is less (2 min) in alkaline medium and produces a stable product. The disadvantage of this reagent is that it needs sample preparation.[80,81]

Other derivatizing reagents such as 3,5-bis-(trifluoromethyl)phenyl isothiocyanate,[82] 2,6-dimethyl-4-quinoline carboxylic acid *N*-hydroxysuccinimide ester (DMQC-OSu), ethyl-acridine-sulfonyl chloride (EAC), naphthyl isothiocyanate (NITC), 2-[(2-dibenzocarbazole)ethoxy]ethyl chloroformate and 2-naphthylenoxycarbonyl chloride are also reported in the literature with

Table 13.2 Stability of the BA derivatized products and their optimum reaction conditions.

Reagent	Stability of product	Optimum reaction conditions	References
DNS-Cl	Stable when kept at 4 °C	pH 9, 40 °C, 45 min	25,26,27,30
DBS-Cl	24 h at 4 °C and 14 days at −20 °C	pH 9.0, 10–15 min, 70 °C	31,32
DEEMM	Proline and hydroxyproline for 24 h, all others, 1 week	pH 9, 30 min, 25 °C	49
Bn-Cl	Stable at 4 °C	pH 8.3, 10 min, 25 °C	51
TMBB-Su	Stable at −40 °C	pH 7.2, 20 min, 20 °C	52
CCR	Stable at room temperature	pH 9.2, 1 min, 25 °C	55
FMOC	Stable at −20 °C	pH 8.5, 2–3 min	87,88
NQS	Unstable after 1 h	pH 10, 5 min, 65 °C	34

validated methods.[83–86] The properties of the important derivatization products are shown in Table 13.2.

13.5 Derivatization Reagents for Gas Chromatography Analysis

To improve the volatility of BAs, which is vital for GC analysis, and to reduce their polarities, derivatization is necessary. Interestingly, not many GC methods are reported in the literature. Isobutyl chloroformate, pentafluoro benzaldehyde and acetic anhydride are the commonly used derivatization agents for GC applications.[89–92]

13.5.1 Isobutyl Chloroformate (IBCF)

Huang *et al.*[93] developed a GC method with SPME sampling for the analysis of non-volatile histamine by the simultaneous extraction and derivatization of BAs from fish samples. IBCF was used as derivatizing reagent and isooctane as the extraction solvent. Thus, the SPME fibre was coated with IBCF derivatizing reagent and extracting solution by dipping into a solution of both (IBCF and isooctane). The derivatized BAs were desorbed from the fibre by injecting directly into a GC injection port.[93]

Cunha *et al.*[94] reported an interesting method for the simultaneous determination of 22 BAs in port wines and grape juices. The sample were derivatized with IBCF and analysed with GC-MS. The advantage of this method was the direct derivatization of the sample with IBCF without any pre-treatment or clean-up process. Besides, this method can be used for the quantification of both volatile and non-volatile BAs. Only a small sample size (1 mL) is required and the analysis is done in about 10 min. The 22 amines found in wines were quantified with this fast and straightforward derivatization and analysis method. A disadvantage was that histamine and tyramine were not stable in the presence of IBCF, so it was analysed after an excess of derivatizing reagent was removed.[94] Wojnowski *et al.* successfully analysed BAs in samples of beef, pork and poultry where the pretreated samples were first extracted with $CHCl_3$ and derivatized with IBCF before their analysis using GC-MS.[95]

Płotka-Wasylka *et al.* determined 13 BAs in homemade wines by simultaneous extraction and derivatization using $CHCl_3$ as the extraction solvent and IBCF as the derivatization reagent. The derivatized samples were analysed using GC-MS.[96] IBCF was also successfully used as the derivatization reagent for GC-MS analysis of 16 biogenic amines in Polish wines.[97]

A total of 18 BAs were determined in 22 beer samples from Portugal using GC-MS. Sample extraction and derivatization were done simultaneously using toluene as the extraction solvent and IBCF as the derivatizing reagent.[21] A similar method of simultaneous extraction and derivatization with toluene and IBCF was used for the determination of BAs in different liqueur samples.[22]

13.5.1.1 Derivatization Procedure

Cunha *et al.*[94] derivatized samples with IBCF by taking 1 mL of the treated sample in a vial containing 1 mL of toluene. Next, 1 mL of 0.5 M of phosphate buffer of pH 12 and 25 µL of derivatizing reagent was added and mixed for 10 min before centrifuging for 5 min. Approximately 500 µL of toluene was separated and mixed with 500 µL of alkaline methanol. The mixture was mixed well and 1.5 mL of 5 M NaOH was added, mixed again and centrifuged. The procedure is suitable for the analysis of all amines except histamine and tyramine.[94] The general reaction of BAs with IBCF is shown in Figure 13.15.

13.5.2 Derivatization with Pentafluoro Benzaldehyde (PFB)

Primary BAs affect sensory properties, so their determination in wine is necessary. Derivatization of primary BAs with PFB followed by GC-MS offered good selectivity and sensitivity compared with HPLC.[98]

13.5.2.1 Derivatization Procedure

For the derivatization procedure, 1 mL of amine (100 mg L^{-1}) was taken in 15% solution of H_2O/C_2H_5 and mixed with 0.5 mL PFB (10 mg mL^{-1}) in a buffer of pH 12. The mixture was mixed manually and then allowed to react

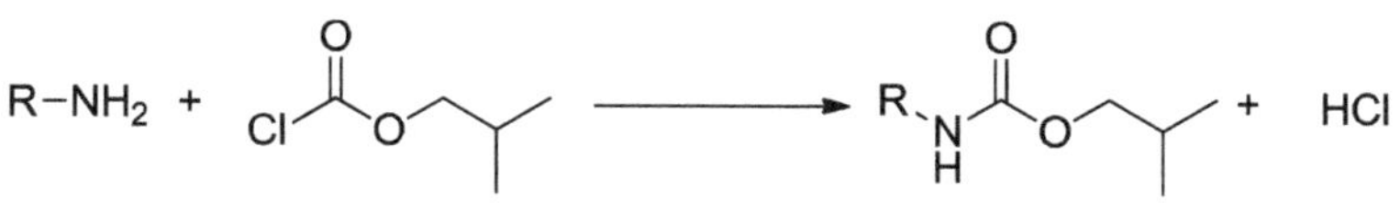

(IBCF)

Figure 13.15 Derivatization of BAs with isobutyl chloroformate.

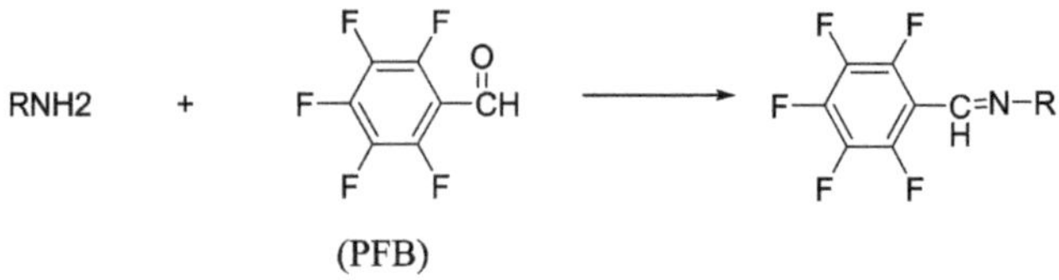

Figure 13.16 Formation of pentafluoro benzyl imine derivatives (PFB imines) from pentafluoro benzaldehyde (PFB) and alkylamines.

Figure 13.17 Derivatization of primary BAs with acetic anhydride.

at 24 °C for 30 min. The derivatized amines were kept in an ice bath for 1–2 min to stop further reaction. Next, 1 mL solution of PFB (10 mg L^{-1}) in hexane and was shaken for 90 s. Finally, 1 mL of 0.1 N solution of NaOH was used to convert the excess derivatization reagent to diol and is extracted with water.[99] The derivatization reaction of BAs with PFB is shown in Figure 13.16.

13.5.3 Derivatization with Acetic Anhydride (AA)

It is rather straightforward to derivatize primary BAs with acetic anhydride in aqueous solution under basic conditions. The derivatization has many advantages, *i.e.*, the derivatives formed are lipophilic and can be easily extracted with organic solvents. Also, the derivative can be further reacted with other acylating reagents and increases their sensitivity for GC-MS[100] analysis. The general derivatization reaction of primary BAs with AA is shown in Figure 13.17.

13.6 Comparison of GC Derivatization Reagents

As mentioned earlier, the list of derivatization reagents for GC analysis is much smaller compared with HPLC. Among them, alkylated chloroformates are predominant since the reaction can be carried out conveniently in aqueous alkaline solutions. Besides, short reaction time and low sample requirement are other notable advantages of using this reagent. It also allows simultaneous extraction and derivatization reactions and for the simultaneous determination of BAs and amino acids in food samples. Furthermore, the derivatives are stable and suitable for GC analysis. The BAs can be extracted using DLLME and SPME.[93]

Another important derivatization reagent for GC is pentafluoro benzaldehyde, which is specific for primary amines. The reaction time is longer than the alkylated derivatizing reagents such as IBCF, but the reaction can

Table 13.3 Determination of BAs using different GC derivatization reagents.[a]

Derivatizing agent	Sample type	Target compound	Optimum conditions	Sample preparation method	Instrument type	LOD/LOQ	References
IBCF	Poultry, pork and beef	β-Phe, Cad, Dmet, His, Put, Sper, Tryp and Tyr	80 µL, 15 min	DLLME	GC-MS	0.003–0.009 $\mu g\ g^{-1}$ and 0.009–0.029 $\mu g\ g^{-1}$	95
IBCF	Wine	Cad, But, His, Eta, Dmet, Det, 2-phenylethylamine, Tyr, Tryp, Sper, Put and Ipa	90 µL, 15 min	DLLME	GC-MS	1.1–4.1 $\mu g\ L^{-1}$/ 3.3–11.7 $\mu g\ L^{-1}$	96
IBCF	Fish	Put, Cad, His and Tyr	20 µL, 1 min	SPME	GC-MS	LOD = 2.98–45.3 $\mu g\ kg^{-1}$	93
IBCF	Wine	Eta, Det, Ipa, but, Ibut, ama, Iama, Hex, Prd, Put, Cad, β-Phe, His and Tyr	Eliminate excess IBCF with solution of alkaline methanol	Centrifugation	GC-MS	1 $\mu g\ L^{-1}$/10 $\mu g\ L^{-1}$	94
HFBA	Wine	Put, Cad, Sper, Spe, β-Phe and Tyr	80 °C and 60 min	Ion pair extraction	GC-MS	LOD = below 10 $mg\ L^{-1}$	102
PFB	Wine	Met, Eta, *n*-Proa, *n*-Buta, *n*-Pena, *n*-Hexa and *n*-Hepa.	pH 12, 24 °C, 30 min and 10 $mg\ mL^{-1}$	Not required	GC-MS	LOQ 0.010 $mg\ mL^{-1}$	99

[a]Abbreviations: LOD, limit of detection; LOQ, limit of quantification.

take place at room temperature. Another derivatization reagent, acetic anhydride, is suitable for primary amines and the stable product can be extracted with organic solvents.[65,96,101] Table 13.3 compares the important characteristics of BA analysis using common GC derivatization reagents. In conclusion, GC can be a suitable method for the analysis of amines, but the stability of derivatized products at low temperature limits the use of this technique.

13.7 Conclusion

Many benefits can be enjoyed when BAs are derivatized: the polarity of BAs is reduced, making it more suitable for reversed phase HPLC separations; improvements in method sensitivity and selectivity; increasing volatility for GC analysis. Different derivatization reagents (*e.g.*, DNS-Cl, dabsyl chloride, benzyl chloride, AQC, NQS) are available. Thus, the selection of these reagents for the intended application becomes necessary. Despite the various benefits of derivatization, analysis time is increased, the procedure become less straightforward and unwanted by-products are sometimes formed. If this happens, laborious clean-up of samples becomes mandatory. Thus, new methods that do not require derivatization, but do not sacrifice the analytical method performance, will be the ultimate goal for obvious reasons.

List of Abbreviations

AQC	6-Amino-quinolyl-*N*-hydroxysuccinimidyl carbamate
ama	Amylamine
BAs	Biogenic amines
Bn-Cl	Benzyl chloride
But	Butylamine
Cad	Cadaverine
CCR	4′-Carbonyl chloride rosamine
CE	Capillary electrophoresis
CEOC	2-(9-Carbazole)ethyl chloroformate
CNBF	2-Chloro-1,3-dinitro-5-(trifluoromethyl)benzene
DBS-Cl	Dabsyl chloride
DEEME	Diethyl ethoxy methylene malonate
Det	Diethylamine
DLLME	Dispersive liquid liquid microextraction
Dmet	Dimethylamine
DNS-Cl	Dansyl chloride
Eta	Ethylamine
FITC	Fluorescein isothiocyanate
FMOC	9-Fluorenylmethoxycarbonyl chloride
GC	Gas chromatography
Hex	Hexylamine
HFBA	Heptafluorobutyric acid

His	Histamine
HPLC	High-performance liquid chromatography
Iama	Isoamylamine
IBCF	Isobutyl chloroformate
Ibut	Isobutylamine
Ipa	Isopropylamine
Ipent	Isopentylamine
LC-MS	Liquid chromatography-mass spectrometry
LOD	Limit of detection
LOQ	Limit of quantification
Met	Methylamine
MS	Mass spectrometry
n-Buta	*n*-Butylamine
n-Hepa	*n*-Heptylamine
n-Hexa	*n*-Hexylamine
n-Pena	*n*-Pentylamine
n-Proa	*n*-Propylamine
NQS	1,2-Naphthoquinone-4-sulfonate
OPA	*o*-Phthaldialdehyde
PFB	Pentafluoro benzaldehyde
Prd	Pyrrolidine
Put	Putrescine
Spe	Spermine
Sper	Spermidine
SPME	Solid phase microextraction
TMBB-Su	1,3,5,7-Tetramethyl-8-(*N*-hydroxysuccinimidyl butyric ester)-difluoroboradiaza-*s*-indacene
Tryp	Tryptamine
Tyr	Tyramine
UHPLC	Ultra-high-performance liquid chromatography
UV/FLD	Ultraviolet with fluorescent detector
UV-VIS	Ultraviolet-visible
β-Phe	2-Phenylethylamine

References

1. A. Halász, Á. Baráth, L. Simon-Sarkadi and W. Holzapfel, Biogenic amines and their production by microorganisms in food, *Trends Food Sci. Technol.*, 1994, **5**, 42–49.
2. L. Beneduce, *et al.*, Biogenic amine in wines, *Ann. Microbiol.*, 2010, **60**, 573–578.
3. M. H. Silla Santos, Biogenic amines: Their importance in foods, *Int. J. Food Microbiol.*, 1996, **29**, 213–231.
4. V. Ladero, *et al.*, Biogenic Amines in Dairy Products, *Microb. Toxins Dairy Prod.*, 2016, **8398**, 94–131.

5. M. Z. Zaman, F. A. Bakar, J. Selamat and J. Bakar, Occurrence of biogenic amines and amines degrading bacteria in fish sauce, *Czech J. Food Sci.*, 2010, **28**, 440–449.
6. A. Peña-Gallego, P. Hernández-Orte, J. Cacho and V. Ferreira, High-performance liquid chromatography analysis of amines in must and wine: A review, *Food Rev. Int.*, 2012, **28**, 71–96.
7. C. Y. Chong, F. A. Bakar, A. R. Russly, B. Jamilah and N. A. Mahyudin, The effects of food processing on biogenic amines formation, *Int. Food Res. J.*, 2011, **18**, 867–876.
8. M. A. Gammone, A. Vicentini, G. Riccioni, M. D. Girolamo, A. D'Aulerio and N. D'Orazio, Food-related atrial fibrillation? The potential role of biogenic amines in "Nutri-Arrhythmias" genesis, *Reports*, 2019, **2**, 1.
9. X. Gong, N. Qi, X. Wang, L. Lin and J. Li, Ultra-performance convergence chromatography (UPC2) method for the analysis of biogenic amines in fermented foods, *Food Chem.*, 2014, **162**, 172–175.
10. N. Benkerroum, Biogenic amines in dairy products: origin, incidence, and control means, *Compr. Rev. Food Sci. Food Saf.*, 2016, **15**, 801–826.
11. G. Sagratini, *et al.*, Simultaneous determination of eight underivatised biogenic amines in fish by solid phase extraction and liquid chromatography-tandem mass spectrometry, *Food Chem.*, 2012, **132**, 537–543.
12. P. W. Stege, L. L. Sombra, G. Messina, L. D. Martinez and M. F. Silva, Determination of melatonin in wine and plant extracts by capillary electrochromatography with immobilized carboxylic multi-walled carbon nanotubes as stationary phase, *Electrophoresis*, 2010, **31**, 2242–2248.
13. G. I. Mohammed, A. S. Bashammakh, A. A. Alsibaai, H. Alwael and M. S. El-Shahawi, A critical overview on the chemistry, clean-up and recent advances in analysis of biogenic amines in foodstuffs, *TrAC, Trends Anal. Chem.*, 2016, **78**, 84–94.
14. F. Bedia Erim, Recent analytical approaches to the analysis of biogenic amines in food samples, *TrAC, Trends Anal. Chem.*, 2013, **52**, 239–247.
15. A. Önal, S. E. K. Tekkeli and C. Önal, A review of the liquid chromatographic methods for the determination of biogenic amines in foods, *Food Chem.*, 2013, **138**, 509–515.
16. Y. Cai, *et al.*, Rapid analysis of biogenic amines from rice wine with isotope-coded derivatization followed by high performance liquid chromatography-tandem mass spectrometry, *Food Chem.*, 2016, **192**, 388–394.
17. H. Wu, *et al.*, Monitoring the contents of six steroidal and phenolic endocrine disrupting chemicals in chicken, fish and aquaculture pond water samples using pre-column derivatization and dispersive liquid-liquid microextraction with the aid of experimental design metho, *Food Chem.*, 2016, **192**, 98–106.
18. F. B. Erim, Recent analytical approaches to the analysis of biogenic amines in food samples, *TrAC, Trends Anal. Chem.*, 2013, **52**, 239–247.

19. S. Sentellas, Ó. Núñez and J. Saurina, Recent advances in the determination of biogenic amines in food samples by (U)HPLC, *J. Agric. Food Chem.*, 2016, **64**, 7667–7678.
20. S. Hernández-Cassou and J. Saurina, Derivatization strategies for the determination of biogenic amines in wines by chromatographic and electrophoretic techniques, *J. Chromatogr. B: Anal. Technol. Biomed. Life Sci.*, 2011, **879**, 1270–1281.
21. S. C. Cunha, R. Lopes and J. O. Fernandes, Biogenic amines in liqueurs: Influence of processing and composition, *J. Food Compos. Anal.*, 2017, **56**, 147–155.
22. C. Almeida, J. O. Fernandes and S. C. Cunha, A novel dispersive liquid-liquid microextraction (DLLME) gas chromatography-mass spectrometry (GC-MS) method for the determination of eighteen biogenic amines in beer, *Food Control*, 2012, **25**, 380–388.
23. F. Mazzotti, *et al.*, *N*-hydroxysuccinimidyl *p*-methoxybenzoate as suitable derivative reagent for isotopic dilution assay of biogenic amines in food, *J. Mass Spectrom.*, 2014, **49**, 802–810.
24. S. J. Liu, J. J. Xu, C. L. Ma and C. F. Guo, A comparative analysis of derivatization strategies for the determination of biogenic amines in sausage and cheese by HPLC, *Food Chem.*, 2018, **266**, 275–283.
25. J. Lee, Y. Kim, J. Her, M. K. Kim and K. Lee, Reduction of biogenic amine contents in fermented soybean paste using food additives, *LWT--Food Sci. Technol.*, 2018, **98**, 470–476.
26. R. Ke, Z. Wei, C. Bogdal, R. K. Göktaş and R. Xiao, Profiling wines in China for the biogenic amines: A nationwide survey and pharmacokinetic fate modelling, *Food Chem.*, 2018, **250**, 268–275.
27. C. Shi, J. Cui, H. Lu, H. Shen and Y. Luo, Changes in biogenic amines of silver carp (Hypophthalmichthys molitrix) fillets stored at different temperatures and their relation to total volatile base nitrogen, microbiological and sensory score, *J. Sci. Food Agric.*, 2012, **92**, 3079–3084.
28. X. Wang, *et al.*, Simultaneous determination of 10 kinds of biogenic amines in rat plasma using high-performance liquid chromatography coupled with fluorescence detection, *Biomed. Chromatogr.*, 2018, **32**, 1–8.
29. G. Vinci, D. Restuccia and R. Antiochia, Determination of biogenic amines in wines by HPLC-UV and LC-ESI-MS, *Chim. Ind.*, 2011, **9**, 128–135.
30. K. Ekici and A. K. Omer, The determination of some biogenic amines in Turkish fermented sausages consumed in Van, *Toxicol. Rep.*, 2018, **5**, 639–643.
31. A. Francioso, *et al.*, HPLC Determination of bioactive sulfur compounds, amino acids and biogenic amines in biological specimens, *Adv. Exp. Med. Biol.*, 2017, **975**, 535–549.
32. F. Zarghampour, Y. Yamini, M. Baharfar and M. Faraji, Electromembrane extraction of biogenic amines in food samples by a

microfluidic-chip system followed by dabsyl derivatization prior to high performance liquid chromatography analysis, *J. Chromatogr. A*, 2018, **1556**, 21–28.
33. R. M. Latorre, Sensitivity enhancement by on-line preconcentration and in-capillary derivatization for the electrophoretic determination of amino acids, *Electrophoresis*, 2001, **22**, 4355–4361.
34. J. Saurina, Chromatographic determination of amino acids by pre-column derivatization using 1, 2-naphthoquinone-4-sulfonate as reagent, *J. Chromatogr. A*, 1996, **740**, 21–30.
35. R. M. Latorre, S. Hernandez-cassou and J. Saurina, Strategies for in-capillary derivatization of amino acids in capillary electrophoresis using 1, 2-naphthoquinone-4-sulfonate as a labeling reagent, *J. Chromatogr. A*, 2001, **934**, 105–112.
36. J. Saurlna and R. Tauler, Multivariate curve resolution applied to reaction between amino acids and 1, 2-naphthoquinone-4-sulfonic acid, *Anal. Chem*, 1995, **67**, 3722–3726.
37. N. Garc, J. Saurina and S. Hern, High-performance liquid chromatographic determination of biogenic amines in wines with an experimental design optimization procedure, *Anal. Chim. Acta*, 2006, **575**, 97–105.
38. H. Rui, X. Zhang, X. Liu, F. Wu, Z. Wang and K. Li, Simultaneous determination of free amino acids and biogenic amines in Chinese rice wine by RP-HPLC with pre-column phenylisothiocyanate (PITC) derivatization, *Food Sci.*, 2016, **37**, 159–163.
39. D. Z. S. Furtado, *et al.*, Profiles of amino acids and biogenic amines in the plasma of Cri-du-Chat patients, *J. Pharm. Biomed. Anal.*, 2017, **140**, 137–145.
40. D. Zildeana, *et al.*, Profiles of amino acids and biogenic amines in the plasma of Cri-du-Chat patients, *J. Pharm. Biomed. Anal.*, 2017, **140**, 137–145.
41. T. Santa, Isothiocyanates as derivatization reagents for amines in liquid chromatography/electrospray ionization – tandem mass spectrometry, *Biomed. Chromatogr.*, 2010, **24**, 915–918.
42. A. Pe, M. J. Ibarz, J. Cacho and V. Ferreira, Determination of the biogenic amines in musts and wines before and after malolactic fermentation using 6-aminoquinolyl-*N*-hydroxysuccinimidyl carbamate as the derivatizing agent, *J. Chromatogr. A*, 2006, **1129**, 160–164.
43. T. S. Weiss, HPLC of biogenic amines as 6-Aminoquinolyl-*N*-hydroxysuccinimidyl Derivatives, *J. Chromatogr. Libr.*, 2005, **70**, 502–523.
44. J. L. Ordóñez, R. M. Callejón, M. L. Morales and M. C. García-Parrilla, A survey of biogenic amines in vinegars, *Food Chem.*, 2013, **141**, 2713–2719.
45. J. L. Ordóñez, R. M. Callejón, A. M. Troncoso and M. C. García-Parrilla, Evaluation of biogenic amines profile in opened wine bottles: Effect of storage conditions, *J. Food Compos. Anal.*, 2017, **63**, 139–147.

46. T. Bauzaa, A. Blake, F. Daumasb and J. C. Cabanis, Determination of biogenic amines and their precursor amino acids in wines of the Vallke du RhSne by high-performance liquid chromatography with precolumn derivatization and fluorimetric detection, *Food Chem.*, 1995, **9673**.
47. A. Jámbor, Amino acid analysis by high-performance liquid chromatography after derivatization with 9-fluorenylmethyloxycarbonyl chloride Literature overview and further study, *J. Chromatogr. A*, 2009, **1216**, 3064–3077.
48. T. Yoshida, A. Uetake, H. Yamaguchi, N. Nimura and T. Kinoshita, New preparation method for 9-anthryldiazomethane (ADAM) as a fluorescent labeling reagent for fatty acids and derivatives, *Anal. Biochem.*, 1988, **174**, 70–74.
49. S. Gómez-Alonso, I. Hermosín-Gutiérrez and E. García-Romero, Simultaneous HPLC analysis of biogenic amines, amino acids, and ammonium ion as aminoenone derivatives in wine and beer samples, *J. Agric. Food Chem.*, 2007, **55**, 608–613.
50. Y. Fu, Z. Zhou, Y. Li, X. Lu, C. Zhao and G. Xu, High-sensitivity detection of biogenic amines with multiple reaction monitoring in fish based on benzoyl chloride derivatization, *J. Chromatogr. A*, 2016, **1465**, 30–37.
51. Y. Fu, *et al.*, High-sensitivity detection of biogenic amines with multiple reaction monitoring in fish based on benzoyl chloride derivatization, *J. Chromatogr. A*, 2016, **1465**, 30–37.
52. P. F. Gao, X. F. Guo, H. Wang and H. S. Zhang, Determination of trace biogenic amines with 1,3,5,7-tetramethyl-8-(N-hydroxysuccinimidyl butyric ester)-difluoroboradiaza-s-indacene derivatization using high-performance liquid chromatography and fluorescence detection, *J. Sep. Sci.*, 2011, **34**, 1383–1390.
53. A. M. Piasta, A. Jastrzebska, M. P. Krzemiński, T. M. Muzioł and E. Szłyk, New procedure of selected biogenic amines determination in wine samples by HPLC, *Anal. Chim. Acta*, 2014, **834**, 58–66.
54. A. Jastrzebska, A. Piasta, S. Kowalska, M. Krzemiński and E. Szłyk, A new derivatization reagent for determination of biogenic amines in wines, *J. Food Compos. Anal.*, 2016, **48**, 111–119.
55. Y. He, *et al.*, Simultaneous determination of food-related biogenic amines and precursor amino acids using *in situ* derivatization ultrasound-assisted dispersive liquid–liquid microextraction by ultra-high-performance liquid chromatography tandem mass spectrometry, *J. Agric. Food Chem.*, 2016, **64**, 8225–8234.
56. J. You and Y. Zhang, Determination of biogenic amines by RPHPLC with fluorescent detection after derivatization with 2-(9-carbazole)ethyl chloroformate (CEOC), *Chromatographia*, 2002, **56**, 43–50.
57. A. Jastrzębska, A. Piasta, M. Krzemiński and E. Szłyk, Application of 3,5-bis-(trifluoromethyl)phenyl isothiocyanate for the determination of selected biogenic amines by LC-tandem mass spectrometry and19F NMR, *Food Chem.*, 2018, **239**, 225–233.

58. J. Y. Lee, Y. G. Kim, J. Y. Her, M. K. Kim and K. G. Lee, Reduction of biogenic amine contents in fermented soybean paste using food additives, *LWT–Food Sci. Technol.*, 2018, **98**, 470–476.
59. C. P. Vieira, *et al.*, Development and validation of RP-HPLC-DAD method for biogenic amines determination in probiotic yogurts, *Arabian J. Chem.*, 2017, DOI: 10.1016/j.arabjc.2017.12.010.
60. M. L. Latorre-Moratalla, *et al.*, Validation of an ultra-high pressure liquid chromatographic method for the determination of biologically active amines in food, *J. Chromatogr. A*, 2009, **1216**, 7715–7720.
61. R. M. Callejón, A. M. Troncoso and M. L. Morales, Determination of amino acids in grape-derived products: A review, *Talanta*, 2010, **81**, 1143–1152.
62. A. Peña-gallego, P. Hernández-orte, J. Cacho and A. Peña-gallego, High-performance liquid chromatography analysis of amines in must and wine, *Food Rev. Int.*, 2012, **28**, 71–96.
63. E. Dadáková, M. Krizek and T. Pelikanova, Determination of biogenic amines in foods using ultra-performance liquid chromatography (UPLC), *Food Chem.*, 2009, **116**, 365–370.
64. C. Proestos, P. Loukatos and M. Komaitis, Determination of biogenic amines in wines by HPLC with precolumn dansylation and fluorimetric detection, *Food Chem.*, 2008, **106**, 1218–1224.
65. H. L. Jiang, *et al.*, Chromatographic determination of biogenic amines in wines after treatment with ionic liquids as novel media, *J. Sep. Sci.*, 2011, **34**, 1055–1062.
66. M. Ivana, L. Ivica, R. Nikolina, P. Ante, A. Ivana and V. Nenad, The content of biogenic amines in Croatian wines of different geographical origins, *Molecules*, 2018, **23**, 2570.
67. R. Romero, D. Gazquez and M. G. Bagur, Optimization of chromatographic parameters for the determination of biogenic amines in wines by reversed-phase high-performance liquid chromatography, *J. Chromatogr. A*, 2000, **871**, 75–83.
68. S. Bover-cid, A. Marine-font, M. C. Vidal-carou and F. Lahoz-portoles, Short communication. Ion-pair high-performance liquid chromatographic determination of biogenic amines and polyamines in wine and other alcoholic beverages, *J. Chromatogr. A*, 2003, **998**, 235–241.
69. D. Kutlan, New aspects of the simultaneous analysis of amino acids and amines as their *o*-phthaldialdehyde derivatives by high-performance liquid chromatography: Analysis of wine, beer and vinegar, *Food Chem.*, 2003, **987**, 311–322.
70. M. T. Kelly, A. Blaise and M. Larroque, Rapid automated high performance liquid chromatography method for simultaneous determination of amino acids and biogenic amines in wine, fruit and honey, *J. Chromatogr. A*, 2010, **1217**, 7385–7392.
71. Ö. Özdestan and A. Üren, A method for benzoyl chloride derivatization of biogenic amines for high performance liquid chromatography, *Talanta*, 2009, **78**, 1321–1326.

72. A. Gonza, N. Jime and C. Ancı, Comparative study of the amine concentration in wines obtained from the traditional fermentation and from a more anaerobic fermentation method, *LWT--Food Sci. Technol.*, 2010, **43**, 771–776.
73. L. Simon-Sarkadi and K. Ganzler, Determination of biogenic amines by capillary electrophoresis, *J. Chromatogr. A*, 1999, **836**, 305–313.
74. F. Chem, Simultaneous HPLC analysis of biogenic amines, amino acids, and ammonium ion as aminoenone derivatives in wine and beer samples, *J. Agric. Food Chem.*, 2007, 608–613.
75. S. Go, *et al.*, Biogenic amine production by *oenococcus oeni* isolates from malolactic fermentation of Tempranillo wine, *J. Food Prot.*, 2009, **72**, 907–910.
76. E. Garcı and P. M. I. Can, Changes in the aromatic composition of Tempranillo wines during spontaneous malolactic fermentation, *J. Food Compos. Anal.*, 2008, **21**, 724–730.
77. M. Valente, C. M. Santos and M. Gonc, Application of gas-diffusion microextraction for high-performance liquid chromatographic analysis of aliphatic amines in fermented beverages, *Anal. Methods*, 2012, 2569–2573.
78. I. Molnár-Perl, Role of chromatography in the analysis of sugars, carboxylic acids and amino acids in food, *J. Chromatogr. A*, 2000, **891**, 1–32.
79. S. R. Hagen, B. Frost and J. Augustin, Precolumn phenylisothiocyanate derivatization and liquid chromatography of amino acids in food, *J. - Assoc. Off. Anal. Chem.*, 1989, **72**, 912–916.
80. A. Villalba-rodr and I. Fern, Study of different biogenic amines in wines from two protected designations of origin (PDOs) in Murcia (Spain) multivariate classification, *J. Agric. Food Chem.*, 2011, 111–119.
81. M. C. Prieto-Blanco, C. Chafer-Pericas, P. Lopez-Mahia and P. Campins-Falco, Micro-extraction assisted derivatization coupled to liquid chromatography for quantifying residual dimethylamine in cationic polymers, *J. Chromatogr. A*, 2008, **1188**, 118–123.
82. E. Szłyk, A. Piasta and M. Krzemin, Application of 3,5-bis-(trifluoromethyl) phenyl isothiocyanate for the determination of selected biogenic amines by LC-tandem mass spectrometry and 19 F NMR, *Food Chem.*, 2018, **239**, 225–233.
83. O. Y. A.-D. Tomofumi Santa Takeshi Fukushima, Derivatization reagents in liquid chromatography/electrospray/rionization tandem mass spectrometry for biomedical analysis, *Drug Discoveries Ther.*, 2007, **1**, 108–118.
84. S. Hernández-Cassou and J. Saurina, Derivatization strategies for the determination of biogenic amines in wines by chromatographic and electrophoretic techniques, *J. Chromatogr. B: Anal. Technol. Biomed. Life Sci.*, 2011, **879**, 1270–1281.
85. M. P. Justyna, C. Morrison, M. Biziuk and J. Namies, Chemical derivatization processes applied to amine determination in samples of different matrix composition, *Chem. Rev.*, 2015, **115**, 4693–4718.

86. L. Ord, A. M. Troncoso, M. D. C. García-parrilla and R. Maria, Recent trends in the determination of biogenic amines in fermented beverages. A review, *Anal. Chim. Acta*, 2016, **939**, 10–25.
87. V. E. S. P. B. Harduf, Z. T. Nir and B. J. Juven, High-performance liquid chromatography amines, derivatized with 9-fluorenylmethyl mate of biogenic, *J. Neurosci. Methods*, 1989, **29**, 27–31.
88. J. Kirschbaum, Pre-column derivatization of biogenic amines and amino acids with 9-fluorenylmethyl chloroformate and heptylamine, *J. Chromatogr. A*, 1994, **661**, 193–194.
89. S. Ronald Jackson, *Wine Science, Principles and applications*, Food Science and Technology, Elsevier, 3rd edn, 2008.
90. B. Pfundstein, A. R. Tricker, E. Theobald, B. Spiegelhalder and R. Preussmann, Mean daily intake of primary and secondary amines from foods and beverages in West Germany in 1989–1990, *Food Chem. Toxicol.*, 1991, **29**, 733–739.
91. A. Y. Smit, W. J. Toit and M. Toit, Biogenic amines in wine: Understanding the headache, *S. Afr. J. Enol. Vitic.*, 2014, **29**, 109–127.
92. D. Torrea and C. Ancín, Content of biogenic amines in a Chardonnay wine obtained through spontaneous and inoculated fermentations, *J. Agric. Food Chem.*, 2002, **50**, 4895–4899.
93. J. Huang, N. Gan, F. Lv, Y. Cao, C. Ou and H. Tang, Environmentally friendly solid-phase microextraction coupled with gas chromatography and mass spectrometry for the determination of biogenic amines in fish samples, *J. Sep. Sci.*, 2016, **39**, 4384–4390.
94. S. C. Cunha, M. A. Faria and J. O. Fernandes, Gas chromatography-mass spectrometry assessment of amines in port wine and grape juice after fast chloroformate extraction/derivatization, *J. Agric. Food Chem.*, 2011, **59**, 8742–8753.
95. W. Wojnowski, J. Namieśnik and J. Płotka-Wasylka, Dispersive liquid-liquid microextraction combined with gas chromatography–mass spectrometry for *in situ* determination of biogenic amines in meat: Estimation of meat's freshness, *Microchem. J.*, 2019, **145**, 130–138.
96. J. Płotka-Wasylka, V. Simeonov and J. Namieśnik, An *in situ* derivatization – dispersive liquid-liquid microextraction combined with gas-chromatography – mass spectrometry for determining biogenic amines in home-made fermented alcoholic drinks, *J. Chromatogr. A*, 2016, **14**, 10–18.
97. M. Papageorgiou, D. Lambropoulou, C. Morrison, J. Namieśnik and J. Płotka-Wasylka, Direct solid phase microextraction combined with gas chromatography – Mass spectrometry for the determination of biogenic amines in wine, *Talanta*, 2018, **183**, 276–282.
98. K. K. Ngim, S. E. Ebeler, M. E. Lew, D. G. Crosby and J. W. Wong, Optimized procedures for analyzing primary alkylamines in wines by pentafluorobenzaldehyde derivatization and GC–MS, *J. Agric. Food Chem.*, 2000, **48**, 3311–3316.

99. K. K. Ngim, S. E. Ebeler, M. E. Lew, D. G. Crosby and J. W. Wong, Optimized procedures for analyzing primary alkylamines in wines by pentafluorobenzaldehyde derivatization and GC–MS, *J. Agric. Food Chem.*, 2000, **48**, 3311–3316.
100. G. B. Baker, R. T. Coutts and A. Holt, Derivatization with acetic anhydride: Applications to the analysis of biogenic amines and psychiatric drugs by gas chromatography and mass spectrometry, *J. Pharmacol. Toxicol. Methods*, 1994, **31**, 141–148.
101. C. Almeida, J. O. Fernandes and S. C. Cunha, A novel dispersive liquid-liquid microextraction (DLLME) gas chromatography-mass spectrometry (GC-MS) method for the determination of eighteen biogenic amines in beer, *Food Control*, 2012, **25**, 380–388.
102. J. O. Fernandes and M. A. Ferreira, Combined ion-pair extraction and gas chromatography-mass spectrometry for the simultaneous determination of diamines, polyamines and aromatic amines in Port wine and grape juice, *J. Chromatogr. A*, 2000, **886**, 183–195.

CHAPTER 14

Chromatographic and Capillary Electrophoretic Methods for the Determination of Biogenic Amines

ABDALGHAFFAR M. OSMAN, CHANBASHA BASHEER* AND MOHD NAIM

Department of Chemistry, King Fahd University of Petroleum and Minerals, Dhahran, 31261, Saudi Arabia
*Email: cbasheer@kfupm.edu.sa

14.1 Introduction

Chromatographic methods are considered among the most common analytical technique used for the quantitation of biogenic amines (BAs). Quantitation at low concentration of BAs from complex sample matrices is a challenging task for the scientific community. In this regard, numerous chromatographic methods have been developed to address the challenges of BAs in a wide range of food and beverages. Most of the chromatographic methods require tedious sample pre-treatment methods before the chromatographic analysis. To achieve better separation and detection of BAs, a selection of suitable column, mobile phase, detector, and optimum operating conditions are needed. In this chapter, various chromatographic and electrophoretic methods used for the analysis of BAs in foodstuffs and

Food Chemistry, Function and Analysis No. 20
Biogenic Amines in Food: Analysis, Occurrence and Toxicity
Edited by Bahruddin Saad and Rosanna Tofalo

Published by the Royal Society of Chemistry, www.rsc.org

beverages and the challenges associated with these chromatography analyses are discussed.

14.1.1 Liquid Chromatography Methods

Reversed-phase high-performance liquid chromatography (RP-HPLC) using C18 column is the most commonly used chromatographic technique for the determination of BAs in food products such as wines and beverages. Both aqueous normal phase liquid chromatography (ANP-LC) and hydrophilic interaction liquid chromatography (HILIC) have been applied for the determination of BAs. Those two modes are between the normal phase (NP) that uses non-polar solvent and polar columns and reversed phase types of LC. HILIC, applied for polar weakly basic compounds, has found great popularity in bioanalytical applications, and it was used for the analysis of polar and charged compounds with unique separation ability.[1–3]

14.1.2 Columns, Mobile Phase, and Detection in LC

Surface-modified silica with C18 is the most common and widely used stationary phase for columns in RP-LC separations for BAs. These types of columns are hydrophobic and non-polar. Therefore, BAs are derivatized to reduce their polarity before separation with this mode. ANP-LC and HILIC separations are performed on a stationary phase based on silica hydride.[4] The particle size of column packing used with ultra-high pressure liquid chromatography (UHPLC) systems is less than 2 µm. In both types, no derivatization of BAs is required before the separation step since the amines are as polar as the stationary phase.

All solvents used in LC are of high purity. In RP-LC, a polar mobile phase is used (typically a mixture of organic solvents such as acetonitrile, methanol, *etc.*). The mobile phase used in both ANP-LC and HILIC is also an organic solvent such as methanol or acetonitrile in addition to including a certain amount of water.[4]

Many factors are considered when choosing a detector for HPLC analysis. These include fast response, sensitivity, linear dynamic range, does not respond to changes in flow rate and mobile phase composition, and others. Various types of HPC detectors were used for the determination of BAs, but the UV detector is the most common because many compounds absorb in the UV region. The fluorescence detector (FLD) is very sensitive to compounds that have fluorescence properties or fluoresce after derivatization. FLD is more sensitive than the UV detector, making it particularly useful for low concentrations of analytes. The mass spectrometry (MS) detector can provide specific structural information and is more sensitive. For BA determination, most of the determinations have been performed using either ultraviolet-visible (UV-vis) or FLD. MS detectors were used mainly with UHPLC.

14.2 RP-LC Separation of BAs

14.2.1 HPLC

As mentioned earlier, BAs must be derivatized and converted into less polar compounds before separation. Besides, most BAs lack chromophores and fluorescent groups required for UV-vis and FLD, respectively. Monitoring of BAs in wines has extensively been conducted using LC techniques, particularly RP-HPLC. BAs in wines have been separated and quantified through different methods of sample pre-treatment, derivatization, and detection.[5–14] The dansyl derivatives of ten BAs were separated and identified using UV-vis detection.[5,6] In other studies, BAs in wine were also converted into derivatives of 2-chloro-1,3-dinitro-5-(trifluoromethyl)-benzene[7] and benzoyl chloride.[8] In the latter method, dispersive solid phase extraction was applied to amines in a wine before the derivatization step to enhance the sensitivity. Dansyl[9,10] and *o*-phthalaldehyde[11,12] derivatives of BAs in wines were also quantified either by UV-vis or FLD. RP-HPLC and RP-UHPLC separations have also been applied in wine to quantify BAs with MS detection.[13–16] RP-HPLC separation and quantitation of BAs have also been applied to a wide range of food types, including dairy products.[17,18] Two extraction methods were applied to preconcentrate the analytes. In the first step, BAs were extracted by solid phase extraction using magnetic mesoporous silica,[17] while precipitation and centrifugation of peptides in the acidic extract were performed to avoid interferences[18] and to enhance the method sensitivity. Dabsyl derivatives of 11 BAs in egg yolk were separated and quantified. This interesting study found a correlation between the content of BAs and shelf life and the storage time.[19]

In order to determine the concentration of BAs in fish, their derivatives were extracted using magnetic ionic liquid (MIL)-dispersive liquid–liquid microextraction before separation.[20] The detection limit for six amines, including histamine, was in the range 3.9–9.6 μg kg^{-1}. Determination of histamine in fish tissue was also achieved without derivatization, but the detection limit was high (1 mg kg^{-1}) compared with the method with derivatization, although the determination was achieved within 5 min.[21]

RP-HPLC has found wide applications for the separation and determination of BAs in fermented foods.[22,23] Soybean products and cheese samples were also analysed for the presence of BAs. The method was highly sensitive, and the separation revealed the presence of seven BAs using FLD. Detection limit for the seven compounds was in the range 12.6–166.6 nmol L^{-1}.[24] BAs were also determined in fermented food (cheese and sausage) using LC coupled with quadrupole time-of-flight mass detector (UHPLC-TOFMS). This method was able to separate and identify the dansylated derivatives of 23 amino acid and amines. The separation and determination were relatively fast (within 25 min). This was attributed to the use of high pressure with sub-2 μm particle size UHPLC column. The detection limit was down to 5 μg L^{-1}.[25] Another sensitive method was developed for the separation and identification of BAs in condiment soy sauce using

UHPLC-MS/MS.[26] Seven BAs were identified and quantified with a linear range of 3.0–600 μg L^{-1}. The limits of detection were in the range 4–8 μg kg^{-1}, while the limits of quantitation ranged from 15 to 30 μg kg^{-1}.[26]

BAs in fruit juices, cocoa, beans,[6] honey,[12] and meat[15,27,28] have also been reported. Generally, the simultaneous determination of both BAs and amino acids needs more time for only BAs. The separation time ranged from 4 min for a few BAs to 85 min for a mixture of 30 BAs and amino acids. More details about the columns and mobile phases used, detection systems, and analytical figures of merits such as sensitivity, linear range and limits of detection (LOD), and quantification (LOQ) are provided in Table 14.1.

14.2.2 UHPLC

Ultra HPLC (UHPLC) sometimes simply referred to as UPLC offers many advantages over HPLC in terms of better resolution, less solvent consumption, shorter analysis time, and lower detection limit because high flow rates and separation columns packed with sub-2 μm particle size[32] are employed. Several determinations of BAs in foods have been performed by UPLC.[33] Eleven amines[6] and a mixture of 24 amines and amino acids[12] were identified and quantified in a total run time of 13 and 40 min, respectively. The simultaneous determination of 23 amino acids and seven BAs was achieved within 25 min.[25] Fast UPLC determination of 12 BAs and polyamines in less than 7 min with FLD[29] and seven BAs within 4 min with LC tandem mass detection (LC-MS/MS)[26] were also reported. More details are provided in Table 14.1.

14.3 ANP-LC and HILIC

HILIC provides an alternative approach for efficient separation of polar compounds such as amines using a stationary phase with higher polarity and relatively polar solvents. It was first proposed for the separation of amino acids and peptides.[34] Both ANP-LC and HILIC modes are between the NP that uses polar columns with non-polar solvent and reversed-phase type of LC. In both types, no derivatization of BAs is required before the separation step since the amines are as polar as the stationary phase. Separation and determination of BAs such as histamine in red wines and some other food products were achieved under ANP conditions, with either UV or MS detection.[35] HPLC-MS/MS method with HILIC column was developed for the determination of serotonin in various types of raw and roasted nuts as well as nut-containing snack bars, chocolates, and spreads,[30] and the other seven BAs in cheese.[31] More details are provided in Table 14.1. A HILIC column structure is represented in Figure 14.1.

HPLC remains the method of choice for the determination of BAs in all type of foods and beverages (alcoholic and others). The reason behind the extensive use of this technique can be attributed to the variety of modes

Table 14.1 HPLC/UHPLC conditions and analytical figures of merits for the determination of BAs.

BAs	Food type	Column (length, ID) particle size	Mobile phase	Separation mode/ detection	Analytical performance: Linear range	Analytical performance: LOD and LOQ	Reference
MEA, ETA, AGM, PEA, PUT, CAD, HIS, IS, SER, TRY, SPD, SPM	Beverages	C18, (100 mm × 4.6 mm), 2.6 µm	(A) acetonitrile (B) water	HPLC–FLD HPLC–UV–Vis	0.01–8 mg L^{-1} 0.06–8 mg L^{-1}	LOD: 0.002–0.023 mg L^{-1} LOQ: 0.006–0.077 mg L^{-1} LOD: 0.013–0.112 mg L^{-1} LOQ: 0.043–0.377 mg L^{-1}	6
HIS, PEA, TRYP, TYR	Wine	C18, (250 mm × 4.6 mm), 5 µm	(A) methanol 80% (B) water 20% (gradient mode)	RP-HPLC–UV–Vis	0.09–9.36 mg L^{-1}	LOD: 0.02–0.03 mg L^{-1} LOQ: 0.05–0.10 mg L^{-1}	7
ETA, PROP, BUT, PUT, CAD, TRY PEA, AML, SPD, HEX, SPM, HIS	Wines	C18, (250 mm × 4.6 mm), 5 µm	(A) sodium tetraborate 5 mM at pH 8 (B) methanol (gradient mode)	RP–HPLC–UV	0.25–20 mg L^{-1}	LOD: 0.133–0.509 mg L^{-1}	8
AGM, MTA, DA, ETA, TRM, PEA, IPA, PUT, CAD, HIM, TYM, SPD, SPM	Wine	C18, (250 mm × 4.6 mm), 5 µm	(A) buffer acetate/ acetonitrile (B) acetonitrile	HPLC–FLD	0.01–7.2 mg L^{-1}	LOD: 0.01–0.06 mg L^{-1} LOQ: 0.01–0.16 mg L^{-1}	9
HIS, CAD, PUT, TYR	Yoghurt	C18, (150 mm × 4.6 mm), 5 µm	(A) acetonitrile 60% (v/v) (B) water 60% (v/v)	HPLC–UV	0.05–120 mg L^{-1}	LOD: 0.008–0.014 mg L^{-1} LOQ: 0.022–0.044 mg L^{-1}	17
PUT, CAD, SPD, SPM, TYR	Yoghurt	C18, (150 mm × 4.6 mm), 5 µm	(A) water (B) acetonitrile	RP-HPLC–DAD	1–50 mg L^{-1}	LOD: 0.18–0.65 mg L^{-1} LOQ: 1.0–5.0 mg L^{-1}	18
TRY, PHE, HIS, TYR, SPD, SPM	Wine, fish	C18, (150 mm × 4.6 mm), 5 µm	(A) water (B) acetonitrile (gradient mode)	HPLC–UV	10.0–1000.0 10.0–1000.0	LOD: 1.3–3.9 µg L^{-1} LOQ: 4.1–9.9 µg L^{-1} LOD: 1.2–2.8 µg kg^{-1} LOQ: 3.9–9.6 µg kg^{-1}	20
HIS	Fish	C18, (250 mm × 4.6 mm), 5 µm	(A) acetonitrile 50% (B) 2,5-dihydroxybenzoic acid 50%	RP–HPLC–UV	1–32 mg kg^{-1}	LOD: 1 mg kg^{-1} LOQ: 3 mg kg^{-1}	21

TYR, TRY, HIS, PUT, CAD, 2-PEA, SPM, SPD	Cheese	C18, (250 mm × 4.6 mm), 5 μm	(A) water (B) acetonitrile (gradient mode)	RP–HPLC–UV	0.25–1250 mg kg^{-1}	LOD: 0.07–0.23 mg kg^{-1} LOQ: 0.20–0.70 mg kg^{-1}	22
TRY, PHEPUT, CAD, HIS, TYR, SPD, SPM,	Fermented radish	C18, (150 m × 4.6 mm), 4 μm	(A) 0.1 M ammonium acetate (B) acetonitrile (gradient mode)	HPLC–UV–vis	—	—	23
TRY, 2-PEA, CAD, PUT, TYR, SPD, HIS	Fish meat, soybean products, cheese	C18, (75 mm × 4.6 mm), 5 μm	(A) acetonitrile 65% (v/v) (B) water 35% (v/v)	HPLC–FLD	0.04–0.81 nmol mL^{-1}	LOD: 0.0126–0.1666 nmol mL^{-1} LOQ: 0.0317–0.7065 nmol mL^{-1}	24
TYR, TRY, HIS, PUT, CAD, 2-PEA, SPD	Cheese, sausage	C18, (100 mm × 2.1 mm), 1.7 μm	(A) water (B) acetonitrile (gradient mode)	UHPLC–TOFMS	20–30 000 μg L^{-1}	LOD: 5–400 μg L^{-1} LOQ: 50–500 μg L^{-1}	25
SPM, SPD, HIS, PUT, TYR, PEA, TRP	Condiment, soy sauce	C18, (100 mm × 2.1 mm), 1.7 μm	(A) 0.1% formic acid in water (B) 0.1% formic acid in acetonitrile	UHPLC–MS/MS	3.0–600 μg L^{-1}	LOD: 4–8 μg kg^{-1} LOQ: 15–30 μg kg^{-1}	26
TRY, BUT, PEA, CAD, PUT, TYR, SPM, SPD	Meat	C18, (250 mm × 4.6 mm), 5 μm	(A) acetonitrile (B) 0.3% ammonium acetate (aqueous)	HPLC–FLD	5.0–800.0 μg L^{-1}	LOD: 0.92–2.57 μg L^{-1}	27
HIS, TYR, CAD	Fish, cheese	C18, (50 mm × 2.1 mm), 1.7 μm	(A) 0.1M sodium acetate/10 mM sodium (B) acetonitrile, 0.2M sodium acetate and 10 mM sodium octanesulphonate	UHPLC–FLD	0.1–50 mg L^{-1}	LOD: $>$ 0.2 mg L^{-1} LOQ: $>$ 0.3 mg L^{-1}	29
5-HT	Nuts, nut-containing products	HILIC, silica (150 × 2.1 mm), 3 μm	(A) 2.5 mM ammonium formate in water (90% v/v) (B) 2.5 mM ammonium formate in acetonitrile/water (10% v/v)	UPLC–MS/MS	—	LOD: 0.4–2.3 ng g^{-1} LOQ: 1.0–7.4 ng g^{-1}	30
CAD, HIS, PUT, SPD, SPM, TRY and TYR	Cheese	HILIC, silica (150 × 2.1 mm), 3 μm	(A) acetonitrile (B) 50 mM ammonium formate	HPLC–MS/MS	6.0–205 μg L^{-1} 10–1000 μg L^{-1}	LOD: 1.1–3.5 μg L^{-1} LOQ: 3–10 μg L^{-1}	31

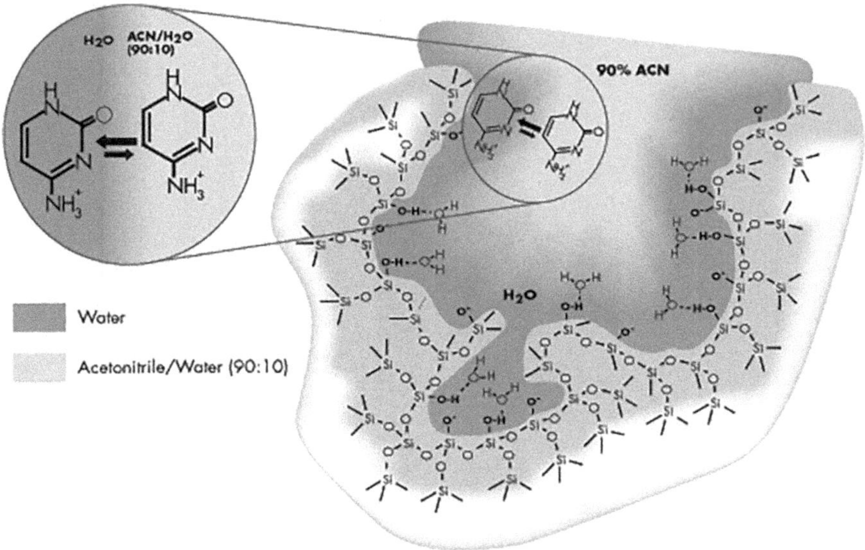

Figure 14.1 Structure of HILIC silica stationary phase and hydrophilic partitioning of a basic analyte. Image has been provided courtesy of Waters Corporation.[36]

that allow the separation of BAs on different types of columns and with the use of a large number of solvents with different compositions. In addition, UHPLC offers highly efficient and fast separation and a low detection limit when it is used with sub-2 μm particle size. Moreover, the coupling of the HPLC with MS increased the analysis sensitivity.

14.3.1 Ion Chromatography

The ion chromatography (IC) method is considered in the family of HPLC techniques. It is used to separate and determine ionic species after converting them into the ionic form[37] if they originally exist in non-ionic form. The stationary phases, also known as ion exchangers, are mainly polymers such as poly(ethylvinylbenzene-divinylbenzene) (EVB-DVB) containing carboxylic groups as a cationic exchanger. Silica gel functionalized with carboxyl groups is also used. Stationary phases containing phosphonic acid have a higher affinity towards the amino groups in BAs higher than those containing carboxylic groups.[38]

Eluents used in IC separation of BAs are of acidic in nature. Many diluted acid solutions are used for this purpose and the concentration of the acid is typically optimized to achieve the best separation, especially when a large number of BAs are to be analysed. Among the eluent solutions, methanesulphonic acid (MSA), sulphuric acid, nitric acid, and formic acid are commonly used. Despite the lower elution power of formic acid compared with

MSA towards some BAs,[39] it is used as an eluent in IC coupled to MS because of its compatibility with MS detection. Elution modes used were either gradient or isocratic. Many detection systems, including conductivity, amperometry, and MS, have been employed for IC determination of BAs. Column dimensions, types of exchangers, eluent composition, and analytical figures of merits of some IC methods are provided in Table 14.2.

IC methods have been used in the separation and determination of BAs in food, wine, and other beverages. Underivatized BAs in fish samples,[40] fish, and beer[41] have been determined using IC with conductometric detection (COND). BAs in various fruit juices and some meat products have been identified and quantified using IC-COND.[42] Recently, IC methods with either MS/MS[39] integrated pulsed amperometric detections[43] have also been applied for the determination of BAs in different cheese samples. The IC-MS/MS method enabled the separation of ten underivatized BAs with high sensitivity.

Despite the extensive use of LC and CE methods for determination of BAs in various types of food, IC methods have the advantage that the determination is achieved without derivatization in most cases.

14.4 Thin Layer Chromatography

Thin layer chromatography (TLC) is one of the first techniques used for separation and identification of BAs.[44–46] It is an economical and straightforward method used for the separation and detection of nanogram amounts of BAs and their precursors and metabolites in biological samples with FLD[47] and without any derivatization.

Separation of BAs are mainly performed on TLC aluminium or glass plates (20 cm × 20 cm) that have been pre-coated with a 0.25 mm layer of silica gel G. BAs were commonly derivatized with dansyl chloride for the visualization on the plate. Different eluting solvent compositions have been used to achieve the separation of BAs derivatives. Examples of the solvent mixtures used were benzene : triethylamine (5 : 1, v/v), benzene : triethylamine : acetone (10 : 2 : 1, v/v), chloroform : trimethylamine (4 : 1, v/v), chloroform : benzene : triethylamine (6 : 4.5 : 1, v/v), and chloroform : diethyl ether : triethylamine (4 : 1 : 1, v/v). The common characteristic of these eluting solvent compositions is that they are of relatively low polarity. The detection technique was the visualization of the separated spots that is normally based on visualization reagents used or by the illumination of the plate with UV light. However, more sophisticated detection where images were captured through an imaging system[48] or by a mobile phone camera[49] and then digitized into optical density with the aid of suitable software have been proposed.

TLC-densitometry was developed and validated for the separation of some BAs in wine and determination of their concentrations over the range 1–20 mg L^{-1}.[48] A novel TLC-image analysis method has been reported for the separation and detection of histamine in fish in less than 15 min[49] using diazotized *p*-nitroaniline reagent for visualization. Quantitation was

Table 14.2 IC conditions and analytical figures of merits for the determination of BAs.

BAs	Food type	Column/ion exchanger (length, ID, particle size)	Eluent	Separation mode/detection	Analytical performance		Reference
					Linear range	LOD and LOQ	
TYR, HIS, AGM, PEA, TRY, PUT, CAD, SPD, SPM	Cheese	Poly(EVB-DVB)-carboxylic (50 mm × 4 mm, 7 μm)	1.0 M formic acid, water (gradient mode)	IC–MS/MS	10–10 000 μg L^{-1}	LOD: 12–46 μg L^{-1} LOQ: 40–153 μg L^{-1}	39
PUT, CAD, SPD, HIS	Fish	Poly(EVB-DVB)-carboxylic (250 mm × 4 mm, 7 μm)	0.1 mol L^{-1} methanesulphonic acid	IC–COND	—	LOD: 0.15–0.50 mg kg^{-1} LOQ: 0.5–1.0 mg kg^{-1}	40
MET, PROP, IBUT BUT, PUT, CAD, HIS, AGM, SPD, SPM	Fish, beers	Poly(EVB-DVB)-carboxylic (250 mm × 4 mm, 7 μm)	Dilute sulphuric acid	IC–COND	0.04–0.1 mM	LOD: 10–20 nM LOQ: 20–65 nM	41
HIS, TYR, PEA, CAD, PUT	Fruit juice, meat products, smoked fish	Silica gel-carboxylic (100 mm × 4 mm, 5 μm)	5 mM nitric acid	IC–COND	0.5–5 mg L^{-1} 5–100 mg L^{-1}	LOD: 0.056–1.63 mg L^{-1} LOQ: 0.19–3.27 mg L^{-1}	42
ETA, PUT, CAD, HIS, SPD, SPM	Cheese	Poly(EVB-DVB)-carboxylic (250 mm × 2 mm, 6 μm)	Methanesulphonic acid	IC–IPAD Ag/AgCl (RE) Au (WE)	—	LOD: 9–09 μg L^{-1} LOQ: 22–278 μg L^{-1}	43

achieved by image analysis with ImageJ software. Another study on the separation and quantification of BAs in fish and meat by TLC-densitometry found that it is comparable with HPLC in terms of precision, accuracy, recovery rate, limits of detection, and quantitation.[49,50] TLC-densitometry is simple, fast, and less expensive than other separation techniques.

Two-dimensional TLC is useful for the separation of mixtures that are not possible using the one-dimension technique. It was reported for the separation and fluorescence-densitometry quantification of dansylated derivatives of BAs in sausage[51] and fish samples.[52] The separation was achieved by the consecutive development using two different solvent systems.[53,54]

In short, TLC offers a valuable alternative to the other advanced LC techniques. Despite the long separation times, TLC is cost-effective, straightforward, and can be carried out by unskilled operators.

14.4.1 Capillary Electrophoresis

Capillary electrophoresis (CE) is an analytical technique that separates ions based on their electrophoretic mobility under the influence of an applied voltage. CE is the second most performed technique applied to the separation and determination of BAs in food, wine, and beverages.[55] It is relatively simple, rapid, reliable, efficient, and offers low reagent consumption. However, a disadvantage of the technique is that it is less sensitive when compared with other methods,[55,56] mainly due to the short sample path length. CE can be divided into the following separation classes: capillary zone electrophoresis (CZE); capillary gel electrophoresis (CGE); capillary isoelectric focusing (CIEF); micellar electrokinetic chromatography (MEKC);[57] and capillary electrochromatography (CEC). Figure 14.2 shows a schematic diagram of a basic CE with different modes of separations.

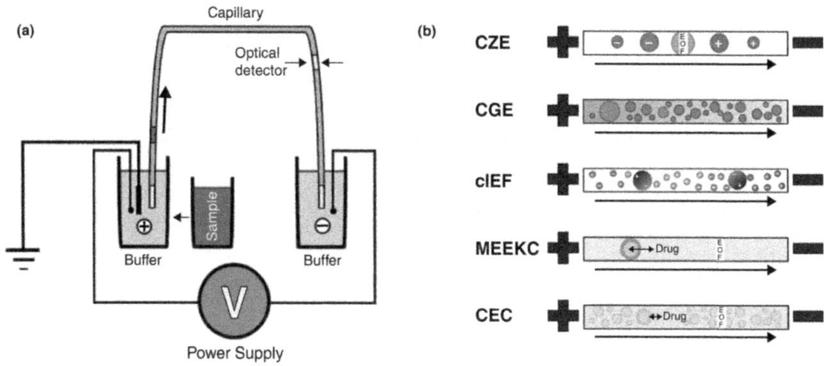

Figure 14.2 (a) Basic setup of a capillary electrophoresis system with a UV detector, and (b) a selection of separation modes employed in CE. Reproduced from ref. 58 with permission from Elsevier, Copyright 2015.

CGE is a robust electrophoresis technique that was used for the analysis of proteins over two decades. In this technique, a gel is packed inside the capillary, and mixing a protein sample with a buffer containing sodium dodecyl sulphate (SDS) is used for separation. Isoelectric focusing (IEF) is a type of CZE where the separation of molecules is based on the differences in their isoelectric points. MEKC is another mode commonly used in the analysis of neutral analytes. A surfactant is added to the buffer solution to interact the micelles with target analytes before they are separated. In CEC, the analytes are separated by the chromatographic electroosmosis effects.

Among the reported CE methods, CZE is the most widely applied mode where the separation of analyte ions is due to zone migration with different velocities through a background electrolyte (BGE) which plays an important role in electrophoretic separation. The primary purpose of using BGE is to maintain the conditions of the separating ions constant. Therefore, these conditions remain independent of the analyte composition.

Under these constant conditions, the sample components are accurately identified by their migration times.[59] The adjustment of the BGE pH ensures constant migration of the analyte ions and the electroosmotic flow.[60]

14.4.2 Capillaries Used in CE

14.4.2.1 Bare-fused Silica Capillary

The most popular capillary used in CE determination of BAs are fused silica capillaries. Fused silica capillaries offer flexibility and excellent thermal and optical properties in the UV region. Bare-fused silica capillaries have been successfully used for the separation of different BAs.[61–63] However, using such capillaries might cause interference of charged molecules with the capillary wall surface and consequently reduces the separation efficiency.

14.4.2.2 Polymer-coated Fused Silica Capillary

The use of polymer-coated silica capillaries suppresses the interferences of the charged molecules with the hydroxyl group on the silica surface and enhances the separation efficiency. Polyvinyl alcohol (PVA)-coated silica[64] and polyimide-coated silica capillaries[65] have been applied for the determination of BAs in wine and beer.

Polymer-based capillaries have been employed in CE analyses as pre-separation and analytical capillaries. Both polytetrafluoroethylene (PTFE)[66] and fluorinated ethylene propylene (FEP)[67] have been used for the CE determination of BAs in wine and other foods. Polymer coating on the capillary or polymer-based capillary is used to eliminate the distortive effects of electroosmotic flow and minimize wall adsorption of the analyte molecules.

14.4.3 CE Detectors

Many systems have been used for the detection and determination of BAs in different types of food and wine using capillary electrophoresis techniques. Those include UV-vis, fluorescence FLD, MS detectors, conductivity, and capacitively coupled contactless conductivity detector, C^4D. Although UV-vis detectors are the most commonly used for CE,[68] methods using fluorescence detection are generally more sensitive.[46] FLD has provided sensitive detection for microchip electrophoresis[69] and has been used with MCE for determination of BAs in fish meat.[70] MS/MS detectors provide more information and more sensitivity.

14.4.4 Some CE Applications

The determination of 8 BAs in wine and beer have been carried out without derivatization using capillary isotachophoresis (CITP) with COND. Results obtained were compared with the HPLC method based on pre-column derivatization. Lower LOD and LOQ have been obtained compared with HPLC analysis.[66] The method was fast and satisfactory in terms of precision and accuracy. BAs in beer and wine have also been analysed by CE-MS/MS,[64] CE-UV,[65] and CE-MS.[61] CE-MS/MS is fast and nine BAs were separated in less than 10 min.[64] MEKC-UV has been applied for the same purpose.[62,71] Quantitative determination of BAs in dairy products (cheese, yoghurt, and kefir) was possible in less than 6 min using CE-C^4D.[63]

Significant increase in sensitivity of the CZE method has been attained by an on-line combination of the CZE and CITP with COND for the determination of histamine in different foodstuffs.[67] The arrangement combines the advantage of the high loading capacity of CITP and high-resolution capacity of CZE. The method showed high sensitivity because of the on-line CITP pre-concentration. The obtained results were compared with HPLC-FLD or HPLC-FLD/MS. Table 14.3 provides some CE methods and their analytical figures of merits.

14.4.5 Microchip Electrophoresis

Microchip electrophoresis (MCE) is a very attractive method for the rapid analysis of BAs in food. In addition to the typical advantages of CE such as high efficiency, low reagent consumption, and low-cost, MCE have been applied for the fast electrophoretic separation with FLD of four BAs within 50 s.[70] This method offers excellent potential for miniaturization and performing fast and real-time detection of BAs for food safety and quality.

14.4.6 Gas Chromatography

GC methods are not widely applied for the determination of BAs in food matrices. Derivatization of BAs is needed prior to the separation to decrease their polarity and increase volatility, which is required for any GC

Table 14.3 CE conditions and analytical figures of merits for determination of BAs.

BAs	Food type	Capillary dimension and length	Run time/min	Separation/detection technique	Background electrolyte	Recovery %	Analytical performance: Linear range	Analytical performance: LOD and LOQ	Reference
SPE, SPM, PUT, HIS, CAD, ETOA, DET, TEA, 2-PE, TETOA, TYR, TRY	Wine	Silica (75 μm ID, 360 μm OD), 100 cm	—	CE–TOFMS	100 mM formic acid	65–117	—	LOD: 25 ng mL^{-1} LOQ: 49.5 ng mL^{-1}	61
HIST, TRY, TYR, 2-BE	Wine	Silica (75 μm ID, 365 μm OD), 40 cm	11	MEKC–UV	5 mM phosphate buffer at pH 3.7 containing 20 mM SDS	98.8–115.6	0.55–10.0 μmol L^{-1}	LOD: 0.11–0.61 μmol L^{-1} LOQ: 49.5 ng mL^{-1}	62
SPD, PUT, HIS, CAD, TYR	Cheese, yoghurt, kefir	Silica (25 μm ID, 360 μm OD), 65 cm	6	CE–C^{4}D	500 mM α-hydroxy isobutyric acid at pH 2.05	—	0.20–0.32 mg L^{-1}	LOD: 0.041–0.098 mg L^{-1} LOQ: 0.14–0.49 mg L^{-1}	63
MET, ETY, TRY, PHE, IAA, PUT, CAD, HIS, TYR, SPM, SPD	Wine, beer	PVA-coated silica (50 μm ID, 360 μm OD), 70 cm	<10	CE–MS/MS (triple-quad)	500 mM acetic acid at pH 2.5	87–113	—	LOD: 1–2 μg L^{-1} LOQ: 3–8 μg L^{-1}	64
SPD, PUT, HA, CAD, PEA, TA, TRY, DA, 5-HT, OA, NE, **α**-CD	Beer	Polyimide-coated silica (75 μm ID), 60 cm	—	CE–UV	12 mM imidazole and 10 mM α-cyclodextrin at pH 4.50	70.4–119.6	—	LOD: 0.36–3.67 μmol L^{-1} LOQ: 1.2–12.2 μmol L^{-1}	65
HIS, TYR, CAD, PUT, TRY, 2-PEA, SPM, SPD	Wine, beer	PTFE pre-separation capillary (0.3 mm ID), 9 cm PTFE analytical capillary (0.3 mm ID), 10 cm	—	CITP–COND	LE: 5 M barium hydroxide, 15 mM valine, and 1% HEC at pH 8.5. TE: 20 mM tris base and 100 mM HCl at pH 8.3	91–101	1–10 mg L^{-1} 5–70 mg L^{-1}	LOD: 0.2–1.4 mg L^{-1} LOQ: 0.5–4.8 mg L^{-1}	66
HIS	Fish, wine, cheese	FEP pre-separation capillary (0.8 mm ID), 9 cm FEP analytical capillary (0.3 mm ID), 16 cm	20	CITP–CZE–COND	15 mM histidine, 5 mM adipic acid, 2 mM sulphuric acid, and 50% MeOH	91 ± 9	20–250 ng mL^{-1}	LOD: 4 ng mL^{-1} LOQ: 12 ng mL^{-1}	67
CAD, PUT, TRY, HIS	Fish, meat	Microchip with 35 μm ID);	0.83	MCE–FLD	sodium polystyrene sulfonate (MW of 500 000)	85.4–112.0	0.01–20 μmol L^{-1}	LOD: 1.3–3.6 nmol L^{-1}	70

separations.[46] Polysiloxanes are the common material used as a stationary phase for the GC columns. Examples of polysiloxanes used in BAs analyses are diphenyl dimethylpolysiloxane, 1,4-bis(dimethylsiloxy)phenylene dimethylpolysiloxane, and cyanopropylmethyl phenylmethyl polysiloxane. The column dimensions and film thickness are mostly (30 m × 0.25 mm ID, 0.25 μm). Helium remains the most commonly used carrier gas in GC analyses. Isobutyl chloroformate was the common derivatizing reagent for BAs because of its ability to form esters.[72] Almost all recent GC determination methods of BAs have been carried out with selective ion monitoring MS detection. Therefore, only GC-MS methods are included in this section.

GC has been applied for the determination of some biogenic diamines aromatic and polyamines in both young and aged port wines and grape juice.[73] Amines were extracted with a chloroform solution of ion-pairing reagent bis-2-ethylhexylphosphate followed by a back-extraction with dilute hydrochloric acid, and then the heptafluorobutyric anhydride derivatized amines were determined. Another faster method for the quantitation of 22 BAs in port wine and grape juice has been developed. Here, derivatization with isobutyl chloroformate in toluene was carried out.[74] While the determination of seven compounds took 18 min in the first method, 22 compounds were quantified within 25 min in this method. A method including simultaneous extraction/derivatization of 18 BAs in beer has been developed using dispersive liquid–liquid microextraction.[75] Both extraction and derivatization were achieved in one step by using extractive/derivatizing reagent consisting of the dispersive and extractive solvents in addition to the derivatizing reagent, isobutyl chloroformate.

Similarly, 13 BAs in homemade wine have been determined through the simultaneous extraction/derivatization with the same derivatizing reagent, but different dispersion and extraction solvents were used.[76] All of the BAs analysed were found in most of the wines. Once again, simultaneous extraction/derivatization has been applied for BAs in a large number of wine samples. This time, the direct immersion solid phase microextraction (DI-SPME) method was employed and different fibre materials were investigated.[77]

Other than wine, GC-MS has been applied to investigate the presence of BAs in fish and meat, and some of their products. Histamine content in fish and fish products were determined without derivatization prior to the separation using CP-SIL 19CB column.[78] Histamine was extracted from fish tissues by methanol with ultrasonication, and the extract was directly injected into the GC unit. Although the determination of a single underivatized BA was conducted in this way, and low LOD (5 μg g^{-1}) was obtained, simultaneous separation of many BAs without derivatization has not been reported using GC. Fast extraction/derivatization has been applied for eight BAs in poultry, pork, and beef samples. The method has shown low limits of detection and quantification and displayed excellent linearity.[79] Recently, a highly sensitive method including both electromembrane extraction and μ-dispersive liquid–liquid microextraction has been applied in the determination of four BAs in canned fish samples. Those include cadaverine,

Table 14.4 GC-MS conditions and analytical figures of merits for determination of BAs.

BAs	Food type	Sample preparation	Derivatization reagent	Capillary column (length, id, particle size)	Linear range	LOD and LOQ	Reference
1,3- DAP, PUT, CAD, SPD SPM, PEA, TYR	Grape juice, wines	Ion-pair extraction	Heptafluorobutyric anhydride	DB-5MS (30 m × 0.25 mm, 0.25 μm)	—	LOD: 10 μg L^{-1}	73
MEA, ETA, DETA, PROP, IPA, BU, IBUT, AML, IML, HEX, PRD, PPD, 1,3-DAP, PUT, VAD, ETA, HIS, TYR	Grape juice, wines	—	Isobutyl chloroformate	HP-5MS (30 m × 0.25 mm, 0.25 μm)	0.010–10 mg L^{-1}	LOD: 1 μg L^{-1} LOQ: 10 μg L^{-1}	74
MET, DMET, ETA, IPA, DETA, IBUT, 2-MBA, PRD, IML, MRF, PPD, AML, 2-PEA, 1,3-DAP, CAD, HIS, PUT, TYR	Beer	DLLME	Isobutyl chloroformate	DB-5MS (20 m × 0.18 mm, 0.18 μm)	0.010–15.0 mg L^{-1}	LOD: 0.3–2.9 μg L^{-1} LOQ: 1.0–9.5 μg L^{-1}	75
BUT, CAD, DET, DMET, ETA, HIS, MET, PROP, PUT, SPER, TEYP, TYR, 2-PEA.	Wine	DLLME	Isobutyl chloroformate	ZB-5MS (30 m × 0.25 mm 0.25 μm)	0.05–1.0 mg L^{-1} and 1.0–10.0 mg L^{-1}	LOD: 1.1–4.1 μg L^{-1} LOQ: 3.3–11.7 μg L^{-1}	76
BUT, CAD, DET, DMET, HEX, HIS, IBUT, IPENT, MET, 2-PE, PROP, PUT, TRYP, TYR	Wine	DI-SPME	Isobutyl chloroformate	ZB-5MS, (30 m × 0.25 mm, 0.25 μm)	1–1000 μg L^{-1}	LOD: 0.009–0.859 μg L^{-1} LOQ: 0.028–1.634 μg L^{-1}	77
HIS	Fish, fish products	Ultrasonic extraction with methanol	Without derivatization	CP-SIL 19CB (30 m × 0.53 mm, 1.5 μm)	4–4000 μg mL^{-1}	LOD: 5 μg g^{-1}	78
DIMET, SPER, 2-PE, PUT, TRY TYR CAD, HIST	Poultry, pork beef	DLLME	Isobutyl chloroformate	ZB-5MS 30 m × 0.25 mm, 0.25 μm)	0.05–10 μg L^{-1}	LOD: 0.003–0.009 μg g^{-1} LOQ: 0.009–0.029 μg g^{-1}	79
PUT, TYR, HIS, CAD	Canned fish	EME and μ-DLLME	Isobutyl chloroformate	HP-5 MS (30 m × 0.25 mm, 0.25 μm)	1–1000 ng g^{-1}	LOD: 0.03–0.09 ng g^{-1} LOQ: 0.09–0.36 ng g^{-1}	80

tyramine, putrescine, and histamine. The LOD and LOQ were very low. The enhanced sensitivity is due to the proper removal of the interferences from the sample matrix. GC-MS separation conditions and analytical figures of merits are provided in Table 14.4.

GC-MS with different sample pre-treatments and extraction procedures has been applied in the determination of BAs in food and beverages. The run time was in the range of 18–25 min. Even though it has proven to be accurate, sensitive and efficient, GC comes after HPLC and CE[46] in terms of number of reports. Derivatization of BAs is inevitable prior to the separation because of their high polarity.

14.5 Conclusion

Different chromatographic methods, in conjunction with various detection systems, were utilized for the determination of many BAs in different types of food, wine, and beverages. Prior to the chromatographic separation, many steps of samples preparation, extraction, and derivatization are involved. Various types of LC columns were applied to perform the separation of BAs in food and wine. The most commonly used LC columns are C18 columns with the reversed mode. The LC methods are coupled with various detectors, with UV/vis and FLD being the most commonly used, and they offered acceptable sensitivity. FLD detectors are generally more sensitive than UV-vis. Mass detectors are generally more sensitive and provide structural information about the analytes. They are useful in the case of very low analyte concentrations. CE is another alternative method for the separation of BAs. Despite its lower sensitivity when compared with other HPLC methods, CE methods are simple, fast, consume lower amounts of samples and reagents, and can be miniaturized and used for the real-time detection of BA. IC methods have the advantage that the separation is achieved without derivatization in most cases. TLC methods are simple and low-cost, but the separation time is rather long. GC-MS were also used in the separation and determination of BAs in foodstuffs. The matrix effect remains a challenge in the determination of BAs in foods, wines, and beverages because of their complex matrices, but numerous methods of extraction and derivatization were applied to overcome this effect.

List of Abbreviations

AGM	Agmatine
AML	Amylamine
Bas	Biogenic amines
BUT	Butylamine
CAD	Cadaverine
CD	Cyclodextrin
CE	Capillary electrophoresis

CIEF	Capillary isoelectric focusing
CITP	Capillary isotachophoresis
COND	Conductometric detector
CZE	Capillary zone electrophoresis
DAD	Diode array detector
DAH	Diamino hexane
DAP	Diamino propane
DET	Diethyl amine
DETA	Diethyl amine
DI-SPME	Direct immersion solid phase microextraction
DLLME	Dispersive liquid–liquid microextraction
DMET	Dimethylamine
ETA	Ethylamine
FLD	Fluorescence detection
HEC	Hydroxyethyl cellulose
HEX	Hexylamine
HIS	Histamine
HISD	Histidine
IBUT	Isobutyl amine
IC	Ion Chromatography
IEC	Ion exchange chromatography
IML	Isoamyl amine
IPA	Isopropyl amine
IPAD	Integrated pulsed amperometric detector
IPENT	Isopentyl amine
LE	leading electrolyte
LIF	Laser induced fluorescence detection
LLE	Liquid–liquid extraction
MBA	Methyl butylamine
MCE	Microchip electrophoresis
MEKC	Micellar electrokinetic chromatography
MET	Methylamine
MRF	Morpholine
MS	Mass spectrometry
MSA	Methane sulfonic acid
PAD	Pulsed amperometric detection
PEA	Phenylethylamine
PEA	Phenylethylamine
PPD	Piperidine
PRD	Pyrrolidine
PROP	Propylamine
PUT	Putrescine
Q-TOFMS	Quadrupole time-of-flight mass spectrometry
RE	reference electrode
SDS	Sodium dodecyl sulphate

SER	Serotonin
SPD	Spermidine
SPER	Spermine
TE	Terminating electrolyte
TRIS base	Tris(hydroxymethyl)aminomethane
TRYP	Tryptamine
TYR	Tyramine
UHPLC	Ultra-high-performance liquid chromatography
UV-vis	Ultraviolet-visible
WE	Working electrode

References

1. A. M. King, *et al.*, Development of a rapid profiling method for the analysis of polar analytes in urine using HILIC–MS and ion mobility enabled HILIC–MS, *Metabolomics*, 2019, **15**, 1–11.
2. T. Ikegami, Hydrophilic interaction chromatography for the analysis of biopharmaceutical drugs and therapeutic peptides: A review based on the separation characteristics of the hydrophilic interaction chromatography phases, *J. Sep. Sci.*, 2019, **42**, 130–213.
3. A. Sentkowska and K. Pyrzynska, HILIC Chromatography: Powerful Technique in the Analysis of Polyphenols, in *Polyphenols in Plants*, Academic Press, 2019, pp. 341–351.
4. S. C. Moldoveanu and V. David, Basic Information about HPLC, in *Essentials in Modern HPLC Separations*, Elsevier Inc., 2013, pp. 1–51.
5. E. H. Soufleros, E. Bouloumpasi, A. Zotou and Z. Loukou, Determination of biogenic amines in Greek wines by HPLC and ultraviolet detection after dansylation and examination of factors affecting their presence and concentration, *Food Chem.*, 2007, **101**, 704–716.
6. R. Preti, M. L. Antonelli, R. Bernacchia and G. Vinci, Fast determination of biogenic amines in beverages by a core–shell particle column, *Food Chem.*, 2015, **187**, 555–562.
7. A. M. Piasta, A. Jastrzębska, M. P. Krzemiński, T. M. Muzioł and E. Szłyk, New procedure of selected biogenic amines determination in wine samples by HPLC, *Anal. Chim. Acta*, 2014, **834**, 58–66.
8. J. Milheiro, L. C. Ferreira, L. Filipe-Ribeiro, F. Cosme and F. M. Nunes, A simple dispersive solid phase extraction clean-up/concentration method for selective and sensitive quantification of biogenic amines in wines using benzoyl chloride derivatisation, *Food Chem.*, 2019, **274**, 110–117.
9. C. I. G. Tuberoso, F. Congiu, G. Serreli and S. Mameli, Determination of dansylated amino acids and biogenic amines in Cannonau and Vermentino wines by HPLC-FLD, *Food Chem.*, 2015, **175**, 29–35.
10. K. Henríquez-Aedo, M. Vega, S. Prieto-Rodríguez and M. Aranda, Evaluation of biogenic amines content in chilean reserve varietal wines, *Food Chem. Toxicol.*, 2012, **50**, 2742–2750.

11. O. Busto, M. Miracle, J. Guasch and F. Borrull, Determination of biogenic amines in wines by high-performance liquid chromatography with on-column fluorescence derivatization, *J. Chromatogr. A*, 1997, **757**, 311–318.
12. M. T. Kelly, A. Blaise and M. Larroque, Rapid automated high performance liquid chromatography method for simultaneous determination of amino acids and biogenic amines in wine, fruit and honey, *J. Chromatogr. A*, 2010, **1217**, 7385–7392.
13. N. García-Villar, S. Hernández-Cassou and J. Saurina, Determination of biogenic amines in wines by pre-column derivatization and high-performance liquid chromatography coupled to mass spectrometry, *J. Chromatogr. A*, 2009, **1216**, 6387–6393.
14. S. Jia, Y. P. Kang, J. H. Park, J. Lee and S. W. Kwon, Determination of biogenic amines in Bokbunja (*Rubus coreanus* Miq.) wines using a novel ultra-performance liquid chromatography coupled with quadrupole-time of flight mass spectrometry, *Food Chem.*, 2012, **132**, 1185–1190.
15. C. M. Mayr and P. Schieberle, Development of stable isotope dilution assays for the simultaneous quantitation of biogenic amines and polyamines in foods by LC-MS/MS, *J. Agric. Food Chem.*, 2012, **60**, 3026–3032.
16. M. Arbulu, M. C. Sampedro, A. Gómez-Caballero, M. A. Goicolea and R. J. Barrio, Untargeted metabolomic analysis using liquid chromatography quadrupole time-of-flight mass spectrometry for non-volatile profiling of wines, *Anal. Chim. Acta*, 2015, **858**, 32–41.
17. R. Molaei, H. Tajik and M. Moradi, Magnetic solid phase extraction based on mesoporous silica-coated iron oxide nanoparticles for simultaneous determination of biogenic amines in an Iranian traditional dairy product; Kashk, *Food Control*, 2019, **101**, 1–8.
18. C. P. Vieira, *et al.*, Development and validation of RP-HPLC-DAD method for biogenic amines determination in probiotic yogurts, *Arabian J. Chem.*, 2017, DOI: 10.1016/j.arabjc.2017.12.010.
19. B. Ramos, O. Pinho and I. M. P. L. V. O. Ferreira, Changes of yolk biogenic amine concentrations during storage of shell hen eggs, *Food Chem.*, 2009, **116**, 340–344.
20. D. Cao, X. Xu, S. Xue, X. Feng and L. Zhang, An *in situ* derivatization combined with magnetic ionic liquid-based fast dispersive liquid-liquid microextraction for determination of biogenic amines in food samples, *Talanta*, 2019, **199**, 212–219.
21. M. Nadeem, T. Naveed, F. Rehman and Z. Xu, Determination of histamine in fish without derivatization by indirect reverse phase-HPLC method, *Microchem. J.*, 2019, **144**, 209–214.
22. C. Zazzu, *et al.*, Biogenic amines in traditional fiore sardo PDO sheep cheese: Assessment, validation and application of an RP-HPLC-DAD-UV method, *Separations*, 2019, **6**, 11.
23. Y. H. Jin, J. H. Lee, Y. K. Park, J.-H. Lee and J.-H. Mah, The occurrence of biogenic amines and determination of biogenic amine-producing

lactic acid bacteria in Kkakdugi and Chonggak Kimchi, *Foods*, 2019, **8**, 73.
24. M. Ishimaru, Y. Muto, A. Nakayama, H. Hatate and R. Tanaka, Determination of biogenic amines in fish meat and fermented foods using column-switching high-performance liquid chromatography with fluorescence detection, *Food Anal. Methods*, 2019, **12**, 166–175.
25. S. Jia, Y. P. Kang, J. H. Park, J. Lee and S. W. Kwon, Simultaneous determination of 23 amino acids and 7 biogenic amines in fermented food samples by liquid chromatography/quadrupole time-of-flight mass spectrometry, *J. Chromatogr. A*, 2011, **1218**, 9174–9182.
26. H. Dong and K. Xiao, Modified QuEChERS combined with ultra high performance liquid chromatography tandem mass spectrometry to determine seven biogenic amines in Chinese traditional condiment soy sauce, *Food Chem.*, 2017, **229**, 502–508.
27. Q. Chang, *et al.*, Use of functionalized covalent organic framework as sorbent for the solid-phase extraction of biogenic amines from meat samples followed by high-performance liquid chromatography, *Food Anal. Methods*, 2019, **12**, 1–11.
28. E. Mazzucco, *et al.*, High-performance liquid chromatography-ultraviolet detection method for the simultaneous determination of typical biogenic amines and precursor amino acids. applications in food chemistry, *J. Agric. Food Chem.*, 2010, **58**, 127–134.
29. M. L. Latorre-Moratalla, *et al.*, Validation of an ultra high pressure liquid chromatographic method for the determination of biologically active amines in food, *J. Chromatogr. A*, 2009, **1216**, 7715–7720.
30. C. Yılmaz, N. G. Taş, T. Kocadağlı and V. Gökmen, Determination of serotonin in nuts and nut containing products by liquid chromatography tandem mass spectrometry, *Food Chem.*, 2019, **272**, 347–353.
31. V. Gianotti, *et al.*, A new hydrophilic interaction liquid chromatography tandem mass spectrometry method for the simultaneous determination of seven biogenic amines in cheese, *J. Chromatogr. A*, 2008, **1185**, 296–300.
32. A.-D. G. Neofotistos, A. S. Tsagkaris, G. P. Danezis and C. Proestos, Emerging Trends in Biogenic Amines Analysis, in *Biogenic Amines*, IntechOpen, 2012, pp. 5–17.
33. J. L. Ordóñez, A. M. Troncoso, M. D. C. García-Parrilla and R. M. Callejón, Recent trends in the determination of biogenic amines in fermented beverages – A review, *Anal. Chim. Acta*, 2016, **939**, 10–25.
34. J. Alpert, Hydrophilic-interaction chromatography for the separation of peptides, nucleic acids and other polar compounds, *J. Chromatogr.*, 1990, **499**, 177–196.
35. A. Dang, J. J. Pesek and M. T. Matyska, The use of aqueous normal phase chromatography as an analytical tool for food analysis: Determination of histamine as a model system, *Food Chem.*, 2013, **141**, 4226–4230.

36. Waters Atlantis Columns, *Application Note*, 2004. Available at https://www.waters.com/webassets/cms/library/docs/720000793en.pdf
37. R. Michalski, Principles and Applications of Ion Chromatography, in *Application of IC-MS and IC-ICP-MS in Environmental Research*, John Wiley & Sons, Inc., 2016, pp. 1–46.
38. Y. Li, P. N. Nesterenko, R. Stanley, B. Paull and M. Macka, Comparison of cation-exchange capillary columns used for ion chromatographic separation of biogenic amines, *J. Chromatogr. A*, 2018, **1571**, 193–200.
39. A. Ščavničar, I. Rogelj, D. Kočar, S. Köse and M. Pompe, Determination of biogenic amines in cheese by ion exchange chromatography, *J. AOAC Int.*, 2018, **101**, 1542–1547.
40. A. Cinquina, *et al.*, Determination of biogenic amines in fish tissues by ion-exchange chromatography with conductivity detection, *J. Chromatogr. A*, 2004, **1032**, 73–77.
41. I. G. Casella, G. A. Palladino and M. Contursi, Determination of aliphatic amines by cation-exchange chromatography with suppressed conductivity detection after solid phase extraction, *J. Sep. Sci.*, 2008, **31**, 3718–3726.
42. A. Jastrzebska, A. Piasta and E. Szłyk, Application of Ion Chromatography for the Determination of Biogenic Amines in Food Samples, *J. Anal. Chem.*, 2015, **70**, 1131–1138.
43. L. Rivoira, M. Castiglioni and M. C. Bruzzoniti, Chromatographic determination of biogenic amines in four typical Italian cheeses for a comprehensive correlation with processing and nutritional characteristics through chemometric approaches, *J. Sci. Food Agric.*, 2019, 9730.
44. N. Seiler, Identification and quantitation of amines by thin-layer chromatography, *J. Chromatogr.*, 1971, **63**, 97–112.
45. A. Costantini, E. Vaudano, L. Pulcini, T. Carafa and E. Garcia-Moruno, An Overview on Biogenic Amines in Wine, *Beverages*, 2019, **5**, 19.
46. J. L. Ordóñez, A. M. Troncoso, M. D. C. García-Parrilla and R. M. Callejón, Recent trends in the determination of biogenic amines in fermented beverages – A review, *Anal. Chim. Acta*, 2016, **939**, 10–25.
47. D. Aures, R. Fleming and R. Hkanson, Separation and detection of biogenic amines by thin-layer chromatography Micro-Analysis of Tissue Amines and of Enzymes Involved in their Metabolism, *J. Chromatogr.*, 1968, **33**, 480–493.
48. A. Romano, *et al.*, Determination of biogenic amines in wine by thin-layer chromatography/densitometry, *Food Chem.*, 2012, **135**, 1392–1396.
49. H. Yu, *et al.*, Rapid determination of histamine in fish by thin-layer chromatography-image analysis method using diazotized visualization reagent prepared with: P-nitroaniline, *Anal. Methods*, 2018, **10**, 3386–3392.
50. M. M. Deabes, K. H. Naguib, A. M. Ayesh, E. M. El-Damaty and G. H. Rowayshed, Comparison of the HPLC and the TLC techniques for

the determination of biogenic amines spiked to sausage and smoked herring samples, *Toxicol. Allied Clin. Pharmacol.*, 2018, **5**, 1–7.
51. A. R. Shalaby, Separation, identification and estimation of biogenic amines in foods by thin-layer chromatography, *Food Chem.*, 1994, **49**, 305–310.
52. K. Naguib, A. M. Ayesh and A. R. Shalaby, Studies on the determination of biogenic amines in Foods. 1. Development of a TLC Method for the determination of eight biogenic amines in fish, *J. Agric. Food Chem.*, 1995, **43**, 134–139.
53. J. Lapa-Guimarães and J. Pickova, New solvent systems for thin-layer chromatographic determination of nine biogenic amines in fish and squid, *J. Chromatogr. A*, 2004, **1045**, 223–232.
54. A. Önal, A review: Current analytical methods for the determination of biogenic amines in foods, *Food Chem.*, 2007, **103**, 1475–1486.
55. M. Papageorgiou, *et al.*, Literature update of analytical methods for biogenic amines determination in food and beverages, *TrAC, Trends Anal. Chem.*, 2018, **98**, 128–142.
56. P. Vanda, A. Miranda, J. M. Leca and J. C Marques, Analytical methodologies for the determination of biogenic amines in wines: An overview of the recent trends, *J. Anal. Bioanal. Sep. Tech.*, 2017, **2**, 52–57.
57. Y. Y. Guo, Y. P. Yang, Q. Peng and Y. Han, Biogenic amines in wine: A review, *Int. J. Food Sci. Technol.*, 2015, **50**, 1523–1532.
58. H. Holtkamp and C. G. Hartinger, Capillary electrophoresis in metallodrug development, *Drug Discovery Today: Technol.*, 2015, **16**, 16–22.
59. W. Kok, Capillary electrophoresis: Instrumentation and operation, *Chromatographia*, 2000, **S51**, S3–S89.
60. J. L. Beckers and P. Boček, The preparation of background electrolytes in capillary zone electrophoresis: Golden rules and pitfalls, *Electrophoresis*, 2003, **24**, 518–535.
61. P. M. Nowak, *et al.*, CE-MS and GC-MS as 'Green' and complementary methods for the analysis of biogenic amines in wine, *Food Anal. Methods*, 2018, **11**, 2614–2627.
62. C.-Y. Huang, Y.-X. Wang, X.-Z. Wang, C.-C. Hu and T.-C. Chiu, Analysis of seven biogenic amines and two amino acids in wines using micellar electrokinetic chromatography, *Appl. Sci.*, 2019, **9**, 1193.
63. V. Adımcılar, N. Öztekin and F. Bedia Erim, A direct and sensitive analysis method for biogenic amines in dairy products by capillary electrophoresis coupled with contactless conductivity detection, *Food Anal. Methods*, 2018, **11**, 1374–1379.
64. D. Daniel, V. B. dos Santos, D. T. R. Vidal and C. L. do Lago, Determination of biogenic amines in beer and wine by capillary electrophoresis-tandem mass spectrometry, *J. Chromatogr. A*, 2015, **1416**, 121–128.
65. L. He, Z. Xu, T. Hirokawa and L. Shen, Simultaneous determination of aliphatic, aromatic and heterocyclic biogenic amines without

derivatization by capillary electrophoresis and application in beer analysis, *J. Chromatogr. A*, 2017, **1482**, 109–114.
66. A. Jastrzebska, A. Piasta and E. Szłyk, Simultaneous determination of selected biogenic amines in alcoholic beverage samples by isotachophoretic and chromatographic methods, *Food Addit. Contam., Part A*, 2013, **31**, 83–92.
67. F. Kvasnička, S. Kavková and A. Honzlová, Electrophoretic determination of histamine, *J. Chromatogr. A*, 2019, **1588**, 180–184.
68. K. Swinney and D. J. Bornhop, Detection in capillary electrophoresis, *Electrophoresis*, 2000, **21**, 1239–1250.
69. S. Thompson and D. Pappas, A fluorescence toolbox: A review of investigation of electrophoretic separations, process, and interfaces, *Electrophoresis*, 2019, **40**, 606–615.
70. J. Guo, *et al.*, A strategy to modulate the electrophoretic behavior in plastic microchips using sodium polystyrene sulfonate, *J. Chromatogr. A*, 2016, **1477**, 132–140.
71. J. Płotka-Wasylka, E. Kłodzińska and J. Namieśnik, Determination of biogenic amines in wine using micellar electrokinetic chromatography, *J. Res. Analytica*, 2017, **3**, 62–66.
72. P. Hušek, Chloroformates in gas chromatography as general purpose derivatizing agents, *J. Chromatogr. B: Biomed. Sci. Appl.*, 1998, **717**, 57–91.
73. J. Fernandes and M. Ferreira, Combined ion-pair extraction and gas chromatography–mass spectrometry for the simultaneous determination of diamines, polyamines and aromatic amines in Port wine and grape juice, *J. Chromatogr. A*, 2000, **886**, 183–195.
74. S. C. Cunha, M. A. Faria and J. O. Fernandes, Gas chromatography-mass spectrometry assessment of amines in port wine and grape juice after fast chloroformate extraction/derivatization, *J. Agric. Food Chem.*, 2011, **59**, 8742–8753.
75. C. Almeida, J. O. Fernandes and S. C. Cunha, A novel dispersive liquid–liquid microextraction (DLLME) gas chromatography-mass spectrometry (GC–MS) method for the determination of eighteen biogenic amines in beer, *Food Control*, 2012, **25**, 380–388.
76. J. Płotka-Wasylka, V. Simeonov and J. Namieśnik, An in situ derivatization – dispersive liquid–liquid microextraction combined with gas-chromatography – mass spectrometry for determining biogenic amines in home-made fermented alcoholic drinks, *J. Chromatogr. A*, 2016, **1453**, 10–18.
77. M. Papageorgiou, D. Lambropoulou, C. Morrison, J. Namieśnik and J. Płotka-Wasylka, Direct solid phase microextraction combined with gas chromatography – Mass spectrometry for the determination of biogenic amines in wine, *Talanta*, 2018, **183**, 276–282.
78. B.-S. Hwang, J.-T. Wang and Y.-M. Choong, A rapid gas chromatographic method for the determination of histamine in fish and fish products, *Food Chem.*, 2003, **82**, 329–334.

79. W. Wojnowski, J. Namieśnik and J. Płotka-Wasylka, Dispersive liquid-liquid microextraction combined with gas chromatography–mass spectrometry for in situ determination of biogenic amines in meat: Estimation of meat's freshness, *Microchem. J.*, 2019, **145**, 130–138.
80. M. Kamankesh, A. Mohammadi, A. Mollahosseini and S. Seidi, Application of novel electromembrane extraction and microextraction method followed by gas chromatography-mass spectrometry to determine biogenic amines in canned fish, *Anal. Methods*, 2019, **11**, 1898–1907.

CHAPTER 15

Miscellaneous Methods for the Detection of Biogenic Amines

D. ZARE* AND M. ENSHAEIEH

Department of Biotechnology, Iranian Research Organization for Science and Technology (IROST), P.O. Box 33535111, Tehran 3313193685, I. R. Iran
*Email: zare@irost.ir

15.1 Introduction

Biogenic amines (BAs) are a group of very important compounds with useful properties for human beings. However, if taken excessively, they can be harmful. The toxicity, carcinogenic and allergic effects of BAs have been reported by many investigators. Controlling BA concentration inside the body has been reported as a useful approach for the suppression of carcinogenic cells.[1] BAs are usually accumulated in different foods and food products in the early days of storage and have been considered as an excellent index for evaluating the freshness and quality of food products. Therefore, simple, fast and precise assessment of these compounds is of paramount importance.[2]

Among the more popular methods for the determination of BAs are high performance liquid chromatography (HPLC), gas chromatography (GC) and capillary electrophoresis (CE), which have been described in Chapter 10. However, most of these analytical methods are time consuming, expensive, require well-trained operators and are not portable.[3–5] Thus, despite the advances in instrumentations, simpler techniques such as colorimetry, fluorimetry and sensors are still required for tasks such as large-scale screening of samples.

Food Chemistry, Function and Analysis No. 20
Biogenic Amines in Food: Analysis, Occurrence and Toxicity
Edited by Bahruddin Saad and Rosanna Tofalo

Published by the Royal Society of Chemistry, www.rsc.org

Colorimetric and fluorometric methods are relatively old methods for the determination of BAs. They remain attractive due to their simplicity, cheapness and user-friendly features. The advances in enzyme technology have opened up new opportunities in sensing and applications through the determination of BAs. The great selectivity of enzymes has been translated into devices such as enzyme-linked immunosorbent assay (ELISA) and biosensors.[5–9]

As microorganisms are the main agent of BA production in food products, detecting and controlling these microorganisms have always been of concern to food specialists. Direct and indirect methods for the detection of BA-producing bacteria have been established and are developing.[10,11]

Thus, in the present chapter, spectroscopic and enzymatic methods for the determination of BAs are discussed. Methods for the detection of BA-producing bacteria are also briefly touched.

15.2 Colorimetric Methods

Colorimetric methods are based on the absorption of radiation in the ultraviolet/visible (UV/Vis) region of the electromagnetic spectrum. However, since BAs lack chromophoric or fluorophoric moieties in their structure, a derivatization step is essential.[12] Generally, the reaction is not specific and is interfered by other BAs and amino acids.[13] To overcome this problem, purification of samples is usually accomplished by cation exchange chromatography. The early versions of colorimetric methods are the association of official analytical chemists (AOAC) official methods 957.07 and 977.13. Basically, these methods include extraction with organic solvents, separation or purification of BAs using carboxylic cation exchange chromatography and derivatization.[14] Selective extraction and chromatographic purification of BA containing samples constitutes the main challenge of colorimetric methods.

Chromophore reagents such as 2,4-dinitroflurobenzene,[5,12,13] pyrylium and coumarin have been introduced for the colorimetric and fluorometric determinations of BAs.[13–15]

The newer colorimetric method for determination of histamine requires saline extraction of histamine followed by centrifugation, *n*-butanol extraction and finally evaporation before the colorimetric reaction with *p*-phenyldiazonium sulfonate is accomplished. The intensity of the pink–orange complex can be measured visually by comparing with the reference colour scale or can be evaluated by a spectrophotometer at 496 nm.[12] The advantage of this method is the fast analysis, requiring approximately 45 min for a single estimation together with sample preparation. However, this assay has not been evaluated in inter-laboratory trials.

Methyl pyridinium derivatives have also been introduced for colorimetric determination of BAs. The compound distinguished histamine, cadaverine and putrescine from interfering amino acids without discriminating between the BAs. The method is appropriate for the fast screening of total BAs in food samples.[15]

15.3 Fluorometric Methods

Certain organic molecules can absorb light energy and accumulate in different interatomic bonds which have been raised to higher energy levels. This energy can be degenerated in various ways such as the emission of light. Fluorescence normally refers to the absorption of light from the ultraviolet (UV) spectrum and its emission in the visible range. Based on this property, fluorometric devices have been constructed. This instrument consists of some important parts: a light source for producing UV radiation with a filter system that allows only a particular wavelength to pass through the sample cuvette. After emission from the sample, a secondary filter system allows only the specified wavelength of light to reach the detector. Figure 15.1 shows the basic components of the fluorescence device.[16]

Since BAs do not have adequate fluorescent activity, they are first transformed into fluorescent derivatives and are sensed as they can quench the fluorescence characteristic of other fluorescent components; 3-(2-Furoyl)-quinoline-2-carboxaldehyde, 5-(4,6-dichloro-*s*-triazin-2-ylamino) fluorescein, fluorescein isothiocyanate, naphthalene-2,3-dicarboxaldehyde and *o*-phthalaldehyde are some fluorescence derivatization reagents for the determination of BAs.[5]

Fluorescence detection has been used for the detection of an extensive spectrum of primary and secondary biological amines. Notable advantages include rational sensitivity and specificity with reasonable cost for a large number of samples without the requirement of experienced personnel. The sensitivity of these methods is greater than 100 times in comparison with colorimetric methods and the limits of detection is down to the nanomolar range. The reagents compared with those required for gas chromatography or mass spectrometry are much cheaper.

Major disadvantages are the different parameters that influence the results so that the method can be complex and time consuming in some cases. Furthermore, most fluorometric reagents react with all or a group of BAs with different emission activity and thus, a separation method should be applied for specific detection. *o*-Phthalaldehyde is the most popular reagent for fluorometric determination of BAs, especially histamine, although it cannot discriminate histidine from histamine.[17] Extraction and

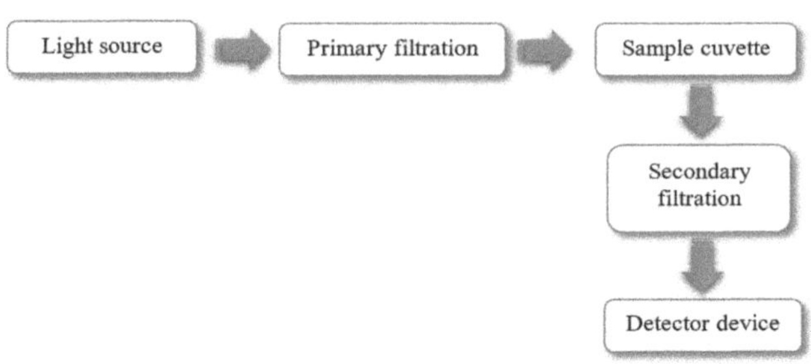

Figure 15.1 Basic instruments for fluorescent measurements.

purification of histamine is accomplished by methanol and an anion exchanger column, respectively. *o*-Phthalaldehyde is then applied for derivatization and measurement is accomplished spectrophotometrically.[18]

The combination of HPLC with fluorometric methods is usually applied for specific detection. Even with the limitations discussed here, fluorescence techniques remain popular for the analysis of BAs in biological samples.

Fluorometric sensing of BAs has been reported using new compounds with aggregation-induced emission. These compounds have indicated suitable emission in aggregated or solid forms.[19] They are valuable alternatives as novel fluorometric methods.[19] The method is based on formation of a hydrogen bond between BAs and receptors. The free receptor does not indicate emission activity while intense emission is observed by the addition of biomolecules and aggregation of the receptor. To avoid interference of water molecules in the formation of hydrogen bonds, the reaction is accomplished in organic solution. Reliable emission activity was reported for some BAs such as putrescine through the interaction with the receptor (carboxylic acid modified tetraphenylethenes) in a dichloromethane solution.[19]

A complex of metal ions and the receptor has been investigated to conduct the reaction in aqueous solution. The new method, introduced by Chopra *et al.*,[20] is based on the nanoaggregation of the complex of Cu^{II} synthetic organic receptor with spermine.[20] The binding selectivity of the complex of Cu^{II}-receptor (dipicolinic acid hydrazide-salicylaldehyde) with different BAs was investigated in the aqueous medium and significant absorbance at 551 nm was observed in the presence of spermine. A recovery of 95–98% has been observed for spermine with 1–2 μM concentration in artificial samples. However, interference of amino acids has not been evaluated and the method has not been validated using real natural samples.

15.4 Enzymatic Methods

As discussed earlier, determination of BAs using chromatographic, colorimetric and fluorometric methods require sample clean up and an extraction step to avoid interaction of derivatization compounds with similar ingredients in food samples. These steps are usually long and decrease the efficiency of the assay. Elimination of sample pretreatment could improve the efficiency and recovery and also increase the rapidity of the method. Enzymatic methods introduced for the determination of BAs are attempts to address this issue. Nowadays, direct enzymatic determination of BAs based on enzyme immunoassays and ELISA and biosensors are being developed.[1,11,21,22]

15.4.1 Direct Enzymatic Determination of BAs

An enzymatic technique for the determination of histamine was developed by Lerke *et al.*[23] in which the enzyme diamine oxidase (DAO) catalyses the

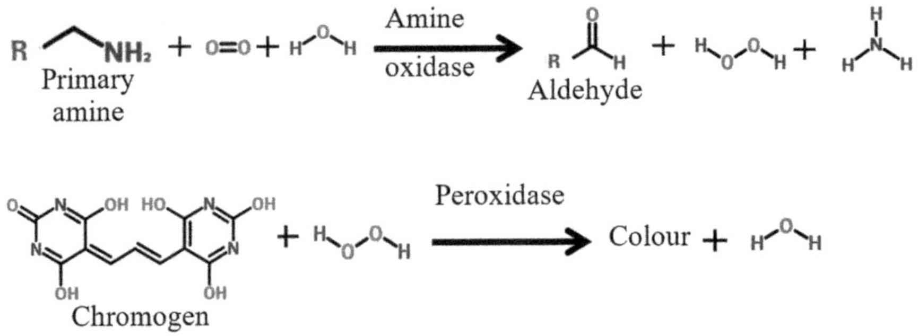

Figure 15.2 Basic reactions for enzymatic determination of biogenic amines.

transformation of histamine to imidazole acetaldehyde and the simultaneous production of hydrogen peroxide. In the presence of horse radish peroxidase, hydrogen peroxide is converted to oxygen and water[7,23] (Figure 15.2).

Parallel to this reaction is the oxidation of a suitable chromogen such as leucocrystal violet (colourless) to crystal violet (purple) which can be readily detected spectrophotometrically for quantitative assays. This reaction is catalysed by the enzyme peroxidase (Figure 15.2). The intensity of the resulting colour is directly proportional to the histamine concentration. The method has been modified[24,25] to measure histamine content using micro plate reader.[26] Detection limit is 0.5 mg kg^{-1} and quantitation limit is 1.5 mg kg^{-1}.[24] This method is rapid with the capability of simultaneous analysis of multiple samples.[25] However, the DAO enzyme can react with some other amines such as putrescine, agmatine and cadaverine when they are present at high levels.[26] For instance, histamine oxidase and histamine dehydrogenase can also react with putrescine and tyramine[27,28] so that it is difficult to identify histamine by these enzymes when some other BA amines are present in foodstuffs.[29]

15.4.2 Immobilized Enzymes

Enzymes are expensive and enzymatic determination of BAs always consumes enzymes. Methods using immobilized enzymes are attractive alternatives. Some enzymes such as polyamine oxidase (PAO) are specific for polyamines (*e.g.*, spermine and spermidine) and DAO are specific for putrescine. These enzymes were used for co-immobilization on nanoparticles composed of chitosan/coconut fibre/zinc oxide (CS/CF/nZnO) for the construction of a polyamine sensing strip.[30] Oxidation of spermine, spermidine or putrescine when present in a sample produces hydrogen peroxide (Figure 15.3). H_2O_2 next reacts with a suitable chromogen, resulting a red colour which absorbs visible light maximally at 520 nm.[30]

The negative point is the susceptibility of interference with electroactive materials that are present in the samples.[4,31] Reusable enzymatic strips

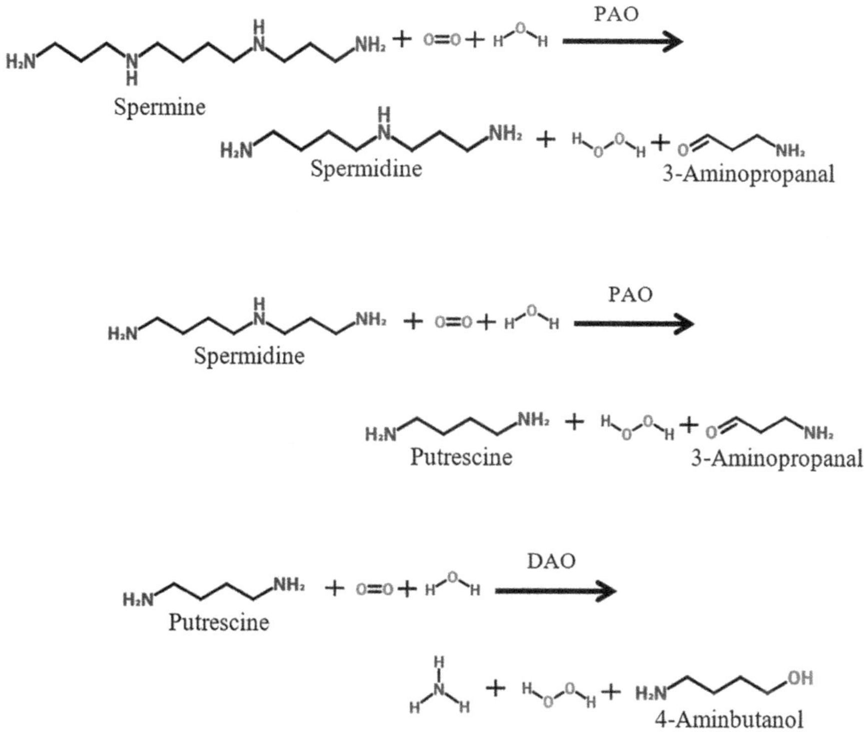

Figure 15.3 Oxidation of spermine, spermidine or putrescine in the presence of oxidases.

with immobilized enzymes on insoluble support are not only sensitive and specific but are also inexpensive. Furthermore, activity and usefulness of the immobilized enzyme can be improved by careful selection of the immobilization support. The basic principles of immobilized enzymes that resulted in biosensor construction will be discussed next.[32,33]

15.4.3 Enzyme-linked Immunosorbent Assay

ELISA is based on the binding of a specific antibody to an antigen. This kind of test has been developed for the detection of many biological molecules. However, the preparation of the specific antibody (monoclonal antibody) with high specificity for the desired antigen is always a challenge. The technique has been developed for the detection of histamine.[34] A variety of commercial kits are also available for the detection of histamine, the most important BA in food samples. The kits are moderately expensive, and a calibration curve is required. However, the simple sample preparation, rapidity and specificity of the technique are commendable.

The basic principle of the histamine test is based on competition between histamine of the sample and the fixed enzyme labelled with histamine for binding to the coated antibody of histamine (antihistamine) on the wells. Histamine in the samples competes with the conjugated histamine (with an enzyme) to the coated antihistamine. Finally, a suitable substrate is added to the well which reacts with the conjugated enzyme to produce or change a detectable colour (Figure 15.4). Although the sample preparation is simple, it should be mentioned that acylation of histamine to form *N*-acylhistamine should be accomplished before running the test. The principle of the kit is competitive ELISA, *i.e.*, the high levels of histamine in the sample decreases the binding probability of labelled histamine to the antibody and thus, decreases the intensity of the colour produced. The detection limit of this immunoassay method is almost 0.8 mg kg^{-1}, which is sensitive enough for the determination of histamine in food samples.[8,17]

The competitive direct ELISA technique has been compared with reversed phase-HPLC for the determination of histamine in wines and good agreement between both methods has been found.[22] Competitive direct ELISA has also been applied for the determination of histamine in fish and other food matrices.[22,35,36]

High precision and sensitivity are some advantages of the ELISA test for histamine determination.[37] However, the test has not been developed for other BAs. Preparation of a monoclonal antibody for BAs such as putrescine is the obstacle. On the other hand, it can be laborious, time consuming and rather expensive, especially when a small number of samples must be evaluated.[37]

15.4.4 Biosensors

The capability of enzymes to undergo specific and fast reaction readily with substrates lends them as a suitable candidate in the detection and determination of biomolecules.

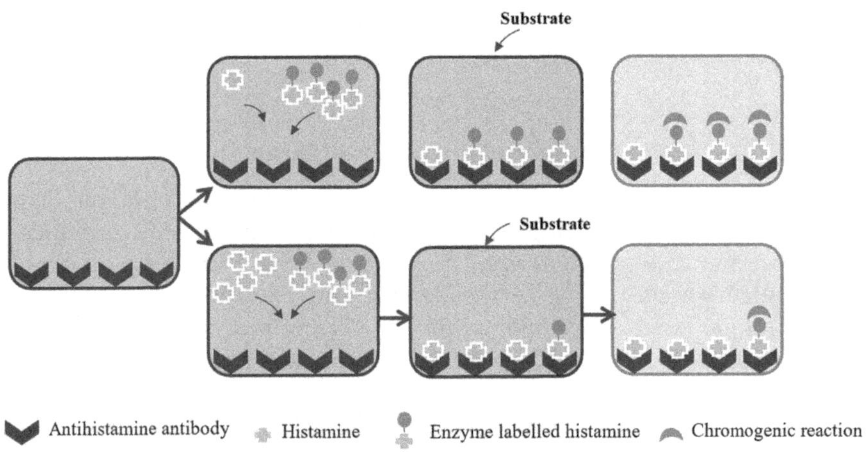

Figure 15.4 A schematic diagram of competitive enzyme-linked immunosorbent assay (ELISA) for the detection of histamine.

Basically, a biosensor can be fabricated from three different parts. In the receptor layer, biomolecules such as enzymes, antibodies, cells or microorganisms are immobilized. These biomolecules are able to react selectively with a target or substrate. As the enzymatic reaction is of high specificity, the clean up or extraction of the analyte of interest is not required. In the second part of the device, the transducer, the product of the enzymatic reaction, is translated into electronic signals. The analyte concentration is often directly proportional to the amount of signal produced. Finally, the third component includes a set of electronic devices that are responsible for amplifying, processing and displaying the recorded signals.[9,38] A schematic diagram of a typical biosensor and its components is shown in Figure 15.5.

Biorecognition of BAs by biosensors comprises a mixture of various enzymes.[9] Amine oxidase enzymes, including monoamine, diamine and polyamine oxidases, are responsible for the oxidation of BAs (Figure 15.4). The main products of these reactions are aldehydes and hydrogen peroxide (HP). HP can be detected in different ways. The reduction of HP, normally catalysed by peroxidase enzyme, releases electrons and the transfer of these free electrons changes the charge of the transducer device. The transducer charge is amplified and displayed by electronic devices. A number of enzymes have been identified for the development of different biosensors and detection of either several or individual BAs. Diamine oxidases (*e.g.*, putrescine and cadaverine oxidase) and polyamine oxidases (*e.g.*, spermine and spermidine oxidases) are the main enzymes applied for the

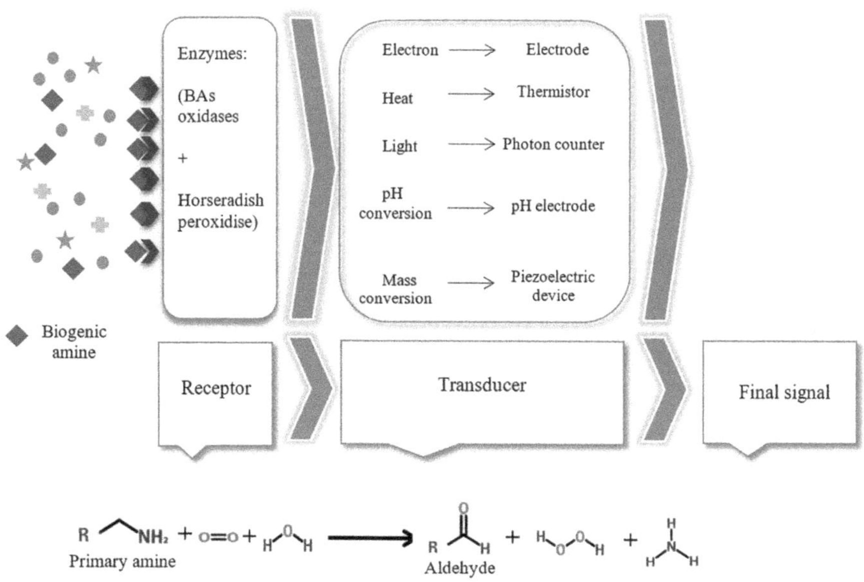

Figure 15.5 Basic principles of biosensors.

construction of biosensors.[9] As food matrices are different, different biosensors have been designed for different foods. For instance, measuring the level of putrescine and cadaverine in fish samples, together with histamine, could be applied as a freshness index while this may not be important in beverages. Furthermore, the design of sensors for liquid samples will be different from that of solid samples such as meat and fish.[39]

As explained earlier, the fundamentals of biosensors are the oxidation and reduction reactions, catalysed by enzymes and oxygen is the electron transfer mediator (Figure 15.6).[42] Thus, oxygen affects the sensitivity and selectivity of these sensors. To improve the selectivity, second-generation sensors wer introduced where a redox molecule was immobilized or added to the solution as mediator which is responsible for the transfer of electrons between the enzyme and the transducer or electrode. Mediators such as ferrocene, quinones, organic salts and ferrocyanide have been introduced.[40] This type of biosensor has been adopted in many commercial biosensors. Although, these mediators improved the selectivity and sensitivity of sensors, it could also increase electron transfer in interfering reactions. Thus, in third-generation biosensors, the mediators are completely eliminated, and the electron transfer is accomplished by the immobilized enzymes (Figure 15.6). The selectivity was further improved as the immobilized enzyme is responsible of electron transfer to the electrode and no intermediate is necessary.[9,40,41]

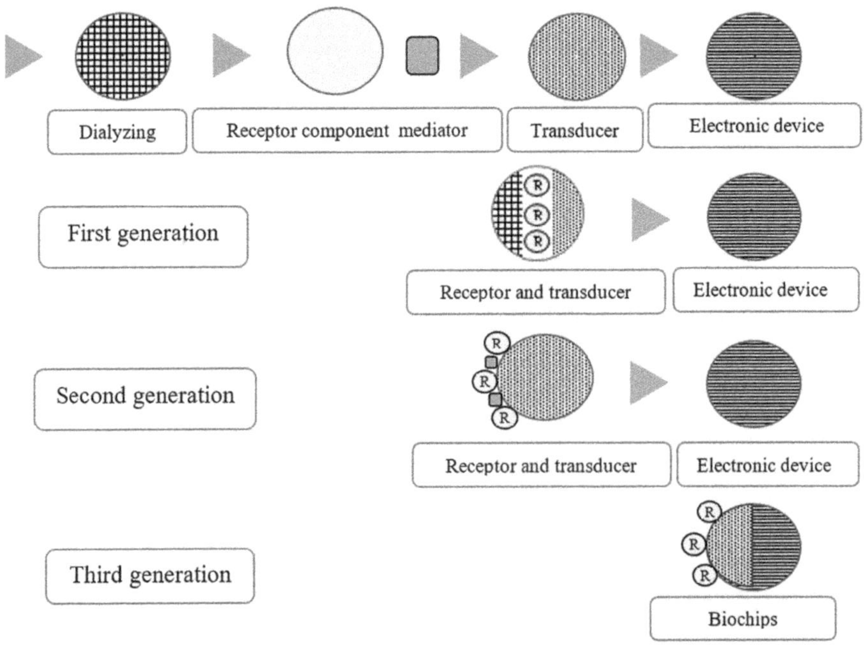

Figure 15.6 Different generations of biosensors.

Several biosensors have been introduced for the determination of BAs in food samples. Biosensors for putrescine and cadaverine have been developed for investigations on the freshness of fish. Other polyamines (PAs) including agmatine, spermidine and spermine are interesting chemical markers for fish freshness as well. Biosensors for these PAs with limit of detection between 10 nM and 98 μM, comparable with colorimetric and chromatography methods, have been reported. The biosensor has also been applied to meat and meat products, dairy products, fruits and alcoholic beverages.[39]

Apart from enzymatic sensors, a number of chemical sensors have also been developed for the determination of BAs. Chemical sensors are usually based on the electro-oxidation capability of biomolecules.[9,43] For instance, a novel enzymeless sensor has been introduced based on the electrochemical activity of putrescine, cadaverine and histamine on a nanomaterial of cetyltrimethylammonium bromide-iron oxide.[43] The electro-oxidation activity when evaluated using a carbon paste electrode produced reasonable LOD and LOQ of 0.003 μM and 0.007 μM, respectively. A boron doped diamond electrode has also been developed for the determination of a series of biomolecules including histamine and polyamines with LOD of 500 nm and 1 μM, respectively.[44] However, unlike biosensors, the specificity of these chemical sensors is not great.[39]

Sensors provide an interesting option for the analysis of BAs with the capability of miniaturization. The method is simple, does not require sample clean up and derivatization and analysis time is short which can be applied for on-site testing. A disadvantage of chemical sensors is that it provides total BA content, rather than the individual amine. Sensors are sensitive to the food matrix and the accuracy of the results must be examined with a confident method before applying on new food matrices.[45,46] Table 15.1 summarizes some chemical and biological sensors for BA determination.

15.5 Flow Injection Analysis

Another alternative for BA analysis is using flow injection analysis (FIA). Conventionally, this technique involves the injection of sample and reagent into a suitable carrier solution under a constant flow condition. Both solutions, when mixed online, pass through a reaction coil. Finally, the products are detected using a suitable detector (*e.g.*, spectrophotometer, spectrofluorometer, electrochemical).[59,60] Figure 15.7 shows the main components of a simple FIA unit.

There is no requirement of pretreating on the sample extract, very low sample and reagent is consumed and the method is miniaturized, cheap and simple. The method is indeed fast as around 60 samples can be screened per hour for detection of histamine.[1] However, the specificity of the method is low when common reagents are applied for derivatization of

Table 15.1 Chemical and biological sensors for biogenic amine (BA) determination.

Foodstuff	Detected biogenic amines	Reacting component in the receptor	Transducer	References
Fish	Histamine	Labelled histamine antibody	Amperometry	47
Fish	Histamine	Diamine oxidase	Amperometry (voltammetry)	48
Fish	Histamine	Copper microparticles and diamine oxidase	Amperometry	49
Fish	Histamine, putrescine, cadaverine, tyramine	Monoamine oxidase, putrescine oxidase	Amperometry	50
Pork, fish	Putrescine, tyramine, tryptamine, phenylethylamine	Monoamine oxidase, diamine oxidase, putrescine oxidase	Amperometry	51
Blood	Spermine, spermidine	Polyamine oxidase, spermine oxidase	Flow injection amperometry	52
Wine	Histamine, putrescine	Putrescine oxidase, histamine dehydrogenase	Amperometry	31
Beer and Wine	Total BAs	Diamine oxidase	Amperometry	53
Beer	Putrescine	Putrescine oxidase	Amperometry	54
Beer	Tyramine	Tyramine oxidase	Amperometry	55
Beer	Tyramine	Tyramine oxidase	Amperometry	56
Beer	Tryptamine	Solid surface-room temperature phosphorescence	Photometry	57
Wine	Histamine	Imprinted nanoparticles	Potentiometry	58
Fish	Histamine, putrescine, cadaverine	Cetyltrimethylammonium bromide and Fe_3O_4 nanoparticles	Amperometry (voltammetry)	43

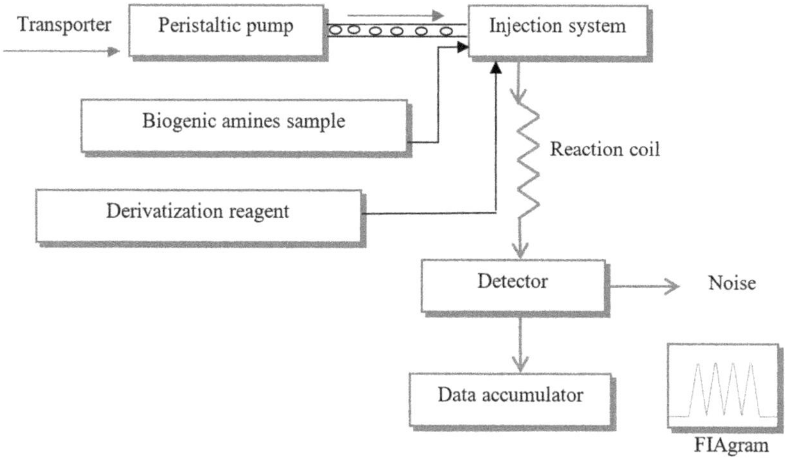

Figure 15.7 Basic components of the flow injection analyser.

BAs and it requires careful selection of reagent and their concentrations, as well as monitoring of pumps for maintaining specificity.[61]

The main challenge in the FIA of BAs is the low specificity of the applied reagents. These reagents are usually do not specify to histamine or any other BAs and interfere with amino acids or some derivatives of amino acids.[59] For instance, in the FIA of histamine, *o*-phthalaldehyde (OPT), a common fluorometric derivatization reagent, reacts with both histamine and histidine. The selectivity of the determination can be improved by the precise control of the flow rate and reaction time.[59]

Alternatively, offline ion exchange separation of the interfering compounds before the FIA injection can be used.[62] In the FIA determination of histamine in commercial red wines, mathematical calibration has been applied to improve selectivity of the FIA and overcome the interference of amino acids or polyphenols by calculations employing partial least squares regression and execution of multivariate calibrations.[62]

As described earlier, biosensors were developed for the analysis of BAs and their selectivity to BAs are reasonable. However, periodic calibration, regeneration and sample manipulation may be required. For instance, dilution and preconcentration may be required when the concentration of the substrate is too high or low, respectively.[63] Thus, it has been tried to overcome the drawbacks of biosensors and improve the selectivity of FIA by coupling these devices.[61] Basically, BA oxidases are immobilized on a suitable support. Then, a suitable flow of substrates (samples) is passed through immobilized enzymes and during oxidation of BAs, the produced hydrogen peroxide is measured by a detector. In some methods, hydrogen peroxide content is evaluated using a reaction with a chromogenic agent through UV/Vis light. It can also be detected using amperometric, spectrophotometric or fluorometric methods.[63] Applying biosensing to FIA system for BA analysis has been successfully evaluated in chicken meat samples and an amperometric biosensor using amine oxidase enzyme as a molecular recognition element was used in combination with FIA.[64]

The hallmarks of the FIA methods are the speed of analysis, reproducibility of results and the ease of automation. The total time required for the analysis of a sample extract in some reports is around one minute. For instance, throughput of 40–60 samples per hour has been reported by some investigators.[61] Table 15.2 indicates some FIA methods applied for BA determination in food samples.

15.6 Detection of Biogenic Amine Producing Bacteria

The main agent for the presence of BAs in food products is bacteria with decarboxylase activity and BAs are the products of decarboxylation of different amino acids. The existence of bacteria with decarboxylase activity in non-fermented foods is usually indicative of mishandled storage status and undesired microbial activity, while the presence in fermented foods is indicative of wrong fermentation conditions.[11] Both Gram-positive and Gram-negative bacteria are found in foods with the capability of BA production. Gram-negative bacteria are usually responsible for the production of BAs in non-fermented foods while Gram positives are usually found in fermented food with the ability of BA production. Thus, the isolation and identification of bacteria from foods can provide valuable information on the background of foods. The bacterial genus including *Enterococcus*, *Lactococcus*, *Lactobacillus*, *Leuconostoc*, *Streptococcus* and the family *Enterobacteriaceae* contain some strains that are the main amine producers in some food products.[11,72]

Among the most common bacteria with decarboxylase activity are the lactic acid bacteria (LAB), *Enterococcus*, *Leuconostoc*, *Streptococcus*, *Pseudomonas*, *Clostridium* and the family *Enterobacteriaceae*. However, the decarboxylase activity is usually strain dependent and the isolation of a suspicious bacterium does not necessary mean that they are a BA producer.[11] Environmental conditions also influence the capability of strains for BA production.[10] The early methods for the detection of BA-producing bacteria were based on direct culture techniques, while a variety of enzymatic and molecular methods were later introduced for the fast and precise detection of these bacteria.[11]

15.6.1 Culture Techniques

Most of the microbiological screening methods commonly involve a differential growth medium which contains a pH indicator.[73] Screening methods based on the application of differential growth media require isolation and growth of the producing microorganism. A suitable growth medium for the screening of BA-producing bacteria usually consists of adequate organic

Table 15.2 Flow injection analysis (FIA) methods applied for biogenic amine (BA) determination in food samples.

FIA construction	Detected BAs	Food tested	Some important characteristics	References
FIA	Histamine	Fish	A range between $1.59 \times 10^{-2} - 1 \times 10^{-5}$ of histamine was tested and the selectivity of flow injection analysis using control of reaction time was improved.	65
FI system based on immobilized diamine oxidase	Putrescine	—	An amperometric detection system for produced hydrogen peroxide was coupled to the system. Linear range was from 10 μM to 1 μM for putrescine sample frequency of ~40 samples per hour was possible.	66
FI system based on immobilized diamine oxidase	Putrescine, cadaverine, histamine	Fish	An amperometric detection system for produced hydrogen peroxide was coupled to the system. Linear response range of up to 400 μM with detection limit of 0.07 μM was observed for putrescine. Detection range of up to 60 μM with a detection limit of 0.5 μM was observed for histamine.	67
FIA	Histamine	Wine and cider	An anion-exchange column was applied for the elimination of interferences. Detection limit: 30 and quantification limit: 101 μg L^{-1}, respectively.	62
FI coupled with enzymeless electrode	Histamine	Fish	Histamine ion-selective electrode has been applied for freshness control. Detection limit was 1.6 ± 10^{-6} M L^{-1}.	68
Amperometric biosensor coupled with FI system	Total BAs	Wine and beer	Fast and sensitive express of total BAs as putrescine in red and white wines. The method underestimated BA content in beer by about 50%.	69
Amperometric biosensor coupled with FI system	Putrescine, cadaverine, tyramine	Chicken meat	Linear range of detection: Putrescine: 5–100 μg ml^{-1} Cadaverine: 5–100 μg ml^{-1} Tyramine 25–1000 μg ml^{-1} Prolonged shelf life for electrode was observed.	64
FIA	Histamine	Wine	Improved selectivity using sample clean up and multivariate calibration by partial least square (PLS) regression.	59
FIA	Histamine	Fish and fish meal	Amberlite resin (weak cation exchanger) was applied for the elimination of interferences. Linear range from 0.5 to 22.0 μg ml^{-1} was observed.	70
FI coupled with amperometric electrochemical sensor	Histamine	Fish	A dynamic range of histamine from 1 to 750 μM with a sensitivity of 15 nA μM^{-1} and a detection limit of 0.33 μM were achieved.	71

nitrogen sources such as yeast extract, meat extract and peptone along with minerals, while glucose is added in some media as a carbon source. Precursor amino acids such as histidine, tyrosine, tryptophan, lysine and ornithine are also added to the culture medium. As the production of BAs in culture medium increases pH, a suitable pH indicator such as bromocresol purple is applied to monitor for pH variations in culture medium and the initial pH of the medium is adjusted to between 5 and 6. After culturing and incubation, changes in the medium colour is indicative of decarboxylase activity and production of BAs. The method is applied in both agar and broth culture media. However, false positive and false negative results are sometimes encountered. Some bacteria utilize amino acids as a carbon source and release amino ($-NH_2$) group in culture media while they are not a BA producer. This phenomenon increases the pH of the medium and is reported as BA producer (false positive). On the other hand, LAB are acid producers, especially in the presence of sugars. For instance, LAB usually need more complicated media and the pH of culture medium decreases due to fermentation activity while they may have positive decarboxylase activity. Enrichment and modification have been applied to enhance the growth of LAB. Addition of metal sulphates, tween 80, citrate, thiamine, pyridoxal-5-phosphate and carbon source has been tried to improve the screening of BA-producing bacteria.[11]

An automated system for microbial metabolic activities detection based on amino acid decarboxylase activity in *Salmonella* spp. and other *Enterobacteriaceae* has also been introduced.[73] In this method, each vial includes liquid growth medium with the appropriate pH indicator while a semi fluid layer has been placed at the bottom of the vials. Decarboxylase activity of bacteria is evaluated by detecting the reflection of colour on the semi fluid layer using a suitable detector while bacteria cells do not penetrate to this layer and thus do not affect the results.[11]

Overall, culture techniques are relatively cheap, easy and do not require complicated instruments, but they are subjected to false positive and false negative reactions.

15.6.2 Enzymatic Techniques

As described earlier, amine oxidases are responsible for the oxidation of BAs and the main by-product of the reaction is hydrogen peroxide. Thus, production of BAs in culture medium is evaluated using diamine oxidase and horseradish peroxidase. In the first reaction, histamine in the culture medium is oxidized by diamine oxidase and hydrogen peroxide is produced. In the second reaction, a colourless chromogen is oxidized by peroxidase to a coloured reagent in the presence of H_2O_2. The method has been successfully applied for the detection of histamine producing bacteria without interference from other BAs such as putrescine, cadaverine and tyramine.[74] However, the turbidity of the culture medium interferes with the colour produced and prevents quantitative assessment. Applying colourless culture media and filtration of culture media after cultivation has been utilized to remove interfering agents and measure quantitative production of histamine.[15] A modified method has been applied for enzymatic determination of histamine where 3,3-diaminobenzidine has been applied as chromogen instead of leuco-crystal violet. This alternative reagent which does not interfere with components of culture medium has been applied for the quantitative determination of histamine in liquid culture media.[11]

Other enzymatic based techniques such as biosensors and ELISA are also useful for rapid assessment of BAs in culture media.[10]

15.6.3 Chromatographic Methods

Chromatographic methods, especially HPLC, are widely used for the determination of BAs produced by bacteria in liquid culture media. However, since the determination of the exact number of BAs is not necessary in the screening of bacteria, especially in small microbiology laboratories, simple, rapid and inexpensive methods are preferred.[10,11,75]

15.6.4 Molecular Techniques

Culture techniques, enzymatic methods and chromatographic techniques are all based on the isolation and growth of suspected bacteria in a suitable medium and detection of BAs produced. Some reports indicated that BA-producing microorganisms may lose this property after frequent cultivation and prolonged storage.[76,77] Furthermore, the methods involve several lengthy isolation and preservation steps of pure culture of microorganisms, cultivation in a suitable culture media (in the presence of BAs amino acids precursors) and finally assessment of the produced BAs.

In molecular methods, time consuming steps of isolation, preservation, cultivation of microorganisms and direct determination of BAs have been omitted. They are also capable of predicting potential BA accumulation in food during storage.

The main molecular methods for assessment of BAs producing capability are polymerase chain reaction (PCR) and DNA hybridization methods.[10]

15.6.5 Polymerase Chain Reaction

Production of BAs in foods is related to amino acid decarboxylases activity which is produced by suspected microorganisms. Tracking of the encoding genes by PCR is a rapid and reliable method to detect BA-producing bacteria. To evaluate the production of BAs by PCR, related oligonucleotide primers are designed based on sequencing of the related BA decarboxylase genes. Since a variety of Gram-positive and Gram-negative bacteria are able to encode BA decarboxylases, different groups of primers have been introduced. According to this method, total DNA from the suspected sample is extracted. The extraction of DNA is made possible using a variety of commercial kits and also different methods. Before the extraction, a suitable method should be applied for lysis of the bacteria cell wall. The extracted DNA is then applied to run PCR. To amplify BA decarboxylase genes, appropriate primers are utilized and the reaction conditions for amplification include denaturation of double strand DNA, annealing and extension is performed according to the adopted method and Taq polymerase. During the running of PCR, BA encoding genes are amplified. The PCR product is then separated using gel electrophoresis where related fragments are detected. The method has been successfully utilized for the screening of Gram-positive and Gram-negative bacteria producing BAs in canned fish, tuna, mackerel, sardine, dairy products and wine.[10,78] However, false negative reports have also been observed[79,80] and many different primers have been introduced to improve capability of the method. It is also noteworthy to mention that the PCR technique is not able to distinguish between viable and non-viable organisms, this being the main disadvantage of the method.[10,81]

15.6.6 DNA Hybridization

The technique of DNA hybridization has also been applied for the screening of BA-producing bacteria. In this method, the PCR product is labelled and applied for DNA/DNA hybridizations. A DNA extract of the microorganisms is immobilized on a membrane surface. Specific DNA which has been generated by amplification of related decarboxylase gene is labelled using a suitable label (*e.g.*, P3′dCTP) and placed on the membranes, including the DNA of the microorganisms. After washing, membrane hybridization is detected using radiography. The method has been applied for the screening of histamine producing microorganisms, but false positive and false negative results have also been reported.[79,81,82]

Molecular methods with their special characteristics like rapidity, sensitivity and prediction of possibility for BA production during storage or fermentation, and also being culture independent, have the potential to solve the deficiencies of traditional methods. Furthermore, the important features of the molecular methods are the detection of potential BA-forming bacteria before the amine is produced, but an intrinsic shortcoming of PCR is the detection of non-viable cells. When PCR is the identification method for risk assessment of BA production in food, it is essential to distinguish between the viable and non-viable organisms.[10,81,82]

15.7 Conclusion

Spectroscopic methods, including colorimetric and fluorometric methods, are moderately cheap, simple and fast methods for BA determination and do not require complicated interpretations and expert operators. The most important challenge of these techniques is low selectivity which makes sample clean up and purification compulsory. Newer rapid, cheap and simple techniques have been introduced to improve the selectivity and accuracy of BA determination. Enzymatic based methods improved selectivity of BA determination while the methods are rapid, moderately cheap and simple. Among the enzymatic methods, ELISA enables the examination of a number of samples simultaneously and biosensors allowed the evaluation of BAs *in situ*. FIA, a miniaturized method to automate the measurement of BA content in samples, and its selectivity have improved when combined with enzymatic methods.

Among methods for the detection of BA-producing bacteria, culture techniques are simple and easy and do not require complicated instruments. On the other hand, instrumental methods such as chromatography and molecular methods are expensive but rapid and precise.

References

1. J. M. Hungerford, *Toxicon*, 2010, **56**, 231–243.
2. B. C. Sara, B. Sava, M. James, P. M. Miguel, R.-G. Elke and S. Giuseppe, *EFSA J.*, 2011, **9**(10), 2393.
3. G. Spano, P. Russo, A. Lonvaud-Funel, P. Lucas, H. Alexandre, C. Grandvalet, E. Coton, M. Coton, L. Barnavon and B. Bach, *Eur. J. Clin. Nutr.*, 2010, **64**, S95.
4. J. L. Ordóñez, A. M. Troncoso, M. D. C. García-Parrilla and R. M. Callejón, *Anal. Chim. Acta*, 2016, **939**, 10–25.
5. G. Mohammed, A. Bashammakh, A. Alsibaai, H. Alwael and M. El-Shahawi, *TrAC, Trends Anal. Chem.*, 2016, **78**, 84–94.
6. A. Önal, *Food Chem.*, 2007, **103**, 1475–1486.
7. K. Punakivi, M. Smolander, M.-L. Niku-Paavola, J. Mattinen and J. Buchert, *Talanta*, 2006, **68**, 1040–1045.
8. D. Chevrier, J.-L. Guesdon, J.-C. Mazie and S. Avrameas, *J. Immunol. Methods*, 1986, **94**, 119–125.
9. K. Kivirand and T. Rinken, *Anal. Lett.*, 2011, **44**, 2821–2833.
10. J. M. Landete, B. de las Rivas, A. Marcobal and R. Muñoz, *Int. J. Food Microbiol.*, 2007, **117**, 258–269.
11. A. Marcobal, B. De las Rivas and R. Muñoz, *J. Verbraucherschutz Lebensmittelsicherh.*, 2006, **1**, 187–196.
12. S. Patange, M. Mukundan and K. A. Kumar, *Food Control*, 2005, **16**, 465–472.
13. P.-Q. Leng, F.-L. Zhao, B.-C. Yin and B.-C. Ye, *Chem. Commun.*, 2015, **51**, 8712–8714.
14. A. Gürses, M. Açıkyıldız, K. Güneş and M. S. Gürses, *Dyes Pigm.*, Springer, 2016, pp. 13–29.
15. B. García-Acosta, M. Comes, J. L. Bricks, M. A. Kudinova, V. V. Kurdyukov, A. I. Tolmachev, A. B. Descalzo, M. D. Marcos, R. Martínez-Máñez and A. Moreno, *Chem. Commun.*, 2006, 2239–2241.
16. J. Lakowicz, *Principles of Fluorescence Spectroscopy*, Springer, New York, 3rd edn, 2006.
17. T. L. Pessatti, J. D. Fontana and M. L. Pessatti, *Public Health Microbiology*, Springer, 2004, pp. 311–316.

18. M. Etienne, Methodology for histamine and biogenic amines analysis, 2006. Available at https://archimer.ifremer.fr/doc/00000/6484/
19. M. Nakamura, T. Sanji and M. Tanaka, *Eur. J. Chem.*, 2011, **17**, 5344–5349.
20. S. Chopra, J. Singh, H. Kaur, A. Singh, N. Singh and N. Kaur, *Eur. J. Inorg. Chem.*, 2015, 4437–4442.
21. T. Dole, S. Koltun, S. M. Baker, R. M. Goodrich-Schneider, M. R. Marshall and P. J. Sarnoski, *J. Aquat. Food Prod. Technol.*, 2017, **26**, 781–789.
22. A. Marcobal, M. Polo, P. Martın-Alvarez and M. Moreno-Arribas, *Food Res. Int.*, 2005, **38**, 387–394.
23. P. A. Lerke, M. N. Porcuna and H. B. Chin, *J. Food Sci.*, 1983, **48**, 155–157.
24. E. López-Sabater, J. Rodríguez-Jerez, A. Roig-Sagues and M. Mora-Ventura, *Food Addit. Contam.*, 1993, **10**, 593–602.
25. B. Ben-Gigirey, C. Craven and H. An, *J. Food Sci.*, 1998, **63**, 210–214.
26. M. Etienne and N. Bregeon, Proceeding of the 22nd annual meeting of the western European fish technologists association, Lisboa, Portugal, 1992.
27. M. Niculescu, C. Nistor, T. Ruzgas, I. Frebort, M. Sebela, P. Pec and E. Csoregi, *Inflammation Res.*, 2001, **50**, 146.
28. Y. Sekiguchi, H. Makita, A. Yamamura and K. Matsumoto, *J. Biosci. Bioeng.*, 2004, **97**, 104–110.
29. S. Moret and L. S. Conte, *J. Chromatogr. A*, 1996, **729**, 363–369.
30. V. Hooda, *Food Chem.*, 2018, **239**, 1100–1109.
31. W. Henao-Escobar, L. Del Torno-de Roman, O. Domínguez-Renedo, M. Alonso-Lomillo and M. Arcos-Martínez, *Food Chem.*, 2016, **190**, 818–823.
32. J. Lange and C. Wittmann, *Anal. Bioanal. Chem.*, 2002, **372**, 276–283.
33. I. Frebort, L. Skoupa and P. Peč, *Food Control*, 2000, **11**, 13–18.
34. J.-L. Guesdon, D. Chevrier, J.-C. Mazie, B. David and S. Avrameas, *J. Immunol. Methods*, 1986, **87**, 69–78.
35. O. Aygün, E. Schneider, R. Scheuer, E. Usleber, M. Gareis and E. Märtlbauer, *J. Agric. Food Chem.*, 1999, **47**, 1961–1964.
36. P. Rogers and W. Staruszkiewicz, *J. Aquat. Food Prod. Technol.*, 2000, **9**, 5–17.
37. R. C. Alves, M. F. Barroso, M. B. González-García, M. B. P. Oliveira and C. Delerue-Matos, *Crit. Rev. Food Sci. Nutr.*, 2016, **56**, 2304–2319.
38. Z. Zhao and H. Jiang, in *Biosensors*, ed. P. A. Serra, Intech, Vukovar, 2010.
39. D. Baratella, E. Bonaiuto, M. Magro, J. de Almeida Roger, Y. Kanamori, G. P. P. Lima, E. Agostinelli and F. Vianello, *Amino Acids*, 2018, 1–17.
40. R. S. Freire, C. A. Pessoa, L. D. Mello and L. T. Kubota, *J. Braz. Chem. Soc.*, 2003, **14**, 230–243.
41. B. D. Meshram, A. K. Agrawal, S. Adil, S. Ranvir and K. K. Sande, *Int. J. Curr. Microbiol. Appl. Sci.*, 2018, **7**, 3305–3324.

42. B. Meshram, A. Agrawal, S. Adil, S. Ranvir and K. Sande, *Int. J. Curr. Microbiol. Appl. Sci.*, 2018, 7, 3305–3324.
43. M. Hasanzadeh, A. Bahrami, M. Alizadeh and N. Shadjou, *RSC Adv.*, 2013, **3**, 24237–24246.
44. J. H. Luong, K. B. Male and J. D. Glennon, *Analyst*, 2009, **134**, 1965–1979.
45. B. Bóka, N. Adányi, D. Virág, I. Frebort and A. Kiss, *Electroanalysis*, 2013, **25**, 237–242.
46. M. Niculescu, T. Ruzgas, C. Nistor, I. Frébort, M. Šebela, P. Peč and E. Csöregi, *Anal. Chem.*, 2000, **72**, 5988–5993.
47. X.-X. Donga, J.-Y. Yanga, L. Luoa, Y.-F. Zhanga, M. C, Y.-M. Suna, H.-T. Leia, Y.-D. Shena, R. C. Beierd and Z.-L. Xua, *Biosens. Bioelectron.*, 2017, **98**, 305–309.
48. A. Karim, A. W. Wahab, I. Raya, H. Natsir and A. R. Arif, The 2nd International Conference on Science (ICOS), Makassar, Indonesia, 2018.
49. D. C. Vanegas, L. Patiño, C. Mendez, D. A. de Oliveira, A. M. Torres, C. L. Gomes and E. S. McLamore, *Biosensors*, 2018, **8**, 1–19.
50. G. C. Chemnitius and U. Bilitewski, *Sens. Actuators, B*, 1996, **32**, 107–113.
51. B. Bóka, N. Adányi, D. Virág, M. Sebela and A. Kiss, *Electroanalysis*, 2012, **24**, 181–186.
52. C. Tortolini, G. Favero and F. Mazzei, in *Polyamines: Methods and Protocols, Methods in Molecular Biology*, ed. R. Alcázar and A. F. Tiburcio, Springer Science+Business Media LLC, New York, 2018, vol. 1694.
53. M. Di Fusco, R. Federico, A. Boffi, A. Macone, G. Favero and F. Mazzei, *Anal. Bioanal. Chem.*, 2011, **401**, 707–716.
54. B. Bóka, N. Adányi, J. Szamos, D. Virág and A. Kiss, *Enzyme Microb. Technol.*, 2012, **51**, 258–262.
55. B. Batra, S. Lata, R. Devi, S. Yadav and C. S. Pundir, *J. Solid State Electrochem.*, 2012, **16**, 3869–3876.
56. S. Lata, S. Yadav, R. Bhardwaj and C. Pundir, *Sens. Instrum. Food Qual. Saf.*, 2011, **5**, 104–110.
57. T. Ramon-Marquez, A. L. Medina-Castillo, A. Fernandez-Gutierrez and J. F. Fernandez-Sanchez, *Biosens. Bioelectron.*, 2016, **79**, 600–607.
58. I. Basozabal, A. Guerreiro, A. Gomez-Caballero, M. A. Goicolea and R. J. Barrio, *Biosens. Bioelectron.*, 2014, **58**, 138–144.
59. S. Hernández-Cassou and J. Saurina, *Anal. Lett.*, 2013, **46**, 1758–1768.
60. C. Ruiz-Capillas and F. Jimenez-Colmenero, *Food Addit. Contam.*, 2008, **25**, 1167–1178.
61. J. M. Hungerford, T. A. Hollingworth and M. M. Wekell, *Anal. Chim. Acta*, 2001, **438**, 123–129.
62. G. del Campo, B. Gallego and I. Berregi, *Talanta*, 2006, **68**, 1126–1134.
63. E. H. Hansen, *J. Mol. Recognit.*, 1996, **9**, 316–325.
64. D. Telsnig, V. Kassarnig, C. Zapf, G. Leitinger, K. Kalcher and A. Ortner, *Int. J. Electrochem. Sci.*, 2012, 7, 10476–10486.

65. J. M. Hungerford, J. Kevin, K. D. Walker, M. M. Wekell, J. E. LaRose and H. R. Throm, *Anal. Chem.*, 1990, **62**, 1971–1976.
66. S. A. M. Marzouka, C. X. Xua, B. R. Cosofreta, R. P. Bucka, S. S. M. Hassanb, M. R. Neumanc and R. H. Sprinkled, *Anal. Chim. Acta*, 1998, **363**, 57–65.
67. M. A. Carsol and M. Mascini, *Talanta*, 1999, **50**, 141–148.
68. C. G. Amorim, R. C. Souza, A. N. Araújo, M. C. B. S. M. Montenegro and V. L. Silva, *Food Chem.*, 2010, **122**, 871–876.
69. M. Di Fusco, R. Federico, A. Boffi, A. Macone, G. Favero and F. Mazzei, *Anal. Bioanal. Chem.*, 2011, **401**, 707–716.
70. M. S. Abdullah and N. A. Fakhri, *J. Univ. Zakho*, 2013, **1**, 261–272.
71. Y.-T. Lin, C.-H. Chen and M. S. Lin, *Sens. Actuator, B*, 2017, **255**, 2838–2843.
72. M. Schirone, R. Tofalo, G. Mazzone, A. Corsetti and G. Suzzi, *Food Microbiol.*, 2011, **28**, 128–136.
73. L. Shelef, A. Surtani, K. Kanagapandian and W. Tan, *Food Microbiol.*, 1998, **15**, 199–205.
74. R. Maijala, *Lett. Appl. Microbiol.*, 1993, **17**, 40–43.
75. S. S. Sumner and S. L. Taylor, *J. Food Prot.*, 1989, **52**, 105–108.
76. E. García-Moruno, A. V. Carrascosa and R. Muñoz, *J. Food Prot.*, 2005, **68**, 625–629.
77. A. Lonvaud-Funel and A. Joyeux, *J. Appl. Bacteriol.*, 1994, **77**, 401–407.
78. M. Izquierdo-Pulido, J.-M. Carceller-Rosa, A. Marine-Font and M. C. Vidal-Carou, *J. Food Prot.*, 1997, **60**, 831–836.
79. R. Alves, A. dos Santos and M. Martins, *Eur. Food Res. Technol.*, 2002, **214**, 178–180.
80. C. Le Jeune, A. Lonvaud-Funel, B. Ten Brink, H. Hofstra and J. Van der Vossen, *J. Appl. Bacteriol.*, 1995, **78**, 316–326.
81. B. De Las Rivas, A. Marcobal, A. V. Carrascosa and R. Munoz, *J. Food Prot.*, 2006, **69**, 2509–2514.
82. J. Landete, S. Ferrer and I. Pardo, *J. Appl. Microbiol.*, 2005, **99**, 580–586.

Subject Index